Solar Heating and Cooling: Engineering, Practical Design, and Economics

Jan F. Kreider
Environmental Consulting Services, Inc.
Boulder, Colorado

Frank Kreith
University of Colorado

and

Environmental Consulting Services, Inc.
Boulder, Colorado

SCRIPTA BOOK COMPANY
a division of
HEMISPHERE PUBLISHING CORPORATION
Washington, D.C.

McGRAW–HILL BOOK COMPANY
New York   St. Louis   San Francisco   Auckland   Düsseldorf   Johannesburg
Kuala Lumpur   London   Mexico   Montreal   New Delhi   Panama
Paris   São Paulo   Singapore   Sydney   Tokyo   Toronto

SOLAR HEATING AND COOLING: Engineering, Practical Design, and Economics

1 2 3 4 5 6 7 8 9 0   K P K P   7 9 8 7 6 5

This book was set in Theme by Hemisphere Publishing Corporation. The editors were Jeremy Robinson, Evelyn Walters Pettit, and Honey Sauberman; the designer was Lilia Guerrero; the production supervisor was Rebekah McKinney; and the compositor was Shirley J. McNett.
The printer and binder was The Kingsport Press.

Library of Congress Cataloging in Publication Data

Kreider, Jan F, date.
    Solar heating and cooling: engineering, practical
design, and economics.

    Includes index.
    1. Solar heating.   2. Air conditioning.
I. Kreith, Frank, joint author. II. Title.
TH7413.K73 1975   697'.78   75-6646

ISBN 0-07-035473-1

Contents

Preface

One of the great maladies of our time is the way sophistication seems to be valued above common sense.

NORMAN COUSINS

By the end of this decade solar-powered systems for heating and cooling of buildings and providing hot service water will be commercially available and will be competitive with fossil fuels and electrical systems in many parts of the United States. By 1985 solar heating and cooling systems could supply about one-third of the 30 percent total United States energy consumption now used for residential and commerical space heating and cooling. By the turn of the century solar energy could provide 20 percent of the energy needs of the United States.

Until a few years ago, proposals for using the sun's energy on a significant scale were likely to be received with skepticism. During the last several years, however, the increasing cost of conventional fuels, the political uncertainties related to the supply of petroleum, and the problems associated with electric-energy generation from nuclear sources have modified this skepticism dramatically.

Although solar energy is the most abundant form of energy available, it is also one of the most dilute and intermittent forms and therefore requires different methods of collection and utilization than forms of energy widely used heretofore. There have been predictions, however, that with current technology the solar energy impinging on 4 percent of the land area of the United States would be sufficient to provide the projected energy needs of the country in the year 2000; by comparison, about 15 percent of the land area of the United States is used today for agricultural purposes. The successful utilization of solar energy on a large scale depends upon the design of systems and a thorough consideration of their economics. In contrast to gas or electric heating systems, which can be easily standardized, no single solar system is satisfactory for all tasks. The design and size of a solar system must, therefore, be matched to its task: It should not only meet technical requirements but should also involve—and resolve satisfactorily—the impact on construction economics, system amortization, and building esthetics.

One significant but generally unrecognized reason the use of solar energy has progressed so slowly is that conventional engineering education has not produced courses that deal with this interdisciplinary field involving principles of physics, chemistry, engineering, architecture, meteorology, and astronomy. Because of this education gap, the authors began to offer solar-energy seminars for architects, engineers, contractors and planners. These seminars presented, in a comprehensive and practical way, the available knowledge of use and economics of solar energy. The seminar material, expanded and supported with examples and related information, forms the basis of this book.

The emphasis in this book is on heating and cooling methods that can be employed economically in the near term. By extension, the principles are also applicable to engineering analysis of solar thermal-power applications, crop drying, and solar distillation. Photochemical, photosynthetic, and photovoltaic processes are not considered in depth; the indirect solar technologies of wind power and ocean-thermal-gradient utilization are likewise not included. This book attempts to describe practical systems, either those of more classical design with long, successful histories of use or those resulting from the latest generation of solar research that seem to have considerable promise because of their use of new concepts, materials, and manufacturing techniques. The reader must recognize that the field is rapidly changing; to remain current one must follow developments on an almost daily basis.

It is recognized that the audience for this book is not homogeneous. Therefore, throughout the book, which may serve as a textbook or a practical handbook, a dual method of presentation is used. Results of analysis are presented in analytical (equation) form and in graphical or tabular form. The seasoned engineer will find the analytical approach more appropriate in his designs by computer, while the architect and builder may find the graphical presentations more useful.

The material is divided into five chapters and eight appendixes. The five chapters contain—in this order—introductory information on solar and conventional energy-use concepts and requirements, fundamental principles of heat transfer and the nature of solar radiation, practical and efficient methods of collecting solar energy, and detailed quantitative descriptions of the practical systems for heating or cooling by means of solar energy together with an analysis of their economics. The appendixes contain extensive tables of reference data, including heat-transfer properties of materials, geographical and seasonal variation of sunshine and climate in the United States, and economic amortization schedules. The appendixes also contain information on the legal implications of solar-energy use for buildings and several typical solar-heated building designs. In keeping with the goal of this book to provide information in a readily usable form, the units of measurement are those endorsed by the American National Standards Institute (ANSI)—those used in practice in the United States and Canada—not those units used in the Système International d'Unités (SI) system.

This book draws on many sources for information, much of which dates from a small group of solar pioneers who were years before their time. Wherever possible, the authors have tried to indicate the specific sources from which the information was derived, but certain information could not be ascribed specifically to a single source. The authors have tried to give credit where credit is due; if the authors have failed to cite an original source, it is an oversight and not an intention. This book is not intended to be a compendium of literature on solar-energy applications; however, a section of general references has been included that can serve as a starting point for a thorough examination of the literature.

The notes upon which this book is based were developed as a part of the program of the Center for Management and Technical Programs at the University of Colorado, Boulder. The authors would like to express their appreciation to the Center for its help and cooperation and to the many authors and publishers who gave permission to use drawings or other materials. The authors wish to acknowledge the patience and helpfulness of Mr. Rolla Rieder during the typing of the manuscript. The authors also wish to acknowledge the contribution of Ms. Honey Sauberman, whose careful editing significantly improved the clarity, continuity, and organization of this book.

JAN F. KREIDER
FRANK KREITH

CHAPTER 1

Introduction: Why Solar Energy?

Eyes, though not ours, shall see
Sky-high a signal flame,
The sun returned to power above
A world but not the same.

CECIL DAY LEWIS

Strictly speaking, all forms of energy are derived from the sun. However, our most common forms of energy—fossil fuels—received their solar input eons ago and have changed their characteristics so that they are now in a highly concentrated form. Since it is apparent that these stored, concentrated energy forms are now being used at such a rapid rate that they will be depleted in the not-too-distant future, we must begin to supply a large portion of our energy needs not from stored, but from incoming solar energy as soon as possible.

ENERGY—CONTEMPORARY SOURCES AND USAGE PATTERNS

Energy is defined in classical thermodynamics as the capacity to do work. From a practical point of view it is the basic ingredient for all industrialized societies. In the United States energy is currently derived from four primary sources: petroleum, natural gas and natural-gas liquids, coal, and wood. The supplies of these common energy sources, except for wood, are finite. Their lifetime is estimated to range from 15 yr for natural gas to 300 yr for coal. As current energy sources become exhausted an energy gap will develop, exacerbated by the synergistic effects of population growth and increased dependence on energy. After nonrenewable energy sources are consumed in what some authors call this "Fossil Fuel Age," mankind must turn to longer-term, permanent energy sources. The two most significant of these are nuclear and solar energy. Nuclear energy requires highly technical and costly means for its safe and reliable utilization and may have undesirable side-effects. Solar

1

energy, on the other hand, shows promise of becoming a dependable energy source without new requirements of a highly technical and specialized nature for its widespread utilization. In addition, there appear to be no significant polluting effects from its use.*

Table 1.1 indicates that of the total national energy consumption, 19 percent is used in residential buildings, 15 percent in commercial buildings, 41 percent in industrial processes, and 25 percent in transportation. The major uses for energy in buildings are in space heating, air conditioning, and service-hot-water supply. Table 1.1 shows this division of energy use as consumption in trillion British thermal units (Btu) and percentage of the national total, as well as in annual growth rate. Space heating of residences accounts for 11 percent of the total national energy consumption, while space heating for commercial occupancy represents an additional 7 percent. Air conditioning represents 2.5 percent of the total national energy consumption but has an annual growth rate of about 16 percent in residences.

The energy losses that determine the effectiveness of heating or air conditioning in buildings are essentially the same. Unnecessary energy losses are caused by inadequate insulation, excessive ventilation, high rates of air infiltration from outside, and excessive fenestration. Building heat losses have recently been recognized by the federal government as a major cause of fuel-resource waste. In the *Minimum Property Standards*† (1965), the Federal Housing Administration (FHA) of the U.S. Department of Housing and Urban Development (HUD) permitted heat loss of 2,000 Btu/(1,000 ft³)(°F) per day in single-family residences. The standard required by HUD Operation Breakthrough in 1970, however, reduced this figure to 1,500, and the newly implemented FHA *Minimum Property Standards* requires these heat losses to be less than 1,000 Btu/(1,000 ft³)(°F) per day. The reduction in energy consumption required by the Standards is to be achieved principally by improved thermal insulation and reduction of air infiltration. Approximately 40 percent of space-heating fuel can be saved through thicker, or more effective insulation and improved draft control in commercial buildings. It appears at the present time technologically and economically feasible to reduce heat losses in buildings to approximately 700 Btu/(1,000 ft³)(°F) per day through the use of proper design, increased insulation, and reduction of unnecessary ventilation and infiltration.[1] Figure 1.1 illustrates the reduction in building heat loss that can be realized by adding storm windows and

---

*Legal questions have been raised concerning possible visual pollution of large solar-collector arrays, questions as yet unanswered. A survey of legal precedents, principles, and devices concerning solar energy—e.g., use in the state of Colorado—is included as Appendix G.
†Standards of construction which must be met to qualify a building for an FHA loan.

TABLE 1.1 Total Fuel Energy Consumption in the United States by End Use*

End use†	Consumption, trillions of Btu		Annual rate of growth %	Percent of national total	
	1960	1968		1960	1968
Residential					
Space heating	4,848	6,675	4.1	11.3	11.0
Water heating	1,159	1,736	5.2	2.7	2.9
Cooking	556	637	1.7	1.3	1.1
Clothes drying	93	208	10.6	0.2	0.3
Refrigeration	369	692	8.2	0.9	1.1
Air conditioning	134	427	15.6	0.3	0.7
Other	809	1,241	5.5	1.9	2.1
Total	7,968	11,616	4.8	18.6	19.2
Commercial					
Space heating	3,111	4,182	3.8	7.2	6.9
Water heating	544	653	2.3	1.3	1.1
Cooking	93	139	4.5	0.2	0.2
Refrigeration	534	670	2.9	1.2	1.1
Air conditioning	576	1,113	8.6	1.3	1.8
Feedstock	734	984	3.7	1.7	1.6
Other	145	1,025	28.0	0.3	1.7
Total	5,742	8,766	5.4	13.2	14.4
Industrial					
Process steam	7,646	10,132	3.6	17.8	16.7
Electric drive	3,170	4,794	5.3	7.4	7.9
Electrolytic processes	486	705	4.8	1.1	1.2
Direct heat	5,550	6,929	2.8	12.9	11.5
Feedstock	1,370	2,202	6.1	3.2	3.6
Other	118	198	6.7	0.3	0.3
Total	18,340	24,960	3.9	42.7	41.2
Transportation					
Fuel	10,873	15,038	4.1	25.2	24.9
Raw materials	141	146	0.4	0.3	0.3
Total	11,014	15,184	4.1	25.5	25.2
National total	43,064	60,526	4.3	100.0	100.0

*Adapted from Stanford Research Institute (1972)[10] by permission of the publishers.
†Electric-utility consumption is allocated to each end use.

increasing ceiling and wall insulation.* Appendix F contains an extensive list of the energy-conservation methods for buildings and

*A quantitative estimate of the savings in energy for home heating, which are technologically feasible, may be obtained as follows. The upper point, A, in (a) and (b) may be assumed to represent the approximate state of insulation and storm window sealing in

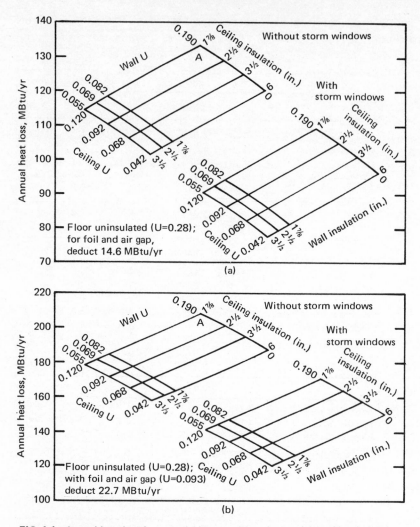

FIG. 1.1 Annual heat loss from model homes with various weights of insulation in (a) New York and (b) Minneapolis (1 MBtu = 10^6 Btu) $U =$ Btu/(hr)(ft²)(°F). (*Adapted with permission from Moyers.*[6])

presents the relative efficiency of each in several climatic zones of the United States.

approximately 90 percent of housing built prior to issuance of minimum property standards. The heat losses from these houses can be reduced by approximately 45 percent by application of heavy ceiling insulation, side wall insulation, and installation of storm windows. Thus, it would not be unreasonable to expect that installation of insulation and storm windows on housing units now in service could reduce the present national demand for fuel consumed in space heating of residences by approximately 40 percent.

Felix[3] has compiled energy-use data for more than 200 countries and colonies; per capita energy-use for a number of these countries is shown in Fig. 1.2. Although it is true that the United States consumes about one-third of the world's energy with less than 6 percent of the world's population, the energy use per dollar of GNP lies on the lower bound of all the world's countries. It seems that below this lower bound, a given level of GNP cannot be sustained. The curve shows that the United States is among the most efficient energy consumers in the world. The present United States standard of living cannot be maintained solely by increased energy conservation; to continue the present standard of living, new sources of supply are required. Solar energy is one such new energy source.

THE NATURE OF SOLAR ENERGY

Solar energy is the world's most abundant permanent source of energy. The amount of solar energy intercepted by the planet earth is 170 trillion kW, an amount 5,000 times greater than the sum of all other inputs (terrestrial nuclear, geothermal, and gravitational energies and lunar gravitational energy). Of this amount 30 percent is reflected to space, 47 percent is converted to low-temperature heat and reradiated to space, and 23 percent powers the evaporation/precipitation cycle of the biosphere; less than $\frac{1}{2}$ percent is represented in the kinetic energy of the wind and waves and in photosynthetic storage in plants. The amount of the sun's energy intercepted by earth is only a tiny fraction—one thousandth of one millionth—of the total released by the conversion of 4 million tons of hydrogen per second to helium in the sun.

Although it is abundant, solar energy impinging on the earth's atmosphere is relatively dilute (approximately 430 Btu/(hr)(ft^2)). Traversing the earth's atmosphere dilutes it further by attenuation, local weather phenomena, and air pollution. Moreover, solar energy is received only intermittently at any point on earth. The solar energy which arrives on the surface of the earth is of two forms: direction radiation and diffuse radiation. *Direct radiation* is collimated and capable of casting a shadow; *diffuse radiation* is dispersed, or reflected, by the atmosphere, and not collimated. In considering how solar energy can best be used, the ratio of direct to diffuse radiation becomes important, as will be shown in Chap. 3. The ratio of direct to diffuse radiation varies with time and location. For example, in the Four Corners states, the ratio of direct to diffuse radiation is on the order of 5, but for a large city the ratio of direct

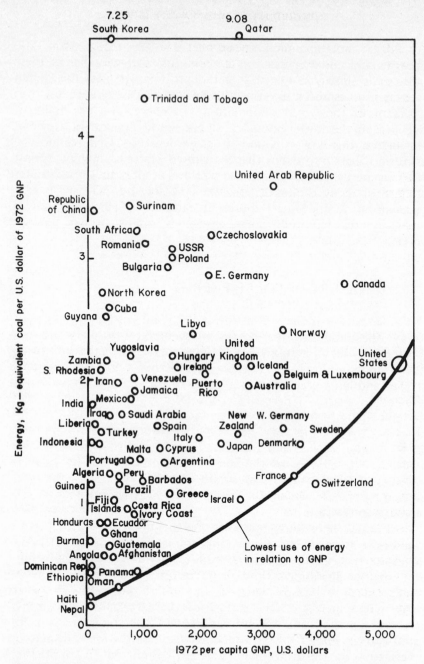

FIG. 1.2 Worldwide energy consumption per unit of GNP compared with the GNP per capita. (*From F. Felix, "Using Electricity More Efficiently," Electrical World, December 1, 1974, p. 59. Copyright 1974, McGraw-Hill, Inc. All rights reserved.*)

to diffuse radiation may only be on the order of 2. The amount of direct radiation diminishes as air pollution increases. The nature of the sun's energy impinging on the surface of the earth and its temporal, seasonal, and geographical variations are described in detail in Chap. 2.

On the average the radiation striking 1 ft^2 of the earth's surface is on the order of 100 to 200 Btu/(hr)(ft^2). Owing to the dilute nature of solar radiation, large collection areas are required, and the initial cost of installing solar heating and cooling equipment is larger than for other contemporary fuels. However, as will be shown in subsequent chapters, if properly designed and constructed, solar heating and cooling of homes and buildings is, at current power costs, less expensive than electric climate control under most conditions and in most locations in the United States.

HISTORICAL PERSPECTIVE

The first person to use the sun's energy on a large scale was Archimedes, who reputedly set fire to an attacking Roman fleet at Syracuse in 212 B.C. He accomplished this "by means of a burning glass composed of small square mirrors moving every way upon hinges . . . so as to reduce it [the Roman fleet] to ashes at the distance of a bowshot." Serious studies of the sun and its potential began in the seventeenth century—when Galileo and Lavoisier utilized the sun in their researches. By 1700 diamonds had been melted and by the early 1800s heat engines were operating with energy supplied by the sun. In the early twentieth century solar energy was used to power water distillation plants in Chile and irrigation pumps in Egypt. By 1930 Robert Goddard had applied for five patents on various solar devices to be used on his project of sending a rocket to the moon. Most of these projects, however, were considered curiosities because they were ahead of their time.

By the 1920s and 1930s a practical use was being made of the sun's energy in California—solar service-hot-water heaters. Devices similar to solar water heaters were also used to heat buildings in the United States. The first building to be practically heated with converted solar service-hot-water heaters was constructed at the Massachusetts Institute of Technology in 1938. Some twenty other experimental building heating projects—an average of about one per year—were completed between 1938 and 1960. Performance data, recorded for a number of these projects, are used in solar-heated building design today.

The space age gave solar energy its first post-war boost. The success of solar cells in powering NASA service modules in terrestrial orbit and in lunar excursions led some engineers to propose other uses for solar energy in the space program. Solar-thermal power plants were designed and prototypes built; other exotic devices were proposed—some even proposed use of the small pressure exerted by sunlight to propel interplanetary vehicles. Although none of these projects was directed at providing large scale, terrestrial solar conversion economically, some of this NASA technology has found use in practical, applied projects for economical solar heating and cooling in the 1970s.

The combined effects of impending shortages of fossil fuels and the government response of initiating federal funding for solar research and development have resulted in the first significant efforts in the United States directed toward practical use of the sun for building heating and cooling. The first federal funding of solar research in the 1970s amounted to 1.7 million dollars in fiscal year 1972. By 1975 the figure was $50 million; by 1976 it was projected to be $200 million. Although these are small amounts in the federal energy research and development budget, they have provided the impetus required to move solar energy from the hobbyist/inventor stage to the first level of practical, widespread use.

CONVERSION OF SOLAR
ENERGY TO HEAT

Solar energy is transmitted from the sun through space to earth by electromagnetic radiation. It must be converted to heat before it can be used in practical heating or cooling systems. Since solar energy is relatively dilute when it reaches the earth, the size of a system used to convert it to heat on a practical scale must be relatively large. Solar-energy collectors, the devices used to convert the sun's radiation to heat, usually consist of a surface that efficiently absorbs radiation and converts this incident flux to heat which raises the temperature of the absorbing material. A part of this energy is then removed from the absorbing surface by means of a heat-transfer fluid that may be either liquid or gaseous.

Since 1900 at least 20 solar-collector designs have been demonstrated as functional. These designs are separated into two generic classes: concentrating and nonconcentrating. *Nonconcentrating, or flat plate, collectors* intercept solar radiation on a metal or glass absorber plate from which heat is transferred and used in the required thermal application. Since the absorber plate has a temperature greater than its environment, unrecoverable heat losses occur from the entire absorbing surface of the collector to the environment. Consequently, 100 percent collector efficiency cannot be realized in practice.

Concentrating collectors attempt to reduce heat loss by using an absorber area smaller than the area that intercepts the suns rays, called the aperture area. This performance improvement is accomplished by reflecting the sun's rays from the large aperture area to the small absorbing area by shaped mirrors or other reflecting surfaces. Since only the direct or collimated portion of solar radiation is amenable to effective concentration, most concentrators must move to track the sun and cannot collect as much diffuse radiation as flat plate collectors. In Chap. 3, where five collector designs are analyzed in detail, it will be shown that these two negative aspects of concentrators can be offset by their greater efficiency, particularly in sunny regions of the world.

Since solar energy is available only during daylight hours and during periods when the sun is not significantly obscured by clouds, a means of providing heat on a continuous basis from this intermittent source is required. In nearly all applications a form of thermal energy storage is used for this purpose. Three practical storage media have shown acceptable performance. Sensible heat storage (storage in a material by virtue of a temperature rise) in pebble or rock beds and water have been the two most widely used means of storage; the first is used with air-cooled solar-conversion devices, the second, with liquid-cooled devices. The third storage method is by the use of latent heat of materials undergoing a phase change. Such materials are capable of larger energy-storage per unit volume than are sensible heat storage materials but to date have shown unacceptable reliability in repeated freeze-thaw cycles. Since the heating and cooling loads on a building do not occur in phase with useful solar collection, all practical, solar building thermal systems require storage. The size and cost of storage are considered in detail in Chaps. 4 and 5.

EXAMPLE OF AN ECONOMIC CASE STUDY: COST OF SOLAR SERVICE-HOT-WATER HEATING

Service-hot-water heating in dwellings accounts for approximately 3 percent of the total national energy consumption. Solar hot-water heaters are commercially available today that could be employed without conflicting with existing plumbing codes. In fact, solar heaters could supply one-half or more of the current, total national energy requirements for hot-water heating.

Although this quantity is only a small part (1.5 percent) of the national energy budget, the cost figures discussed below will indicate why solar service-water heating could—and should—have immediate economic application. To assess the viability of solar hot-water heating, the capital investment required to supply an increment of 1 percent of the national energy supply by solar and by two conventional means will be compared. This 1 percent increment represents one-third of the energy presently used for water heating.

The solar method

The cost to the consumer of installing a solar device can be compared with the total savings the consumer would realize in reduced expenditure for fuel or electrical power. This approach will be taken in a subsequent part of this book, but there is another way to consider the cost of solar energy: i.e., to compare the net capital outlay required on a national scale to increase the national energy supply by 1 percent.

The cost of hot-water solar collectors with a 1-m^2 surface and the capacity to retain an average of 3.5 kw-hr per day of solar energy in the form of low-temperature heat would be approximately $19/$m^2$. This figure corresponds to a capital cost of 1.5 cents to collect 1 kw-hr/yr. A typical dwelling in the United States uses approximately 10,000 kw-hr/yr for hot-water heating. A collector to provide one-half this energy annually would have an area of 4 m^2 and would cost approximately $76. The cost of implementing solar, service-hot-water heating in 60 million dwellings to effect a 1.5 percent reduction in national demand upon conventional fuels would be approximately $4.5 billion. Thus, it is reasonable to use *$3 billion as the estimated capital required to increase the national energy supply by 1 percent, through implementation of solar energy* to provide low-temperature heat.

Conventional methods

As an alternative to solar heating, it would be reasonable to consider the possibility of expanding the national capacity to supply natural gas by 1 percent. Because domestic supplies are unavailable, however, an expansion of the gas supply would entail the importation of liquefied natural gas (LNG). Current estimates for the capital cost of LNG plants indicate that a plant capable of delivering 100 million ft^3 of gas per day will cost between $200 and $300 million to build. If the heating value of the gas is assumed to be approximately 1,000 Btu/ft^3, the capital costs of producing LNG are between $160 and $240/kW. Assuming that such a plant can operate for 8,000 hr/yr, the capital cost of liquefication may be estimated as 2 to 3 cents to increase the national energy supply by the LNG equivalent of 1 kw-hr/yr. If the costs of transportation and storage systems are included, the total cost of delivering 1 kw-hr/yr through liquefication would be approximately 5 cents. Thus, *increasing the present capacity of the national energy supply by 1 percent through gas liquefication would cost approximately $10 billion.*

Another reasonable method of increasing the national energy supply would be to utilize electricity for hot-water heating. A modern power plant costs between $200 and $300/kW to build, and an additional $100/kW capital requirement is necessary for distribution systems. Thus, the capital cost of increasing electrical power capacity may be taken as approximately $400/kW. Assuming the plant averages 65 percent of peak operating capacity, the capital cost of increasing the capacity of the national electric supply by 1 kw-hr/yr is found to be approximately 7 cents. Therefore, *to increase the capacity of the national energy supply by 1 percent through expansion of electrical power would cost approximately $14 billion.*

Cost comparison

One reason the projected 1 percent expansion of energy supply by use of solar energy for low-temperature heat costs less than the corresponding expansion of capacity to produce electric power is that the latter requires energy in a high-quality, concentrated form, i.e., fuels for high-temperature combustion. High-temperature energy is readily convertible to work but is generally expensive to generate; on the other hand low-temperature heat, which is not readily convertible to work, is usually relatively inexpensive to generate. In fact, low-quality energy is often discarded in power plants.

The difference in capital requirement between solar and conventional methods for increasing the national energy supply by 1 percent ranges from $7 to $11 billion. Even though some details of these estimates may be debatable—and they should be considered more as orders of magnitude than as definitive figures—they do provide a basis for judging whether local use of solar energy for low-temperature heat offers a viable alternative

> energy supply source in the context of national energy planning. Based on the data and the calculations in the preceding subsections, it is reasonable to conclude that solar energy for hot-water supply is economically viable.

SUMMARY

Solar energy currently represents the only inexhaustible energy resource that could be used economically to supply man's increasing energy demands—demands increasingly more difficult to meet by means and sources used previously. Although comparatively dilute, the energy reaching the earth from the sun far exceeds the energy requirements of all the world's population. The first economical, large-scale thermal application of solar energy in this century will be heating and cooling of residential and commercial buildings. It is the purpose of this book to provide the architect, engineer, and builder with the tools required to design and construct properly engineered heating and cooling systems.

TERMINOLOGY

Terms and basic concepts used throughout the book are defined below:

Insolation: Radiation from the sun received by a surface.

Infrared: Portion of the electromagnetic spectrum of radiant energy that occupies the band of wavelengths longer than those in the adjacent visible spectrum band; associated with thermal radiation at common terrestrial temperatures.

Spectrum: Format of wavelengths which encompass all types of electromagnetic radiation, including thermal radiation.

Solar constant: Average intensity of solar radiation in near-earth space; its value is 429 Btu/(hr)(ft^2) ($\pm$ 1.5%).

Laws of thermodynamics
  First law: In nonrelativistic processes energy is neither created nor destroyed; it is only transformed from one form to another.
  Second law: When the free exchange of heat takes place between two bodies, it is always transferred from the warmer to the cooler body.

Heat: In thermodynamics heat is defined as energy in transition due to a temperature difference. In common language heat also denotes energy stored in a body by virtue of its temperature.

Other special terms and concepts are defined as they are used in the text.

REFERENCES

1. Berg, C. A.: Energy Conservation through Effective Utilization, *Science,* vol. 181, p. 128, 1973.
2. Daniels, F.: "Direct Use of the Sun's Energy," Yale University Press, New Haven, 1964.
3. Felix, F.: U.S. Using Electricity More Efficiently, *Electrical World,* p. 59, Dec. 1, 1974.
4. Krenz, J. H.: "Energy Conversion and Utilization," Allyn and Bacon, Boston, in press.
5. Löf, G. O. G.: World Distribution of Solar Radiation, *University of Wisconsin, Dept.* 21, Madison, 1966.
6. Moyers, J. C.: *The Value of Thermal Insulation in Residential Construction: Economics and Conservation of Energy,* Oak Ridge Nat'l. Lab., Oak Ridge, Tenn. 1971.
7. NSF/NASA Solar Energy Panel: An Assessment of Solar Energy as a National Energy Source, University of Maryland, College Park, 1972.
8. *Proc. U.N. Conf. New Sources of Energy (Rome, 1961).* United Nations, New York, 1964.
9. *Proc. World Symp. Appl. Solar Energy (Phoenix, 1955),* Stanford Research Inst., Menlo Park, Calif., 1956.
10. Stanford Research Inst.: *Patterns of Energy Consumption in the United States,* U.S. Gov't. Printing Off., Washington, 1972.
11. *Trans. Conf. Use of Solar Energy,* University of Arizona Press, Tucson, 1958.
12. Zarem, A. M., and D. D. Erway: "Introduction to the Utilization of Solar Energy," McGraw-Hill Book Company, New York, 1963.

CHAPTER 2

Fundamentals of Heat Transfer

The sun is satisfied with days.

ROBERT FROST

To estimate the size, the efficiency, and the cost of equipment necessary to transfer a specified amount of heat in a given time, a heat-transfer analysis must be made. The dimensions of a solar collector, a heat exchanger, or a refrigerator depend not so much on the amount of heat to be transmitted but rather on the *rate* at which heat is to be transferred under given external conditions. From an engineering viewpoint the determination of the rate of heat transfer at a specified temperature difference is the key problem in sizing a solar collector to provide a given temperature in a home or building.

PRINCIPLES OF CONDUCTION, RADIATION, AND CONVECTION

Heat is transmitted by three distinct modes:

1. *Conduction,* or heat transfer by molecular scale vibration and rotation
2. *Radiation,* or heat transfer by photons
3. *Convection,* or heat transfer by large-scale fluid motion

In nature, heat usually flows not by one but by several mechanisms acting simultaneously; however, the three modes of heat transfer are initially analyzed separately in the discussion that follows.

Conduction

Conduction is a process by which heat flows from a region of higher temperature to a region of lower temperature within a solid, liquid, or gaseous medium or between different media in direct physical contact. Of the three processes conduction is the only mechanism by which heat can flow in opaque solids. The basic relations for heat transfer by conduction state that the rate of heat flow (flux) by conduction q_k in a material is equal to the product of the following three quantities:

1. k, the thermal conductivity of the material, in Btu per hour per square foot per degree Fahrenheit per foot.
2. A, the area of the medium e.g., a building wall, through which heat flows by conduction, in square feet.
3. dT/dx, the temperature gradient through the medium, in degree Fahrenheit per foot.

In equation form we have

$$q_k = -kA \frac{dT}{dx} \tag{2.1}$$

Equation (2.1) can be integrated for heat flow through a wall of thickness W having a temperature T_1 over one surface and T_2 over the other to yield the relation

$$q_k = \frac{kA}{W}(T_1 - T_2) \tag{2.2}$$

In Eq. (2.2) the temperature difference between the higher temperature T_1 and the lower temperature T_2 is the driving potential which causes the flow of heat. The value W/Ak is equivalent to a thermal "resistance" R_k the wall offers to the flow heat by conduction, and therefore

$$q_k = \frac{T_1 - T_2}{R_k} \tag{2.3}$$

The subscript k indicates the transfer mechanism is conduction. The concept of resistance will be found helpful in the analysis of thermal systems where several modes of heat transfer occur simultaneously.

Radiation

Radiation is a process by which heat flows from a body at higher temperature to a body at a lower temperature when the bodies are separated in space or even when a vacuum exists between them. The term radiation is generally applied to all kinds of electromagnetic wave phenomena, but in heat transfer only those phenomena that are the result of temperature and can transport energy through a medium such as air or space are of interest. The energy transmitted in this fashion is called *radiant heat*. Radiation is the mode of heat transfer by which the sun transfers energy to the earth.

The quantity of energy leaving a surface as radiant heat depends on the absolute temperature and the nature of the surface. The absolute temperature equals the temperature in degrees Fahrenheit plus 460, the temperature difference between absolute zero ($-459.69°F$) and $0°F$. A perfect radiator, a so-called *black body*, emits radiant energy from its surface at a rate q_r given by

$$q_r = A\sigma T^4 \tag{2.4}$$

The heat-flow rate q_r will be in Btu per hour if A is the surface area in square feet, T is the surface temperature in degrees Rankine, and σ is a dimensional constant with a value of 0.1713×10^{-8} Btu/(hr) $(ft^2)(°R^4)$.

Real bodies do not meet the specifications of an ideal radiator but emit radiation at a lower rate than do black bodies. The ratio of the radiation emission of a real body to the radiation emission of a black body at the same temperature is called the *emittance*. Thus a real body emits radiation at a rate

$$q_r = \bar{\epsilon} A\sigma T^4 \tag{2.5}$$

where $\bar{\epsilon}$ is the average emittance of the surface.

If radiation exchange takes place between two gray bodies, e.g., two large, parallel plates with areas A_1 and A_2, the rate of heat transfer between them can be written in the form

$$q_{r,net} = A_1 \mathcal{F}_{1-2}\sigma (T_1^4 - T_2^4) \tag{2.6}$$

where $\mathcal{F}_{1-2}$ is a quantity which depends only on surface properties, orientation, and shape.

For many purposes it is convenient to write the equation for radiation heat transfer (Eq. (2.6)) in the same form as that for conduction heat transfer (Eq. (2.3)):

$$q_{r,net} = \frac{T_1 - T_2}{R_r} \tag{2.7}$$

where the resistance to radiation heat transfer R_r is

$$R_r = \frac{1}{A_1 \mathcal{F}_{1-2}\sigma (T_1^2 + T_2^2)(T_1 + T_2)} \tag{2.8}$$

For example, for infrared radiation between two parallel, large, flat plates of the same area with emittances $\bar{\epsilon}_{ir1}$ and $\bar{\epsilon}_{ir2}$,

$$\mathcal{F}_{1-2} = \frac{1}{[(1/\bar{\epsilon}_{ir1}) + (1/\bar{\epsilon}_{ir2})] - 1}$$

Other expressions and the method of calculation of $\mathcal{F}_{1-2}$ are given in heat-transfer texts.[2]

Convection

Convection is a process that transfers heat from one region to another by motion of a fluid. The rate of heat transferred by convection q_c between a surface and a fluid can be calculated from the relation

$$q_c = h_c A (T_s - T_f) \qquad (2.9)$$

where q_c = rate of heat flow by convection, Btu/(hr)
      A = base area of heat transfer by convection, ft^2
      T_s = surface temperature, °F
      T_f = fluid temperature, °F
      h_c = convection heat-transfer coefficient, Btu/(hr)(ft^2)(°F)
The thermal resistance to convective heat transfer R_c is given by

$$R_c = \frac{1}{h_c A} \qquad (2.10)$$

Table 2.1 lists orders of magnitudes of thermal conductivities, and Table 2.2 gives convective heat-transfer coefficients. Equations for calculating convective heat-transfer coefficients are treated in detail in heat-transfer texts; App. C presents a summary of useful equations for calculating h_c; Table A.12 contains conductivities.

TABLE 2.1   Order of Magnitude of Thermal Conductivity k

	Thermal conductivity k	
Substance	Btu/(hr) (ft) (°F)	W/(m) (°K)
Gases at atmospheric pressure	0.004–0.10	0.0069–0.17
Insulating materials	0.02–0.12	0.034–0.21
Nonmetallic liquids	0.05–0.40	0.086–0.69
Nonmetallic solids		
(e.g., brick, stone, cement)	0.02–1.5	0.034–2.6
Liquid metals	5.0–45	8.6–76
Alloys	8.0–70	14–120
Pure metals	30–240	52–410

TABLE 2.2 Order of Magnitude of Convective
Heat-transfer Coefficients h_c

Substance	Heat-transfer coefficient h_c	
	Btu/(hr) (ft²) (°F)	W/(m²) (°K)
Air, free convection	1–5	6–30
Superheated steam or air, forced convection	5–50	30–300
Oil, forced convection	10–300	60–1,800
Water, forced convection	50–2,000	300–12,000
Water, boiling	500–10,000	3,000–60,000
Steam, condensing	1,000–20,000	6,000–120,000

COMBINED HEAT-TRANSFER MECHANISMS

In the preceding section the three mechanisms of heat transfer have been considered separately. In most engineering situations, however, heat is transferred by two or three of the mechanisms acting simultaneously. For example, the surface of a solar-energy collector loses heat simultaneously by convection as well as by radiation to the environment. For heat loss through a combination of radiation and convection, the total rate of heat transfer q from the surface of the collector is given by the equation

$$q = q_r + q_c = \frac{T_{coll} - T_{out}}{R_{cr}} \qquad (2.11)$$

where R_{cr} = combined resistance for the two mechanisms, convection and radiation acting in parallel
  T_{coll} = average collector-surface temperature
  T_{out} = outside air temperature

As another illustration of combined heat transfer, consider the heat loss through the wall of a building in winter. Suppose the interior temperature of the building is kept at a temperature T_{in} and heat is transferred to the interior surface of the building by convection. In steady state the heat transferred by convection to the interior wall is then conducted through the wall to its exterior surface and from the exterior surface, by the combined action of conduction and radiation, to the environment. The equations for the heat flow can be written as follows:

$$q_c = h_{c,in} A(T_{in} - T_{s,in}) = \frac{T_{in} - T_{s,in}}{R_1} \qquad (2.12a)$$

$$q_k = \frac{kA}{W}(T_{s,\,in} - T_{s,\,out}) = \frac{T_{s,\,in} - T_{s,\,out}}{R_2} \qquad (2.12b)$$

$$q = q_c + q_r = \frac{T_{s,\,out} - T_{out}}{R_3} \qquad (2.12c)$$

where

T_{in} = inside air temperature
T_{out} = outside air temperature
$T_{s,\,in}$ = interior-wall-surface temperature
$T_{s,out}$ = exterior-wall-surface temperature
R_1, R_2, R_3 = three resistances acting in series

$$R_3 = \frac{1}{(1/R_c) + (1/R_r)} \qquad (2.13)$$

as in the previous case.

Often in practice only the temperatures inside and outside the building are known. The intermediate temperatures—interior and exterior wall temperatures—can be eliminated by algebraic addition of Eq. (2.12a), (2.12b), and (2.12c). This operation yields

TABLE 2.3 U-factor Values for Typical Residential Wall Construction*

Wall No.	Construction component	Component thickness, in.	U value Btu/(hr) (ft^2) ($^\circ$F)
1	Face brick	4	
	Block	8	
	Firring space	¾	0.09
	Urethane board	½	
	Plasterboard	½	
2	Stucco	½–1	
	Block	8–12	
	Firring space	¾	0.13
	Urethane board	½	
	Plasterboard	½	
3	Wood siding	½–¾	
	Building paper		
	Sheathing	⅝	0.13
	Stud/space (with insulation)	3⅝	
	Plasterboard	½	
4	Glass, double-plate	¼	
	Insulated window (separated by ½-in. air space)		0.60

*It is assumed the windows are double-glazed, and the glass-to-wall ratio is 20%.

$$q = \frac{T_{in} - T_{out}}{R_1 + R_2 + R_3} \qquad (2.14)$$

where the thermal resistances of the three series-connected heat-flow steps in the systems are defined above.

In Eq. (2.14) the rate of heat flow q is expressed only in terms of an over-all temperature potential and the resistances of the individual sections in the heat-flow path. These values can be combined into what is generally called an *overall transmittance*, or *overall heat-transfer coefficient U*, as shown below:

$$UA = \frac{1}{R_1 + R_2 + R_3} \qquad (2.15)$$

Values of U factors for typical, residential wall construction are shown in Table 2.3.

EXAMPLE OF HEAT TRANSFER FOR TWO TYPES OF WALL CONSTRUCTION

Compare the rate of heat transfer through the wall of a building constructed by two alternate methods. In method 1 the wall consists of $\frac{1}{4}$-in. plasterboard on the interior with 2 x 4 in. wood supports spaced 12 in. center-to-center and an external layer of shingles $\frac{1}{4}$ in. thick. In the alternate method of construction, method 2, the wall consists of the same external and interior material but support consists of 2 x 6 in. wooden studs spaced 16 in. apart. In both methods 1 and 2 transfer coefficient by convection on the interior is 1.0 Btu/(hr)(ft^2), and on the exterior the combined heat transfer coefficient is 2.5 Btu/(hr)(ft^2)($^\circ$F).

SOLUTION

The thermal circuits for both types of wall construction are identical, consisting of 2 series resistance elements at the interior and exterior, with two parallel thermal resistances between them, as shown in Fig. 2.1.

The definitions for the example solution are as follows:

A = Unit wall surface area
b_1 = Air space thickness
b_2 = Wood stud thickness
$h_{c,out}$ = Thermal convection coefficient, outside wall
$h_{c,in}$ = Thermal convection coefficient, inside wall
k_a = Thermal conductance, air space

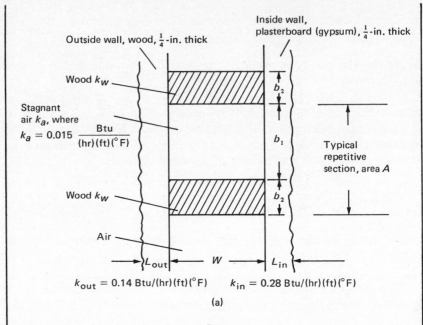

$k_{out} = 0.14$ Btu/(hr)(ft)(°F) $k_{in} = 0.28$ Btu/(hr)(ft)(°F)

(a)

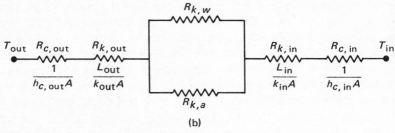

(b)

FIG. 2.1 (a) Schematic of wall; (b) thermal circuit.

k_{in} = Thermal conductance, inside wall
k_{out} = Thermal conductance, outside wall
k_w = Thermal conductance, wood studs
L_{in} = Inside wall thickness
L_{out} = Outside wall thickness
$R_{c,in}$ = Thermal resistance to convection, inside wall
$R_{c,out}$ = Thermal resistance to convection, outside wall
$R_{k,a}$ = Thermal resistance to conduction, air space
$R_{k,in}$ = Thermal resistance to conduction, inside wall
$R_{k,out}$ = Thermal resistance to conduction, outside wall
$R_{k,w}$ = Thermal resistance to conduction, wood studs
$R_{k,wa}$ = Combined resistance to conduction, wood studs plus air space
T_{in} = Building inside temperature, °F

T_{out} = Outside ambient temperature, °F

W = Distance between inside and outside walls

If it is assumed the air between the wooden supports is stagnant so that heat is primarily transferred by conduction, the effective resistance $R_{k,wa}$ of the wood-support resistance and the air resistance for a wall section $(b_1 + b_2)$ wide and 1 ft high is given by the equation

$$R_{k,\text{wa}} = \frac{R_{k,w} \, R_{k,a}}{R_{k,w} + R_{k,a}}$$

Method 1: 2 x 4 in. supports, 12 in. apart

$$R_{k,w} = \frac{W}{k_w b_2} = \frac{4}{0.12 \times 2} = 16.7 \ (\text{hr})(\text{ft}^2)(°\text{F})/\text{Btu}$$

$$R_{k,a} = \frac{W}{k_a b_1} = \frac{4}{0.015 \times 10} = 26.7 \ (\text{hr})(\text{ft}^2)(°\text{F})/\text{Btu}$$

$$R_{k,\text{wa}} = \frac{16.7 \times 26.7}{16.7 + 26.7} \times 1.0 = 10.3 \ (\text{hr})(\text{ft}^2)(°\text{F})/\text{Btu}$$

Method 2: 2 x 6 in. supports, 16 in. apart

The section area A is $1 \times (b_1 + b_2) = 1.33 \ \text{ft}^2$.

$$R_{k,w} = \frac{W}{k_w b_2} = \frac{6}{0.12 \times 2} = 25 \ (\text{hr})(\text{ft}^2)(°\text{F})/\text{Btu}$$

$$R_{k,a} = \frac{W}{k_a b_1} = \frac{6}{0.015 \times 14} = 28.6 \ (\text{hr})(\text{ft}^2)(°\text{F})/\text{Btu}$$

$$R_{k,\text{wa}} = \frac{25 \times 28.6}{25 + 28.6} \times 1.33 = 17.8 \ (\text{hr})(\text{ft}^2)(°\text{F})/\text{Btu}$$

For both methods

$$R_{c,\text{out}} + R_{k,\text{out}} = \frac{1}{2.5} + \frac{0.25/12}{0.14} = 0.4 + 0.15$$

$$= 0.55 \ (\text{hr})(\text{ft}^2)(°\text{F})/\text{Btu}$$

$$R_{c,\text{in}} + R_{k,\text{in}} = \frac{1}{1.0} + \frac{0.25/12}{0.28} = 1.0 + 0.074$$

$$= 1.07 \ (\text{hr})(\text{ft}^2)(°\text{F})/\text{Btu}$$

Thus, in method 1 the total thermal resistance is

$$0.55 + 10.3 + 1.07 = 11.92 \ (hr)(ft^2)(°F)/Btu$$

while in method 2 the total thermal resistance is

$$0.55 + 17.8 + 1.07 = 19.42 \ (hr)(ft^2)(°F)/Btu.$$

The corresponding U values for methods 1 and 2 are, respectively,

$$U_1 = \frac{1}{11.92} = 0.084 \ Btu/(hr)(ft^2)(°F)$$

$$U_2 = \frac{1}{19.42} = 0.051 \ Btu/(hr)(ft^2)(°F)$$

which means that construction method 2 reduces the heat loss by 40 percent.

CONCLUSION

This analysis could be refined by considering the free convection between the interior and exterior walls or the addition of insulation, but the essential conclusion would remain the same: a small change in some building codes could reduce the heat loss substantially by reducing the U value without reducing structural strength.

EXAMPLE OF ECONOMIC ANALYSIS AND CALCULATION OF HEAT LOSS

In this example a single-plate, double-strength window is compared to a double-plate window to determine the

1. Heat losses per square foot through both types of window
2. Cost per square foot of heat losses for both types of window
3. Investment cost per year for both types of window
4. Comparison of over-all costs for both types of window

Computation of heat losses

Single-plate, double-strength window. A window consisting of a single plate of glass $\frac{1}{8}$ in. thick is shown with its equivalent thermal circuit in Fig. 2.2(a), where

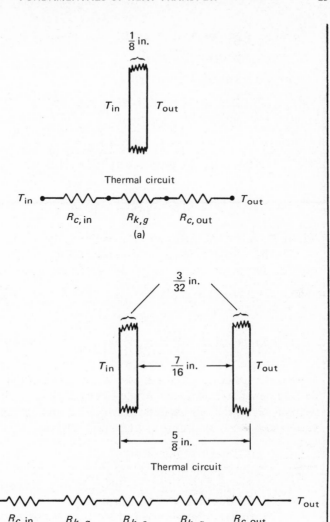

FIG. 2.2 Schematic of window and thermal circuit for (a) single- and (b) double-plate windows.

T_{in} = inside air temperature
T_{out} = outside air temperature
$R_{c,in}$ = thermal resistance to heat transfer by convection on the inside surface of the glass pane
$R_{c,out}$ = thermal resistance to heat transfer by convection on the outside surface of the glass pane

and
$$h_{c,in} = 1 \text{ Btu/(hr)(ft}^2)(^\circ F)$$
$$h_{c,out} = 4 \text{ Btu/(hr)(ft}^2)(^\circ F)$$

On the basis of 1 ft^2 of window

$$R_{c,in} = \frac{1}{h_{c,in}A} = \frac{1}{[1 \text{ Btu/(hr)(ft}^2)(^\circ F)] \text{ ft}^2} = 1 \text{ (hr)}(^\circ F)/\text{Btu}$$

$$R_{k,g} = \frac{w}{k_g A} = \frac{1/8 \times (1 \text{ ft/12 in.})}{[0.45 \text{ Btu/(hr)(ft}^2)(^\circ F)] \text{ ft}^2} = 0.023 \text{ (hr)}(^\circ F)/\text{Btu}$$

$$R_{c,out} = \frac{1}{h_{c,out}A} = \frac{1}{[4 \text{ Btu/(hr)(ft}^2)(^\circ F)] \text{ ft}^2} = 0.25 \text{ (hr)}(^\circ F)/\text{Btu}$$

where k_g is the thermal conductivity of glass.

If the average inside air temperature T_{in} is 68°F and the outside air temperature T_{out} is 43°F, the heat loss q is

$$q = \frac{T_{in} - T_{out}}{R_{c,in} + R_{k,g} + R_{c,out}}$$

$$= \frac{25^\circ F}{(1 + 0.023 + 0.25)(\text{hr})(^\circ F)/\text{Btu}} = 19.6 \text{ Btu/(hr)(ft}^2)$$

Double-plate (insulated) window. A double-plate window consisting of two plates of single-strength glass, each 3/32-in. thick with a 7/16-in. air space between them, is shown with its equivalent thermal circuit in Fig. 2.2(*b*), where

T_{in} = inside air temperature
T_{out} = outside air temperature
$R_{c,in}$ = thermal resistance to heat transfer by convection on the inside surface of the glass pane
$R_{c,out}$ = thermal resistance to heat transfer by convection on the outside surface of the glass pane
$R_{k,g}$ = thermal resistance to heat transfer by convection on the inside surface of the glass pane
$R_{k,a}$ = thermal resistance to heat transfer by conduction in an air gap
$R_{c,in}$ = 1 (hr)(°F)/Btu

$$R_{k,g} = \frac{w}{k_g A} = \frac{3/32 \times (1 \text{ ft/12 in.})}{0.45 \text{ Btu/(hr)(ft)}(^\circ F)} = 0.017 \text{ (hr)}(^\circ F)/\text{Btu}$$

$$R_{k,a} = \frac{w}{k_a A} = \frac{7/16 \times (1 \text{ ft/12 in.})}{0.014 \text{ Btu/(hr)(ft)}(^\circ F)} = 2.6 \text{ (hr)}(^\circ F)/\text{Btu}$$

$R_{c,out}$ = 0.25 (hr)(°F)/Btu

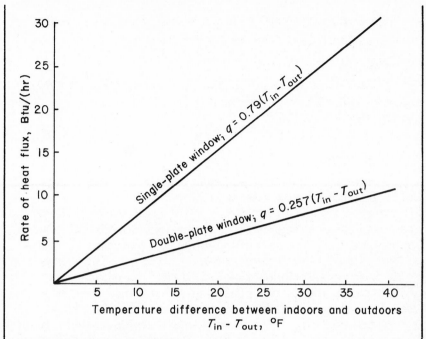

FIG. 2.3 Effect of various temperature differences $T_{in} - T_{out}$ for single- and double-plate windows on rate of heat flux q.

where k_a is the thermal conductivity of heat through the air gap, and the heat loss q is calculated as follows:

$$q = \frac{T_{in} - T_{out}}{R_{c,in} + 2R_{k,g} + R_{k,a} + R_{c,out}}$$

$$= \frac{25°F}{(1 + 0.034 + 2.6 + 0.25)(hr)(°F)/Btu}$$

$$= 6.4 \; Btu/(hr)(ft^2)$$

The effect of temperature difference on heat loss for the two types of window is shown in Fig. 2.3.

Cost of heat losses (natural-gas heating)

Single-plate window. The energy content of natural gas for Boulder, Colo., is 840 Btu/ft³. If the heating unit is 80 percent efficient, the heat delivered by 1 ft³ of gas is 0.8 × 840 Btu/ft³ or 672 Btu/ft³. If the cost is assumed to be \$0.072/100 ft³, the cost of net energy is \$1.07 per million Btu.

Since the heat loss q for the single-plate window is 19.6 Btu/(hr), the cost of this loss per hour per square foot of area is 19.6 Btu per hr × $1.07 per million Btu, or ($2.10 × 10⁻⁵)/(hr)(ft²).

Double-plate window. If the heat loss q for a double-plate window is 6.4 Btu/(hr)(ft²), then the cost of this loss per hour per square-foot area is ($6.85 × 10⁻⁶)/(hr)(ft²).

Cost analysis

Investment cost for a single-plate window. The average cost of single-plate window glass is $1.25/ft², based on prices of 3 × 4 ft plates. If the life of a plate of glass is assumed to be 20 yr, the cost (at 8 percent interest) is approximately $0.125/(ft²)(yr), i.e., approximately 10 percent of the initial cost per year.

Investment cost for a double-plate window. The average cost of a double-plate window glass is $3.12/ft², based on prices of 3 × 4 ft plates. If the life of a plate of glass is assumed to be 20 yr, the cost (at 8 percent interest) is approximately $0.312/(ft²)(yr).

Total cost (C_{wi1}) *for a single-plate window.* If the hourly cost of heat loss is ($2.10 × 10⁻⁵)/(hr)(ft²), then the annual cost of heat loss C_{hl} is

$$C_{hl} = (\$2.10 \times 10^{-5})/(hr)(ft^2) \times \frac{24 \text{ hr}}{\text{day}}$$

$$\times \frac{30 \text{ days}}{\text{month}} \times \frac{8 \text{ heating months}}{\text{year}} = \$0.12/(ft^2)(yr)$$

and the total cost C_{wi1} is

$$C_{wi1} = (\$0.12 + \$0.125)/(ft^2)(yr) = \$0.245/(ft^2)(yr)$$

Total cost (C_{wi2}) *for a double-plate window.* If the hourly cost of heat loss is ($6.85 × 10⁻⁶)/(hr)(ft²), then the annual cost of heat loss C_{hl} is

$$C_{hl} = (\$6.85 \times 10^{-6})/(hr)(ft^2) \times \frac{24 \text{ hr}}{\text{day}}$$

$$\times \frac{30 \text{ days}}{\text{month}} \times \frac{8 \text{ heating months}}{\text{year}} = \$0.04/(ft^2)(yr)$$

and the total cost C_{wi2} is

$C_{wi2} = (\$0.04 + \$0.312)/(ft^2)(yr) = \$0.352/(ft^2)(yr)$

The increased capital cost of double-plate compared to single-plate window construction is $\$0.312 - \$0.125 = \$0.187/(ft^2)(yr)$. The benefit, i.e., the value of heat saved, is $\$0.08/(ft^2)(yr)$. If electricity must be used, the cost saving (at a unit heating cost of $6 per million Btu) increases to $\$0.48/(ft^2)(yr)$.

CONCLUSIONS

Three conclusions can be drawn from the calculations shown in this example:

1. Double-plate windows will prove economical when the price of natural gas exceeds $\$0.168/100 ft^3$.
2. If a building is heated electrically, the use of double-plate windows can reduce heating costs appreciably with no increase in energy price.
3. Energy-conservation measures can be analyzed by classic techniques of economics to determine if investment is warranted for a given energy source and projected savings through conservation.

RADIATION CALCULATIONS

To understand the methods of operation of various solar-energy collectors it is necessary to examine the nature of radiation in some detail. Radiation can be envisioned as a transport of energy by electromagnetic waves, or it can be viewed as heat transport by photons, or quanta, of energy. Neither viewpoint by itself completely describes all the phenomena observed in nature, but for analysis of radiation heat transfer, it is most convenient to consider radiation transfer in terms of its wavelength, which is designated by the Greek symbol lambda. Wavelength (λ) is related to the propagation velocity of light C and the frequency ν by the relation

$$\lambda = \frac{C}{\nu}$$

The unit of wavelength used in engineering practice is the micron μ (10^{-6} m), which is equal to approximately 4×10^{-5} in.

The Spectral Nature of Radiation

Thermal radiation is defined as the radiant energy emitted by a medium by virtue of its temperature. Thus, the emission of thermal radiation is governed by the temperature of the emitting body. The wavelengths encompassed by thermal radiation range approximately between 0.1 and 100 μ. The sun has an effective surface temperature of about 10,000°R and emits most of its energy between 0.1 and 3 μ, while a building wall that emits radiation at a temperature of about 550°R, that is, approximately between 60 and 100°F, emits most of the radiation at wavelengths above 3 μ, in the infrared spectral range.

To understand how a solar collector operates, it is necessary to consider the radiant energy emission of a black body, per unit time

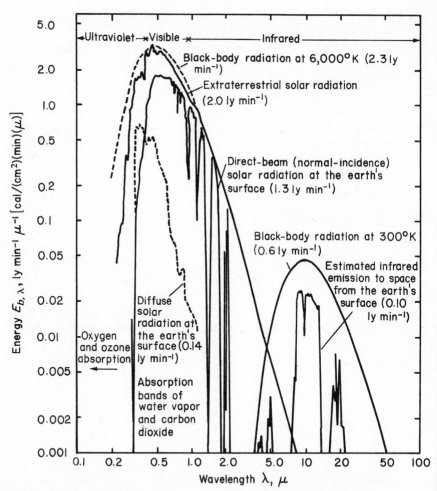

FIG. 2.4 Electromagnetic spectra of solar and terrestrial radiation. The black-body radiation at 6000 °K is reduced by the square of the ratio of the sun's radius to the average distance between the sun and the earth to give the flux that would be incident on the top of the atmosphere.

per unit area at a wavelength λ. This quantity $E_{b,\lambda}$ is called the *spectral emissive power*. The adjective "spectral" denotes that the radiation depends on the wavelength spectrum. If the total radiation spectrum is divided into small wavelength bands of width $\Delta\lambda$, the quantity $E_{b,\lambda}\,\Delta\lambda$ denotes the amount of radiation emitted in the waveband $\Delta\lambda$. Figure 2.4 shows the variation of the spectral emissive power at a given temperature as a function of λ for black bodies at solar and building temperatures.

For engineering purposes it is usually easiest to calculate the amount of radiation emitted in a given wavelength from a tabulation such as that shown in Table 2.4. The first column in this table gives the product of λ and T, the wavelength in microns and the absolute temperature in degrees Rankine. The second column gives the spectral emissive power $E_{b,\lambda}$, divided by σT^5; this ratio is useful in bandwidth calculations as shown in App. B. The third column gives the amount of radiation in the wavelength interval between zero and the value λT divided by σT^4 $(= \int_0^\infty E_{b,\lambda}d\lambda)$ to make the quantity dimensionless. The use of Table 2.4 is illustrated in the following example.

EXAMPLE

Silica glass transmits 92 percent of the incident radiation in the wavelength range between 0.35 and 2.7 μ and is essentially opaque at wavelengths not in this range. Determine the percent of the total solar radiation glass will transmit; assume the sun is a black body at 10,000°R.

SOLUTION

For the wavelength range within which the glass is transparent, the lower end of the spectrum corresponds to $\lambda T = 3,500$ and the upper limit to $\lambda T = 27,000$. Table 2.4 shows that the wavelength band between 0 and 0.35 μ ($\lambda T = 3,500$) contains 6 percent of the total emissive power at 10,000°R; the band between 0 and 2.7 μ ($\lambda T = 27,000$) contains 96.9 percent. Therefore, 96.9 percent minus 6.0 percent or 90.9 percent of the total radiant energy incident on the glass from the sun is in the transparent wavelength range between 0.35 and 2.7 μ, and 90.9 × 0.92, or 83.6 percent, of this solar radiation is transmitted through the glass. Figure 2.5 shows the spectral transmittance.

EXAMPLE

Determine the percent of energy from a thermal-radiation source at 200°F (660°R) that would be transmitted through a piece of silica glass (calculations left as an exercise).

TABLE 2.4 Radiation Functions*

λT	$\dfrac{E_{b,\lambda} \times 10^3}{\sigma T^5}$	$\dfrac{E_0 - \lambda T}{\sigma T^4}$	λT	$\dfrac{E_{b,\lambda} \times 10^3}{\sigma T^5}$	$\dfrac{E_0 - \lambda T}{\sigma T^4}$	λT	$\dfrac{E_{b,\lambda} \times 10^3}{\sigma T^5}$	$\dfrac{E_0 - \lambda T}{\sigma T^4}$
1,000	0.0000394	0	7,200	10.089	0.4809	13,400	2.714	0.8317
1,200	0.001184	0	7,400	9.723	0.5007	13,600	2.605	0.8370
1,400	0.01194	0	7,600	9.357	0.5199	13,800	2.502	0.8421
1,600	0.0618	0.0001	7,800	8.997	0.5381	14,000	2.416	0.8470
1,800	0.2070	0.0003	8,000	8.642	0.5558	14,200	2.309	0.8517
2,000	0.5151	0.0009	8,200	8.293	0.5727	14,400	2.219	0.8563
2,200	1.0384	0.0025	8,400	7.954	0.5890	14,600	2.134	0.8606
2,400	1.791	0.0053	8,600	7.624	0.6045	14,800	2.052	0.8648
2,600	2.753	0.0098	8,800	7.304	0.6195	15,000	1.972	0.8688
2,800	3.872	0.0164	9,000	6.995	0.6337	16,000	1.633	0.8868
3,000	5.081	0.0254	9,200	6.697	0.6474	17,000	1.360	0.9017
3,200	6.312	0.0368	9,400	6.411	0.6606	18,000	1.140	0.9142
3,400	7.506	0.0506	9,600	6.136	0.6731	19,000	0.962	0.9247
3,600	8.613	0.0667	9,800	5.872	0.6851	20,000	0.817	0.9335
3,800	9.601	0.0850	10,000	5.619	0.6966	21,000	0.702	0.9411
4,000	10.450	0.1051	10,200	5.378	0.7076	22,000	0.599	0.9475
4,200	11.151	0.1267	10,400	5.146	0.7181	23,000	0.516	0.9531
4,400	11.704	0.1496	10,600	4.925	0.7282	24,000	0.448	0.9589
4,600	12.114	0.1734	10,800	4.714	0.7378	25,000	0.390	0.9621
4,800	12.392	0.1979	11,000	4.512	0.7474	26,000	0.341	0.9657
5,000	12.556	0.2229	11,200	4.320	0.7559	27,000	0.300	0.9689
5,200	12.607	0.2481	11,400	4.137	0.7643	28,000	0.265	0.9718
5,400	12.571	0.2733	11,600	3.962	0.7724	29,000	0.234	0.9742
4,600	12.458	0.2983	11,800	3.795	0.7802	30,000	0.208	0.9765
5,800	12.282	0.3230	12,000	3.637	0.7876	40,000	0.0741	0.9881
6,000	12.053	0.3474	12,200	3.485	0.7947	50,000	0.0326	0.9941
6,200	11.783	0.3712	12,400	3.341	0.8015	60,000	0.0165	0.9963
6,400	11.480	0.3945	12,600	3.203	0.8081	70,000	0.0092	0.9981
6,600	11.152	0.4171	12,800	3.071	0.8144	80,000	0.0055	0.9987
6,800	10.808	0.4391	13,000	2.947	0.8204	90,000	0.0035	0.9990
7,000	10.451	0.4604	13,200	2.827	0.8262	100,000	0.0023	0.9992
							0	1.0000

*Adapted from Kreith (1973)[2] by permission of the publishers.

Since only 0.3 percent of the total radiation is contained in the wavelength band in which the glass is transparent (0.35 to 2.7 μ), 99.7 percent of the radiation from the source is prevented from passing through the glass. This phenomenon is the principle of the greenhouse: as long as the temperature differences are small, a surface of glass will effectively trap radiation emitted at infrared temperatures. However, heat can escape through the glass by conduction and then from the glass surface by convection and by reradiation.

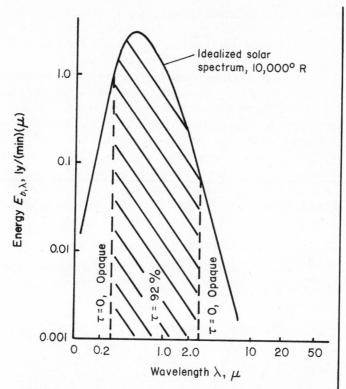

FIG. 2.5 Bandwidth transmittance values for the example problem.

Appendix B contains a detailed example of the method used to determine spectral radiation properties for opaque surfaces, and App. C contains extensive tables of radiation properties of about 100 engineering materials. If the reader requires more detailed data, Touloukian, *et al.*,[7] have prepared one of the most extensive compilations of radiation data in existence.

Absorptance, Reflectance, and Transmittance

When solar radiation impinges on the surface of a body, it is partially absorbed, partially reflected, and, if the body is transparent, partially transmitted. According to the law of conservation of energy, the relationship between the absorbed, reflected, and transmitted energy is

$$\bar{\alpha}_s + \bar{\rho}_s + \bar{\tau}_s = 1$$

where $\bar{\alpha}_s$ = solar absorptance, i.e., the fraction of the incident solar radiation absorbed by a substance

$\bar{\rho}_s$ = solar reflectance, i.e., the fraction of the incident solar radiation reflected by a surface

$\bar{\tau}_s$ = solar transmittance, i.e., the fraction of the incident solar radiation transmitted through a non-opaque substance

The relative magnitudes of $\bar{\alpha}_s$, $\bar{\rho}_s$, and $\bar{\tau}_s$ not only vary with the temperature, the surface characteristics, body geometry, and the material but also vary with wavelength. Solids and liquids are usually opaque in most engineering applications, and transmittance $\bar{\tau}_s$ for these types of matter is zero. Gases, on the other hand, reflect very little, and $\bar{\rho}_s$ can therefore be neglected in a majority of problems. The meanings of $\bar{\alpha}_s$, $\bar{\rho}_s$, and $\bar{\tau}_s$ are shown graphically in Fig. 2.6.

Specular and Diffuse Radiation

The reflection of radiation from a surface may be *specular* or *diffuse*. In *specular*, or *regular*, *radiation* the angle of incidence of

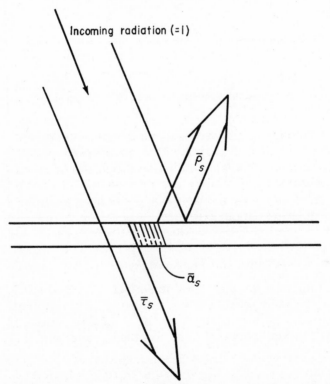

Incoming radiation (=1)

$\bar{\rho}_s$

$\bar{\alpha}_s$

$\bar{\tau}_s$

FIG. 2.6 Schematic representation of transmittance $\bar{\tau}_s$, absorptance $\bar{\alpha}_s$, and reflectance $\bar{\rho}_s$.

a radiation ray is equal to the angle of reflection, a situation analogous to the reflection of light from a mirror surface. Regular, or nearly total, reflection occurs only on highly polished surfaces; however, most materials used in engineering practice are "rough," and the reflection of an incident bundle of rays on most materials is nearly isotropic or diffuse; that is, the reflected radiation is distributed approximately uniformly into all directions.

The analysis of radiation problems is considerably simplified when limited to *diffuse radiation,* and in the following presentation, unless otherwise noted, it is assumed that reflection and emission are diffuse. This assumption is satisfactory for most problems, a notable exception being the polished-metal coatings used on solar reflectors and some satellite surfaces which are specular.

SOLAR-RADIATION FUNDAMENTALS

This section examines the nature of solar radiation emanating from the sun and the characteristics of this energy as it reaches the surface of the earth.

The sun is a star around which the earth revolves. It is a quite ordinary star, about average in size, mass, and brightness, and it is thought to be in a stable state of evolution; according to astronomers it has changed little during the past 3 billion yr and is not expected to change much in the next 3 billion yr. Consequently, its radiation may be considered an inexhaustible source of energy for terrestrial mankind.

The sun is a sphere of intensely hot gases, with a diameter of 8.6×10^5 mi and a mass of 2.2×10^{27} tons (about 334,000 times the mass of the earth). To an observer from the earth, the sun appears to rotate on its axis about once every 4 weeks, with an average distance of 9.3×10^7 mi from the earth. The sun emits energy in the forms of light and thermal radiation at the rate of 5×10^{23} hp. The temperatures in the central interior regions of the sun are on the order of 36×10^{6} °R, but as seen by an observer from the earth, the effective surface temperature of the sun is 5,760°K, according to recent measurements.

In oversimplified terms, the sun is a continuous fusion reactor in which the constituent gases are contained by gravitational forces. According to one prevalent theory, the most important energy generation process is the combination of four hydrogen atoms to form helium. Since the mass of helium is less than that of hydrogen, some mass is lost in the reaction. According to Einstein's hypothesis on the equivalence of mass and energy, this change in mass is

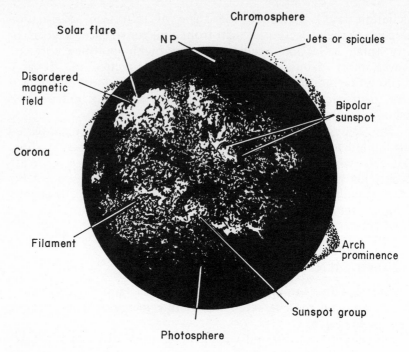

FIG. 2.7 Adaptation of a spectroheliogram taken by Lockheed Solar Observatory showing solar flare (prominences added by the artist); NP = the sun's north pole.

converted to energy and then radiated into space. Figure 2.7 shows an artist's conception of the sun with a typical solar flare shown on its surface.

The most familiar feature of the sun is its outer layer, called the photosphere, a bright surface that can be seen through an ordinary telescope. More complete examination shows that this photosphere is covered by a mosaic of small, bright polygonal cells, called *granules*. According to the most widely accepted astrophysical model, the sun's thermonuclear reaction, which converts hydrogen into helium, takes place in a core of very high density (about 100 times that of water); the energy so produced is transmitted outward by radiation for some distance from the core, but nearer the photosphere the gas becomes more opaque and convection takes place. The photosphere is a thin layer, less than 100 or 200 mi deep, in which the temperature decreases from about 11,000 to 8,000°F. The solar radiation received at surface of the earth emanates mostly from this layer.

Outside the photosphere is a transparent solar atmosphere, which can be observed during a total solar eclipse. This solar atmosphere consists of a layer of cooler gases, called the *reversing layer,* and outside the reversing layer is a layer called the *chromo-sphere.* The solar flares that erupt sporadically from the chromo-sphere layer can disrupt shortwave communication on earth. The outermost layer is the *corona,* a gas body of low density and high temperature that extends far into the solar system.

Despite the extremely complicated structure of the sun, for engineering purposes on earth it is acceptable to treat the radiation emanating from this star as though it came from a "black-body" radiator at approximately 5760°K, or 10,400°R. For more detailed calculation, the reader may refer to measurements of solar radiation made outside the atmosphere and compiled by NASA in 1971.[6]

The Solar Constant and Spectral Distribution of Solar Radiation

Figure 2.8 shows the sun-earth relationship. The earth's orbit around the sun is elliptical, with an eccentricity of 3 percent. At the mean distance between the earth and the sun, the sun subtends an angle of about 32 minutes. The amount of solar energy received by a unit area of surface placed perpendicular to the sun's rays in near-earth space at the earth's mean distance from the sun is called the *solar constant* (I_{sc}). The solar constant has been measured from rockets and satellites and an analysis of available data suggests that its value is approximately 1,353 watts/m^2, 1.94 cal/(cm^2)(min), or 429 Btu/(hr)(ft^2). Owing to the small variation in distance between the earth and the sun, the solar constant varies with the time of the year. The effect of the time of the year on the solar constant is shown in Fig. 2.9.

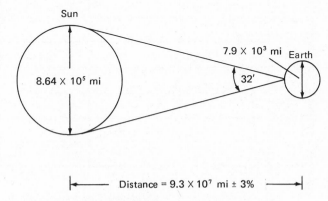

Sun

7.9 × 10^3 mi Earth

8.64 × 10^5 mi   32'

Distance = 9.3 × 10^7 mi ± 3%

FIG. 2.8 Geometric relationship of the earth and sun; the approximate value of the solar constant is 1.940 ly/min (1.940 cal/(cm^2)(min)), 1,353 watts/m^2, or 429 Btu/(hr)(ft^2).

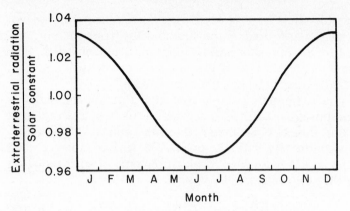

FIG. 2.9   Effect of the time of year on the ratio of extraterrestrial
radiation to the nominal solar constant.

The solar constant represents the total energy in the solar spectrum. This quantity, however, is not sufficient for most engineering calculations as discussed in the preceding section, and it is necessary to examine the distribution of energy within the spectrum. Figure 2.10 shows the spectral irradiance at the mean sun-earth distance for a solar constant of 1,353 watts/m^2 as a function of wavelength according to the standard spectrum data published by NASA in 1971.[6] The same data are also presented in Table 2.5, and their use is illustrated in the following example.

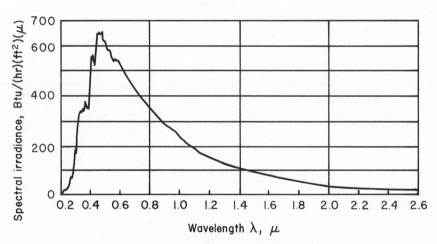

FIG. 2.10   Extraterrestrial spectral distribution of radiation at the average distance.
(*Adapted by permission of the publishers from M. P. Thekaekara.*[6])

TABLE 2.5 Extraterrestrial Solar Irradiance*,†

λ μ	E_λ‡ Btu/(hr) $(ft^2)(\mu)$	D_λ§ %	λ μ	E_λ Btu/(hr) $(ft^2)(\mu)$	D_λ %	λ μ	E_λ Btu/(hr) $(ft^2)(\mu)$	D_λ %
0.115	0.002	1×10^{-4}	0.43	520	12.47	0.90	283	63.37
0.14	0.010	5×10^{-4}	0.44	574	13.73	1.00	237	69.49
0.16	0.073	6×10^{-4}	0.45	636	15.14	1.2	154	78.40
0.18	0.397	1.6×10^{-3}	0.46	655	16.65	1.4	107	84.33
0.20	3.39	8.1×10^{-3}	0.47	645	18.17	1.6	77.7	88.61
0.22	18.2	0.05	0.48	658	19.68	1.8	50.4	91.59
0.23	21.2	0.10	0.49	619	21.15	2.0	32.7	93.49
0.24	20.0	0.14	0.50	616	22.60	2.2	25.1	94.83
0.25	22.5	0.19	0.51	597	24.01	2.4	19.7	95.86
0.26	41.2	0.27	0.52	581	25.38	2.6	15.2	96.67
0.27	73.6	0.41	0.53	584	26.74	2.8	12.4	97.31
0.28	70.4	0.56	0.54	566	28.08	3.0	9.83	97.83
0.29	153	0.81	0.55	547	29.38	3.2	7.17	98.22
0.30	163	1.21	0.56	538	30.65	3.4	5.27	98.50
0.31	219	1.66	0.57	543	31.91	3.6	4.28	98.72
0.32	263	2.22	0.58	544	33.18	3.8	3.52	98.91
0.33	336	2.93	0.59	539	34.44	4.0	3.01	99.06
0.34	341	3.72	0.60	528	35.68	4.5	1.87	99.34
0.35	347	4.52	0.62	508	38.10	5.0	1.21	99.51
0.36	339	5.32	0.64	490	40.42	6.0	0.57	99.72
0.37	375	6.15	0.66	471	42.66	7.0	0.32	99.82
0.38	355	7.00	0.68	453	44.81	8.0	0.19	99.88
0.39	348	7.82	0.70	434	46.88	10.0	0.076	99.94
0.40	453	8.73	0.72	417	48.86	15.0	0.015	99.98
0.41	555	9.92	0.75	392	51.69	20.0	0.005	99.99
0.42	554	11.22	0.80	352	56.02	50.0	0.0001	100.00

*Adapted from Thekaekara (1973)[6] by permission of the publishers.
†Solar constant = 429 Btu/(hr)(ft²).
‡E_λ is the solar spectral irradiance averaged over a small bandwidth centered at λ.
§D_λ is the percentage of the solar constant associated with wavelengths shorter than λ.

EXAMPLE

Calculate the fraction of solar radiation within the visible part of the spectrum, i.e., between 0.40 and 0.75 μ.

SOLUTION

The first column in Table 2.5 gives the wavelength. The second column gives the averaged solar spectral irradiance in a band centered at the wavelength in the first column. The third column, D_λ, gives the percentage of solar total radiation at wavelengths shorter than the value of λ in the first column. At a value of .40 μ, 8.7 percent of the total radiation occurs at shorter wavelength. At a wavelength of .75 μ, 51.7 percent of the radiation occurs at shorter wavelengths. Consequently, 43 percent of the total radiation lies within the band between 0.40 and 0.75 μ, and the total energy received outside the earth's atmosphere within that spectral range is 184 Btu/(hr)(ft^2).

In addition to spectral variations, the amount of solar radiation incident on a surface depends on its location, orientation, the time of year, and the time of day. The solar radiation incident on a surface is called *insolation* (*I*) and will be considered in detail later in this chapter under "Seasonal Variation of Insolation."

Solar Radiation at the Earth's Surface

The radiation impinging upon the outer fringes of the earth's atmosphere is attenuated by several constituents of the earth's atmosphere—H_2O, CO_2, O_3, and O_2. For engineering purposes, it is usually not convenient to calculate the changes extraterrestrial radiation undergoes in passing through the atmosphere, and meteorological measurements made at the surface of the earth are used whenever they are available. Meteorological measurements cannot always be obtained, however, and if they cannot be obtained, it is important to understand the properties that affect the solar radiation passing through the atmosphere.

The radiation at the outer fringes of the atmosphere is collimated and is called *beam radiation.* Owing to reflection and scattering processes within the atmosphere, a portion of the total collimated radiation is changed from direct-beam into *diffuse,* or nondirectional, radiation. The principal factor that determines the amount and degree of attenuation is the distance through which radiation travels from the outer surface of the atmosphere to the receiving point on the earth. At sea level, when the sun is directly overhead, this distance is the shortest and is defined as *1 air mass* (*m*). The distance of atmosphere, or the air mass, traversed changes as the sun moves through its virtual daily motion; when the sun reaches an angle of 60° between the zenith and the line of sight, the distance through which the beam travels before it reaches the earth has doubled, and the air mass is equal to 2.

In general, atmospheric transmittance $\bar{\tau}_{atm}$ is given to 3 percent accuracy by the relation

$$\bar{\tau}_{atm} = 0.5\left(e^{-0.65m\,(z,\alpha)} + e^{-0.095m\,(z,\alpha)}\right) \tag{2.16}$$

where α is the solar altitude angle (see p. 47), and the air mass m at an altitude z above sea level is

$$m(z,\alpha) = m(0,\alpha)\,\frac{p(z)}{p(0)} \tag{2.17}$$

where p is the atmospheric pressure and the air mass $m(0,\alpha)$ at sea level (altitude 0) is given by

$$m(0,\alpha) = [1229 + (614 \sin \alpha)^2]^{1/2} - 614 \sin \alpha \tag{2.18}$$

The surface beam radiation I_b is given by

$$I_b = I_{sc}\,\tau_{atm} \tag{2.19}$$

where I_{sc} is the solar constant. To date, no reliable way is available to predict the diffuse component of radiation from the clear-sky value. However, if the insolation at noon is known, the collimated part of the solar irradiation at any time of day and season can be calculated, given similar atmospheric conditions of turbidity and moisture content, by Eqs. (2.16) through (2.19). A discussion of methods of calculating atmospheric attenuation of solar energy is presented in numerous textbooks on meteorology. In addition, a tentative equation for the diffuse component is given in Eq. (3.15).

PRINCIPLES OF SOLAR COLLECTOR WITH WAVELENGTH SELECTIVE SURFACE

To evaluate quantitatively the performance of a flat-plate, or planar, solar collector, consider the simplified model shown in Fig. 2.11. The surface of the collector consists of a flat plate, oriented normal to the direction of the sun's rays. It is assumed that the collector temperature is uniform and constant and that the useful energy is removed from the rear of the surface by means of an appropriate circulating fluid. An energy balance per unit surface area of collector has the form

$$\bar{\alpha}_s I = \bar{\epsilon}_{ir}\sigma T_{coll}^4 + h_{c1}\,(T_{coll} - T_{out}) + q_f - \bar{\alpha}_{sky}\,\sigma T_{coll}^4 \tag{2.20}$$

Since the value of a solar collector consists in its ability to collect and retain as great an amount as possible of the insolation it receives, *collector efficiency* can be defined as the ratio of collected energy q_f to insolation I, i.e., the ratio of solar output to insolation.

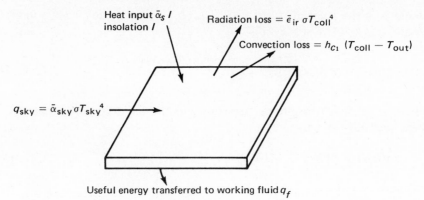

FIG. 2.11 Energy balance of a flat-plate solar collector, where the efficiency $\eta = q_f/I$, and $\bar{\alpha}_s$ = solar absorptance, I = insolation, $\bar{\alpha}_{sky}$ = surface absorptance of infrared sky radiation, σ = Boltzmann constant (that is, $(0.1713 \times 10^{-8} \text{ Btu})/(\text{hr})(\text{ft}^2)(^\circ \text{R}^4))$), T_{sky} = sky temperature, $\bar{\epsilon}_{ir}$ = average surface infrared emittance, h_{c_1} = coefficient of heat transfer from the outer collector cover, T_{coll} = average collector-surface temperature, and T_{out} = outside ambient temperature.

$$\eta = \frac{q_f}{I} \tag{2.21}$$

Combining Eqs. (2.20) and (2.21) gives the following expression for collector efficiency:

$$\eta = \bar{\alpha}_s - \frac{\bar{\epsilon}_{ir}\sigma T_{coll}^4}{I} - \frac{h_{c_1}(T_{coll} - T_{out})}{I} + \frac{\bar{\alpha}_{sky}\sigma T_{sky}^4}{I} \tag{2.22}$$

Equation (2.22) indicates that the collector efficiency increases if either the absorptance $\bar{\alpha}_s$ increases or the convection transfer coefficient h_{c_1} or emittance $\bar{\epsilon}_{ir}$ decreases. A maximum collector efficiency is reached when the latter quantities go to zero. A *selective surface* is a surface for which $\bar{\alpha}_s \neq \bar{\epsilon}_{ir}$ ($\bar{\alpha}_s > \bar{\epsilon}_{ir}$ for most solar-energy applications), while a *nonselective surface* is a surface for which $\bar{\alpha}_s = \bar{\epsilon}_{ir}$.

An interesting comparison of collector surfaces with selective coatings can be made with Eq. (2.22). If I and T_{out} are specified collector efficiency can be related to the plate temperature. The results for a number of practical surfaces where sky radiation is ignored are shown in Fig. 2.12 (the data used in Fig. 2.12 are shown in Table 2.6). The figure shows that black paint (frequently used in solar collectors) is not the best surface, particularly at high

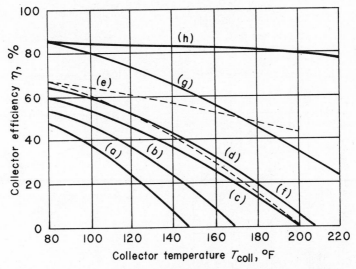

FIG. 2.12 Effect of collector temperature T_{coll} for a number of surfaces on collector efficiency η under the following conditions: Insolation $I =$ 300/Btu/(hr)(ft^2), air (ambient) temperature $T_{out} = 80°$F, and coefficient of heat transfer from the absorber to the outer environment $h_{c_1} = 0.22$ $(T_{coll} - T_{out})/^1/_3$. (a) Lamp-black paint; (b) graphite; (c) Tyler screen, 20 × 350; (d) Tyler screen, 20 × 200; (e) Tyler screen with no convection, 20 × 200; (f) Tyler screen, 28 × 500; (g) coated surface; (h) coated surface with no convection.[1]

temperatures. The best types of surfaces, in order of increasing efficiency, are

1. Graphite
2. Tyler screens
3. Coated surfaces

The differences among the various surfaces are significant: For instance, at a collector temperature of 120°F, black paint has an efficiency of 24 percent, a Tyler 28 × 500 screen an efficiency of 51

TABLE 2.6 Data Used for Calculating Curves in Fig. 2.12[1]

	Material					
	Tyler-screen mesh size				Lamp-black	Coated
Parameter	20 × 350	20 × 200	28 × 500	Graphite	paint	surface
$\bar{\alpha}_s$	0.73	0.77	0.86	0.85	0.97	0.90
$\bar{\epsilon}_{ir}$	0.26	0.25	0.38	0.60	0.97	0.10

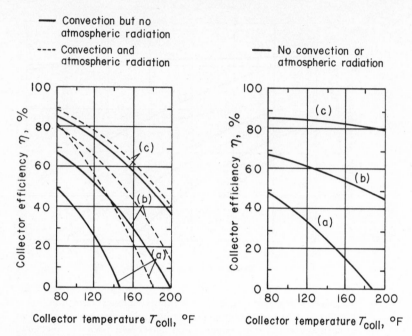

FIG. 2.13   Effect of collector temperature on collector efficiency. (a) Lamp-black paint; (b) Tyler screen, 28 × 500; (c) coated surface.[1]

percent.* If sky radiation is included, collector efficiency is greater, as shown in Fig. 2.13.

Placing a glass plate over a surface reduces the value of the over-all heat-transfer coefficient by introducing additional resistances for the convection heat loss. At the same time, the glass also reduces the incoming solar radiation because approximately 90 percent of the insolation penetrates the glass and reaches the surface. Moreover, the installation of the glass surface increases the cost. These matters will be considered in detail later in this book.

EXAMPLE

Calculate efficiency η and useful energy transferred to the working fluid q_f from the following data:

$$I = 200 \text{ Btu/(hr)(ft}^2)$$
$$T_{out} = 40°F = 500°R$$

---

*The experimentally determined values of the infrared emittance and solar absorptance of the Tyler surfaces were taken from measurements made at the University of Minnesota, Minneapolis.[5]

$T_{sky} = 20°F = 480\,°R$
$T_{coll} = 140°F = 600°R$
$h_{c1} = 0.22\,(T_{coll} - T_{out})^{1/3}$

Case 1: Nonselective surface

For a nonselective surface $\bar{\alpha}_s = \bar{\epsilon}_{ir} = \bar{\alpha}_{sky} = 0.90$, where ϵ_{ir} is the collector-absorber-plate infrared emittance on a nonselective surface.

Case 2: Selective surface

For a selective surface $\bar{\alpha}_s = 0.90$ and $\bar{\epsilon}_{ir} = \bar{\alpha}_{sky} = 0.30$

SOLUTION

$h_{c1} = 0.22 \times 100^{1/3} = 1.0\,\text{Btu/(hr)(ft}^2)(°F)$

Case 1: Nonselective surface

$$\eta = 0.90 - \left\{ \left[\frac{0.90 \times 0.1713 \times 6^4 + 1 \times (140 - 40)}{200} \right] - \frac{0.90 \times (0.1713 \times 4.8^4)}{200} \right\}$$

$\eta = 0.90 - [(1.01 + 0.50) - 0.41] = -0.2$

Since η is less than 0 ($\eta = -0.2$), no useful energy is transferred to the working fluid.

Case 2: Selective surface

$\eta = 0.90 - [(0.34 + 0.50) - 0.14] = 0.20$

$q_f = \eta \times I = 0.20 \times 200 = 40\,\text{Btu/(hr)(ft}^2)$

SEASONAL VARIATION OF INSOLATION

Horizontal Surfaces

The amount of solar radiation incident at the outer edge of the atmosphere depends on the time of year, the time of day, and the

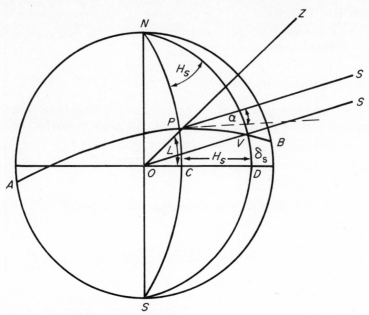

FIG. 2.14 Definition of solar-hour angle H_S (CND), solar declination δ_S (VOD), and latitude L (POC).

location, expressed by the latitude. Fig. 2.14 shows that the solar irradiation on a horizontal surface is directly proportional to the sine of the angle of solar altitude α. This angle is not directly known and must therefore be expressed in terms of known quantities such as latitude, the time of day expressible as the hour angle, and the solar declination, which can be obtained from a book of astronomical data, such as the *Air Almanac* or *The American Ephemeris.* *

To obtain an expression for the sine of the solar altitude angle, let P be the point of observation on earth, with its zenith, i.e., overhead point, at Z, and let the sun be in the direction OS. A plane passed between the two lines OZ and OS intersects the surface of the earth in a great circle, and the angle, ZOS, measured by the arc PV on the circle, equals the sun's zenith angle, which is the complement of the sun's altitude angle as shown in Fig. 2.14. In the spherical triangle NPV, the angle H_S at N, measured by the arc CD on the equator, is the hour angle defined as the angle through which the earth would turn to bring the meridian of P directly under the sun. Applying the cosine law of spherical trigonometry to the triangle NPV leads to the expression

*The *Air Almanac* is published annually by the U.S. Nautical Almanac Office, Washington, D.C.; the *American Ephemeris* is published annually by the Superintendant of Documents, Washington, D.C.

$$\sin \alpha = \sin L \sin \delta_s + \cos L \cos \delta_s \cos H_s \tag{2.23}$$

where L = latitude of the point P

       δ_s = solar declination

       H_s = local solar hour angle measured west from solar noon

The solar declination δ_s can be obtained from Fig. 2.15, and the hour angle H_s is equal to 15 times the number of hours from local solar noon. For example, 10 A.M. local solar time corresponds to a local solar-hour angle of 30°. The method for calculating local solar time from clock time is given in App. C in the text describing Figs. C.1(a) to C.1(g).

To determine the total amount of insolation I_{tot} during a day at a given point on a horizontal plane outside the earth's atmosphere the following relation is used:

$$I_{tot} = \frac{2I_{sc}(H_m \sin L \sin \delta_s + \cos L \cos \delta_s \sin H_m)}{\Omega} \tag{2.24}$$

where Ω = angular velocity of the earth, equal to $\pi/12$ rad/hr

       I_{sc} = solar constant

       H_m = hour angle between sunrise and noon (which can be obtained from the information presented in Fig. 2.15)

H_m is a function of the latitude and the solar declination angle, which depends on the time of year. The solar declination δ_s may be considered constant during a given day, however.

At sunrise (when $\alpha = 0$),

$$\sin \alpha = 0 = \sin L \sin \delta_s + \cos L \cos \delta_s \cos H_m \tag{2.25}$$

which yields

$$\cos H_m = -\tan L \tan \delta_s \tag{2.26}$$

Combining Eq. (2.26) with Eq. (2.24) yields

$$I_{tot} = \frac{2I_{sc} \sin L \sin \delta_s (H_m - \tan H_m)}{\Omega}$$

$$= \frac{24}{\pi} I_{sc} \sin L \sin \delta_s (H_m - \tan H_m) \tag{2.27}$$

If it is assumed that a constant rate of solar radiation is emitted by the sun, relative amounts of insolation delivered to the earth's

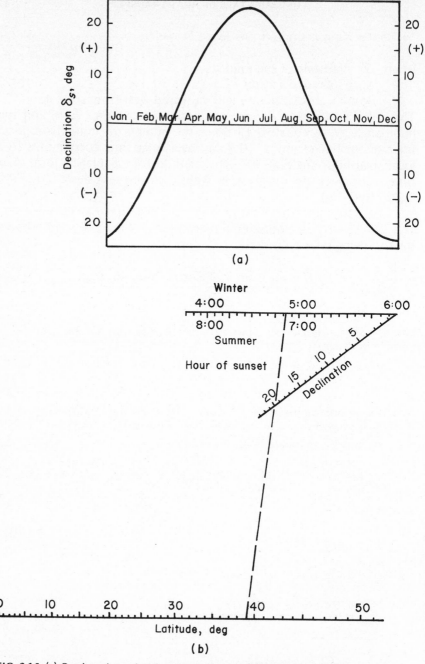

FIG. 2.15 (a) Graph to determine the solar declination.[8] (b) Sunset nomograph example showing determination of sunset time for summer (7:08 P.M.) and winter (4:52 P.M.) when the latitude is 39°N and the solar declination angle is 20°. (*From A. Whillier, "Solar Radiation Graphs," Solar Energy, vol. 9, p. 164. Copyright 1965 by Pergamon Press. By permission of the publishers.*)

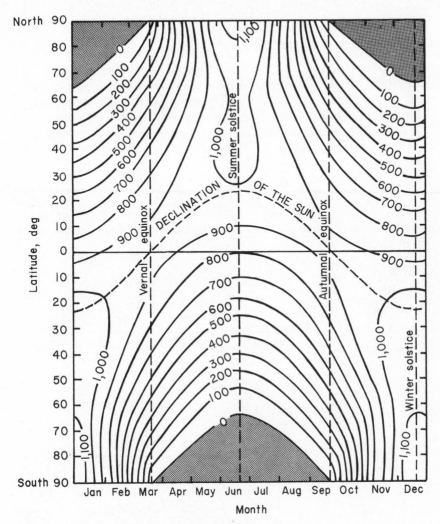

FIG. 2.16 The daily variation of the solar radiation at the top of the atmosphere as a function of latitude. The values are measured in langleys per day.

outer atmosphere at different latitudes and different times of year may be computed.*

Figure 2.16 shows the variation in the daily insolation on a horitontal surface outside the atmosphere with seasons and latitude, as computed by Eq. (2.27). The analysis in this section is also applicable to solar radiation at the earth's surface if the solar constant I_{sc} is replaced by the beam component of radiation at the surface I_b in Eqs. (2.24) and (2.27).

*Values of H_m and δ_s can also be obtained from *The American Ephemeris.*

South-facing Tilted Surfaces
(Northern Hemisphere)

For a surface tilted at an angle β from the horizontal the insolation can be divided into two components, one perpendicular and one parallel to the tilted surface. Only the perpendicular component impinges on the surface, and if the incidence angle between the surface normal to the collector and the direction of the sun is i, as shown in Fig. 2.17, the effective component of insolation $I_{b,\text{coll}}$ is given by

$$I_{b,\text{coll}} = I_b \cos i \tag{2.28}$$

where I_b is the beam, or direct, component of insolation at the earth's surface and

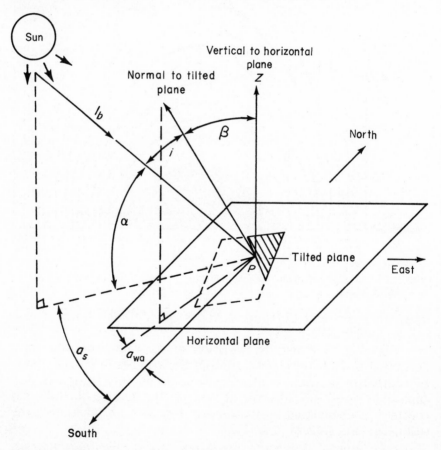

FIG. 2.17 Definition of solar and surface angles for a non-south-facing tilted surface.

$$\cos i = \sin \delta_s \sin (L - \beta) + \cos \delta_s \cos (L - \beta) \cos H_s \qquad (2.29)$$

where L is the latitude.

Non-south-facing Tilted Surfaces (Northern Hemisphere)

If a tilted wall faces a direction other than due south, Eqs. (2.30) and (2.31) are used to calculate the incidence angle i:

$$\cos i = \cos (a_s - a_{wa}) \cos \alpha \sin \beta + \sin \alpha \cos \beta \qquad (2.30)$$

where a_s = azimuth of the sun, i.e., the angle of the sun measured in the horizontal plane westward from due south

a_{wa} = azimuth angle of the normal to the insolated surface, measured in the same way as solar azimuth

and

$$a_s = \sin^{-1} \frac{\cos \delta_s \sin H_s}{\cos \alpha} \qquad (2.31)$$

A convenient way to visualize the results of the preceding equations is to refer to a series of sun-path diagrams such as those presented in the *Smithsonian Meteorological Tables*. Two of these diagrams are reproduced in Fig. 2.18, and several are included in App. C, Figs. C.1(a) to C.1(g). An application of these diagrams is given in the following example.

EXAMPLE

Find the zenith angle and the azimuth at 11 A.M. solar time of February 21st at 32°N latitude.

SOLUTION

Follow path line 5 in Fig. 2.18(a) to its intersection with the 11 A.M. line. At this point read the zenith angle (45°) and the azimuth (21° east of south).

Sun-path diagrams can be used to construct shading maps to determine the seasonal and diurnal shading characteristics of trees or existing adjacent buildings. Figure 2.19 shows how an existing building can shade the proposed collector. The method is quite simple: A point C along the lower edge of the collector is selected

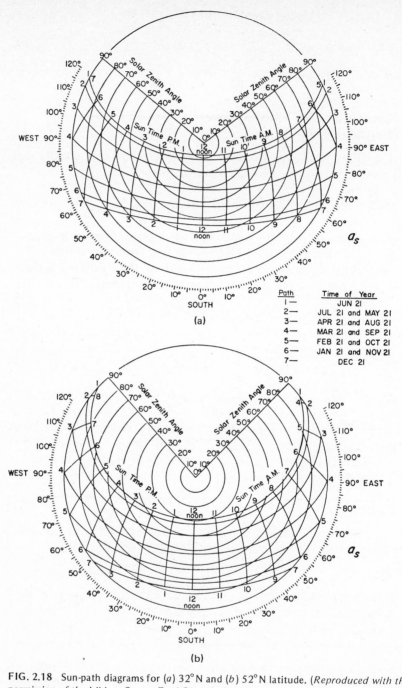

Path	Time of Year
1—	JUN 21
2—	JUL 21 and MAY 21
3—	APR 21 and AUG 21
4—	MAR 21 and SEP 21
5—	FEB 21 and OCT 21
6—	JAN 21 and NOV 21
7—	DEC 21

(a)

(b)

FIG. 2.18 Sun-path diagrams for (a) 32°N and (b) 52°N latitude. (*Reproduced with the permission of the Libbey-Owens-Ford Glass Co.*)

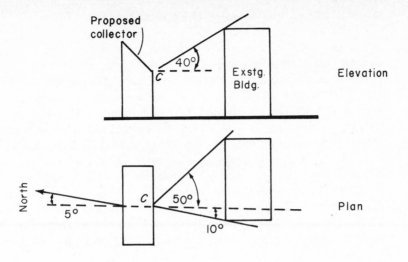

Proposed
collector

40°

c̄

Exstg.
Bldg.

Elevation

North

5°

C

50°

10°

Plan

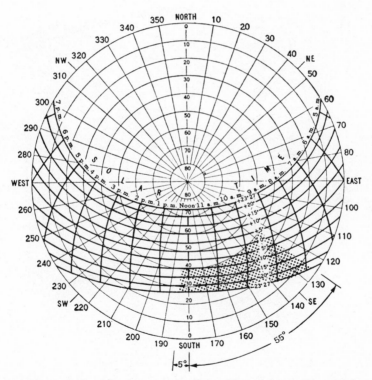

FIG. 2.19 Example of shade-map construction showing effect of plan and elevation location of existing structure on winter collector shading.

and a plan and elevation view drawn from which the critical solar altitude and azimuth angles can be measured. The critical points of azimuth and altitude are plotted on a sun-path diagram as shown in Fig. 2.19(b); the effect of the path of the sun on a non-south-facing wall is also shown in Fig. 2.19(b). The shaded area represents the portion of the year during which the collector will be at least partly shaded. The shading map shown in Fig. 2.19(b) is approximate since it shows the solar altitude at a uniform value of 40°. The value of 40° is the maximum value corresponding to a sun position along the solid line in the elevation view of Fig. 2.19(a). However, for solar altitudes of 40° or less and solar azimuth angles ±40° from due south, the error is not large using this approximation. For strictly correct shading maps, the so-called "shadow-angle protractor" is

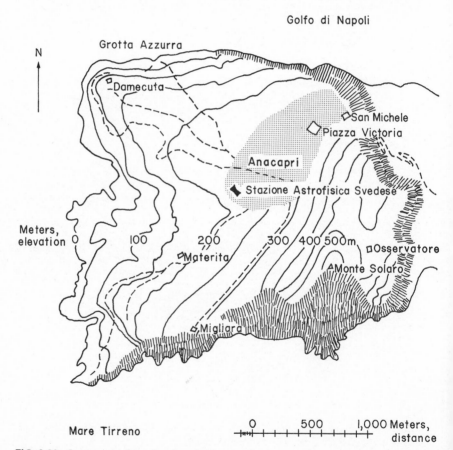

FIG. 2.20 Contour map showing elevations and other factors that would affect the placement of a collector at the Stazione Astrofisica Svedese on the Isle of Capri (40° 6′ N latitude, 14° 3′ E longitude).[4]

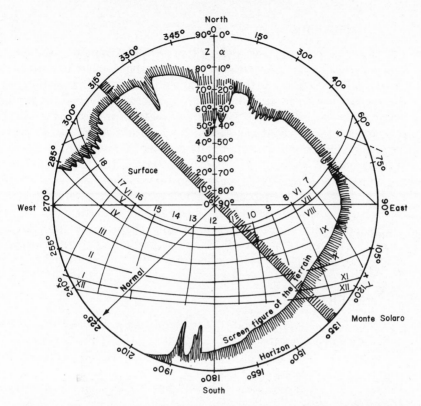

FIG. 2.21 Solar chart for the Stazione Astrofisica Svedese showing the profile of the horizon as seen from the site of a proposed solar collector; minimum solar obstruction orientation is at 225°.[4]

used. Since such protractors are not generally available and the approximate method is quite adequate for the northern and midlatitude area of the United States, only its use is discussed in this text.

An alternative to the protractor approach for complex terrain features that may shade the collector is to "plot" the terrain on a sun-path diagram. An example of terrain plotted on a sun-path diagram is demonstrated in Fig. 2.20, which shows the location of the Swedish Astrophysical Station (Stazione Astrofisica Svedese) on a partial contour map of the Isle of Capri in the Tyrrhenian Sea, and Fig. 2.21 is a plot of the horizon as viewed from the station.[4] It is apparent from this terrain plot that a solar collector should not face due south because if it did, the winter sun in the east would be obscured by the terrain. The best orientation for this particular collector is toward the southwest, where the terrain shadowing will be less in winter.

SUMMARY

In this chapter the principles of heat transfer used in the rational design of solar-energy collectors have been quantitatively discussed; by extension, these same principles of heat transfer govern the methods used in designing building energy conservation systems. The nature of the sun and its energy creation process, as well as the properties of insolation in near-earth space and at the earth's surface have been described in detail. The efficient conversion of the sun's radiant energy to heat in solar collectors is fundamentally determined by the interplay of heat transfer, optical, and astronomical phenomena; in this chapter these principles have been described in qualitative physical terms as well as in analytical form. The fundamental principles analyzed in this chapter are integrated in Chap. 3 to provide a rational means of predicting thermal performance of several types of solar collectors.

REFERENCES

1. Irvine, T. F., Jr., *et al.*: Solar Collector Surfaces with Wavelength Selective Radiation Characteristics, *Solar Energy*, vol. 2, nos. 2-3, p. 12, 1958.
2. Kreith, F.: "Principles of Heat Transfer," 3d ed., Intext Educational Publishers, Inc. New York, 1973.
3. Kreith, F.: "Radiation Heat Transfer," Intext Educational Publishers, Inc. New York, 1964.
4. *Proc. U.N. Conf. New Sources of Energy (Rome, 1961),* United Nations, New York, vol. 5, p. 207, 1964.
5. Sellers, W. D.: "Physical Climatology," University of Chicago Press, Chicago, 1967.
6. Thekaekara, M. P.: Solar Energy Outside the Earth's Atmosphere, *Solar Energy*, vol. 14, p. 109, 1973.
7. Touloukian, Y. S., *et al.*: "Thermophysical Properties of Matter," vols. 7-9, Plenum Data Corp., New York, 1970, 1972.
8. Whillier, A.: Solar Radiation Graphs, *Solar Energy*, vol. 9, p. 165, 1965.

CHAPTER 3

Methods of Solar-energy Collection and Use

He had been eight years upon a project for
extracting sunbeams out of cucumbers, which were
put in phials hermetically sealed, and let out
to warm the air in raw inclement summers.

JONATHAN SWIFT

Methods of solar-energy collection range from active, mechanical, solar thermal collectors to photoelectrical devices to windmills to fully passive collection systems. In this chapter primary emphasis is placed on the performance characteristics of solar thermal collectors that can be used for building heating and cooling. Several viable systems will be described in detail and performance comparisons will be made.

FLAT-PLATE-COLLECTOR FUNDAMENTALS

A typical flat-plate collector consists of one or more transparent flat front plates, one or more insulating zones bounded by the covers, and an absorbing rear plate. Heat is removed from the rear plate by a gaseous or liquid heat-transfer fluid, such as air or water. Thermal insulation is usually placed behind the absorber to prevent heat losses from the rear surface. The front covers are generally glass that is transparent to incoming solar radiation and opaque to the infrared reradiation from the absorber. The glass covers act as a convection shield to reduce losses from the absorber plate beneath.

For water systems the absorber plate can be any metal, plastic, or rubber sheet that incorporates water channels, while for air systems the space above the collector plate can also serve as the conduit. Many metal products, such as the aluminum, copper, and steel absorbers shown in Fig. 3.1(*a*), are suitable for use in water or air systems. Nonmetal absorbers are currently available, but their construction is different from that of the metal absorbers. Because

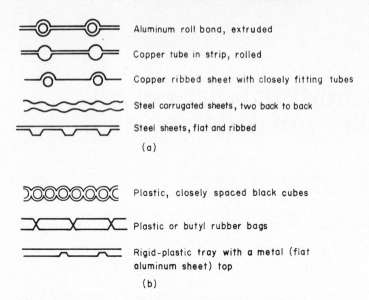

FIG. 3.1 Examples of currently available (*a*) metal and (*b*) nonmetal
products that can be used as the absorber plate in a flat-plate collector.

the thermal conductivity of plastics and rubber is much less than that
of metals, plastic or rubber absorbers require closer contact between
the surface and the liquid; plastic and rubber collectors are only
suitable for lower temperature uses. Typical nonmetal absorber
shapes are shown in Fig. 3.1(*b*).

Heat from the absorber is removed by water channels connected
at top and bottom by a manifold; examples of various types of
manifolds are shown in Fig. 3.2. To ensure a steady, balanced flow,
the header (the conduit into which water flows from the absorber
rear-plate water channels) should have a cross-section area larger than
the area of the water channels served.*

The surface finish of the absorber plate may be a flat black
paint with an appropriate primer. The primer coat should preferably
be thin since a thick undercoat of paint would increase the resistance
to heat transfer. The primer should be of the self-etching type. If the
primer is not a self-etching type, the repeated thermal expansion and
contraction of the plate may cause the paint to peel after a year or
so. Several types of baked-on finishes are also available.

The most favorable orientation of a heating-only solar collector
is facing due south at an inclination angle to the horizontal equal to
the latitude plus 15°. However, throughout most of the continental

*Table C.12 in App. C lists a number of manufacturers who can provide commercial
quantities of flat-plate collectors.

United States, placing a collector in a vertical position causes a performance reduction of only about 15 percent from the optimum arrangement, reduces undesirable summer heating, and permits much easier incorporation of the collector into the wall of the building.

In general, the thermal conversion performance of the solar collector can be improved by increasing the transmission of energy through the collector to the working fluid and by reducing thermal losses. The following items govern the amount of solar energy collected.

1. The transmittance of the transparent covers (ideally, transmittance should exceed 90 percent in the solar spectrum)
2. The absorptance of the absorber plate to the incident radiation (absorptances should exceed 95 percent, a level that can be readily obtained by appropriate coating)
3. Thermal resistance between the absorber surface and the heat-transfer fluid (this resistance, which should be as small as possible, depends largely on the thermal conduction through the solid absorber plate, the nature of the convection in the flow channel, and the nature of the heat-transfer medium)
4. Emittance of the absorber plate surface in the infrared spectrum

Thermal losses can be separated into three components:

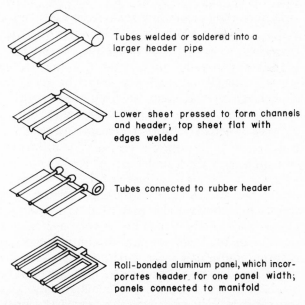

Tubes welded or soldered into a larger header pipe

Lower sheet pressed to form channels and header; top sheet flat with edges welded

Tubes connected to rubber header

Roll-bonded aluminum panel, which incorporates header for one panel width; panels connected to manifold

FIG. 3.2 Examples of manifolds used in typical flat-plate collectors.

Conductive Losses

Conduction through the back and the sides of a collector is usually negligible if the back and sides of the collector are well insulated. An overall heat transfer coefficient (U) value of less than 0.04 Btu/(hr)(ft^2)($^\circ$F) is suggested to minimize back losses.

Convective Losses

Convective losses occur from the absorber plate to the environment through intermediate convection exchanges between the air enclosed in each insulating zone and the boundaries of each zone—the collector covers. In the absence of wind, external convection loss from the outermost cover is by the mechanism of natural convection; but even in low winds, forced convection occurs and increases the loss substantially. (Natural convection occurs without an imposed external flow whereas forced convection occurs in the presence of an external flow.) As shown later, sizing the air gap between the collector covers at 0.4 to 0.75 in. reduces internal convective losses to the minimum possible level.

Convection loss between glass plates can also be inhibited if a honeycomb-type, cellular structure is placed between the absorber and the outer window plate. However, in addition to the increase in cost, a cellular structure also reflects a part of the incoming radiation, thus preventing solar radiation from reaching the absorber plate. A cellular structure also increases the thermal conductivity of the space between the absorber and the outer air. A honeycomb that transmits solar radiation, is opaque in the infrared spectrum, and has a low thermal conductivity could be ideal for a solar collector. Evacuation of the space between the absorber and the outer cover has been proposed to reduce internal convection and conduction, but the cost of added supports and maintenance of a vacuum are excessive.

Radiative Losses

Radiative losses from the absorber can be reduced by the use of spectrally selective absorber coatings. Such coatings have a high absorptance of about 0.9 in the solar spectrum and a low emittance, usually on the order of 0.2, in the infrared spectrum in which the absorber radiates to the environment. Selective absorber coatings therefore decrease heat losses and increase collector efficiency. Selective-black coatings are commercially available from a few sources, but their cost, stability, and directional radiation properties should be carefully checked before using them on a large scale.

FLAT-PLATE SOLAR-
COLLECTOR PERFORMANCE
(VERTICAL CONFIGURATION)

The material presented in this section is derived from a report prepared at the University of Pennsylvania's Center for Energy Management and Power.[1,7] The assumptions made in the analysis are as follows:

1. The collector is liquid cooled and in the steady state.
2. The temperature gradients in the glass cover plates are negligible.
3. Thermal and radiation properties of the collector materials are independent of temperature.
4. Edge heat losses are negligible compared with heat losses through the collector plate.
5. All collector plates are vertical and south facing.

The analytical results in Fig. 3.3 show the influence of spacing between glass plates for a collector consisting of two glass plates with

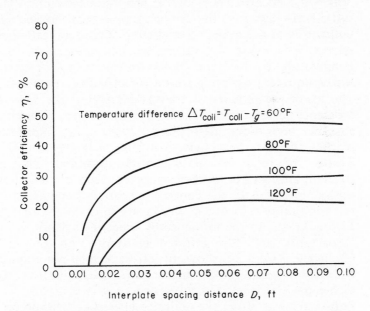

FIG. 3.3 Effect of various temperature differences ΔT_{coll} and interplate-spacing distances D on the efficiency η of a double-plate collector under the following conditions: Collector absorber-plate temperature $T_{coll} = 130°F$, collector outer-plate temperature T_g ranges from 10 to 70°F, insolation $I = 150$ Btu/(hr)(ft^2) and solar absorptance $\bar{\alpha}_s$ and infrared emittance $\bar{\epsilon}_{ir} = 0.95$. (*Adapted with permission from M. Altman.*[1])

a flat-black absorber surface of roll-bonded flow channels. When the distance between plates is too small, heat loss occurs by conduction through the air gap. The efficiency increases with increasing air gap but reaches an asymptote at an interplate spacing of about 0.6 in. A widely recommended optimum value for the interplate spacing is 0.5 in. and the results of Fig. 3.3 essentially confirm this rule-of-thumb. The results in Fig. 3.3 represent a solar-energy intensity (insolation I) on the collector plate of 150 Btu/(hr)(ft^2), a condition that represents typical winter insolation on a vertical collector facing south between 10 A.M. and 2 P.M. on a clear day at 40°N latitude. The absorber plate was assumed to be covered by a black coating having a solar absorptance of 0.95 and an infrared emittance of 0.95. The two glass plates in the solar collector were assumed to have a 0.90 transmittance, 0.05 absorptance and reflectance in the solar spectrum, and 0.05, 0.90, and 0.05, respectively, in the infrared spectrum.

The collector efficiencies are strongly affected by the difference between the temperature of the absorber plate T_{coll} and the outer glass plate T_g. It is desirable to hold the absorber temperature low to obtain a high efficiency, but at the same time one must obtain a sufficiently high absorber temperature so that the heat supply to the building is at a practical temperature for space heating and storage. As a compromise between these two conflicting demands an absorber temperature of 130°F was chosen for the study. The effect of using a single, double, or triple plate on the efficiency is shown in Fig. 3.4. The lower the number of transparent plates, the more sharply the efficiency drops with increasing temperature difference between absorber and outer glass plate, $(T_{coll} - T_g)$. More solar energy reaches the absorber plate when only a single cover plate is used because the addition of a second glass plate reduces the net transmittance of the glass-plate system from 0.90 to 0.81. At high values of $T_{coll} - T_g$ the convective heat loss becomes dominant, and the addition of a second or third plate substantially reduces this loss. Figure 3.5 shows the influence of selective coatings, with high solar absorptance and low infrared emittance, on the solar-collector efficiency. Figure 3.6 shows the effect of solar-radiation intensity on collector efficiency: the higher the solar-radiation intensity, the higher the efficiency at a given operating temperature. For example, if $T_{coll} - T_g$ is 80°F, the collector efficiency is 10 percent when I is equal to 100 Btu/(hr)(ft^2); when I is equal to 300 Btu/(hr)(ft^2), the efficiency rises to 57 percent. Trebling the incident solar radiation increases the amount of heat collected sixfold. The large increase in heat collected when incident radiation is trebled is because the convective heat loss is a function of $T_{coll} - T_g$ only; radiative heat losses are affected mainly by the absorber temperature, which was

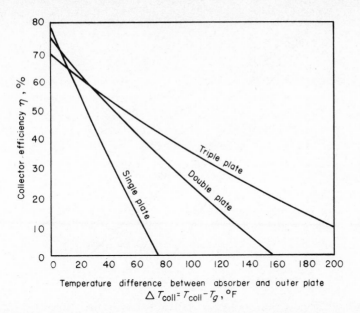

FIG. 3.4 Effect of various temperature differences ΔT_{coll} and number of glass plates on collector efficiency η under the following conditions: Collector absorber-plate temperature $T_{coll} = 130°F$, interplate spacing $D = 3/8$ in., insolation $I = 150$ Btu/(hr)(ft²), and solar absorptance $\bar{\alpha}_s$ and infrared emittance $\bar{\epsilon}_{ir} = 0.95$. (*Adapted with permission from M. Altman.*[1])

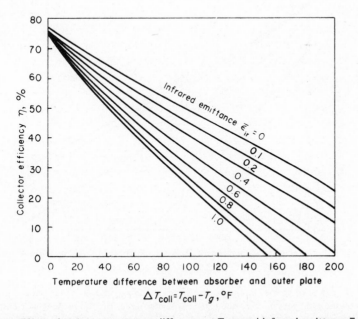

FIG. 3.5 Effect of various temperature differences ΔT_{coll} and infrared emittance $\bar{\epsilon}_{ir}$ values on the efficiency η of a double-plate collector under the following conditions: Collector absorber-plate temperature $T_{coll} = 130°F$, interplate spacing $D = 3/8$ in., insolation $I = 150$ Btu/(hr)(ft²), and solar absorptance $\bar{\alpha}_s = 0.95$. (*Adapted with permission from M. Altman.*[1])

63

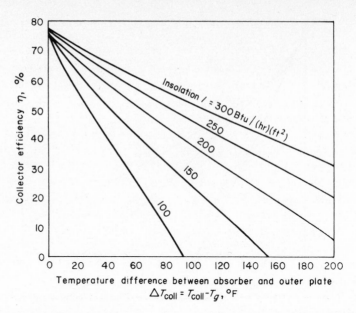

FIG. 3.6 Effect of various temperature differences ΔT_{coll} and insolation I values on the efficiency η of a double-plate collector under the following conditions: Collector absorber-plate temperature $T_{coll} = 130°$ F, interplate spacing $D = 3/8$ in., and solar absorptance $\bar{\alpha}_s$ and infrared emittance $\bar{\epsilon}_{ir} = 0.95$. (*Adapted with permission from M. Altman.*[1])

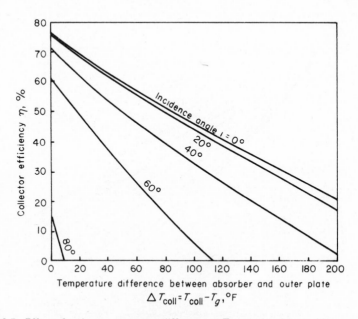

FIG. 3.7 Effect of various temperature differences ΔT_{coll} and various incidence angle i values on the efficiency η of a double-plate collector under the following conditions: Collector absorber-plate temperature $T_{coll} = 130°$F, interplate spacing $D = 3/8$ in., insolation $I = 250$ Btu/(hr)(ft^2), and solar absorptance $\bar{\alpha}_s$ and infrared emittance $\bar{\epsilon}_{ir} = 0.95$. (*Adapted with permission from M. Altman.*[1])

assumed constant in both cases. Consequently, most of the increase in solar incidence radiation shows up as collected useful energy.

Figure 3.7 illustrates the effect of solar-radiation incidence angle (angle between the surface normal and the sun's rays) on collector efficiency for an insolation I of 250 Btu(hr)(ft^2).

The effect of angle of incidence on efficiency is twofold: first, as the incidence angle increases, the amount of solar radiation impinging on the collector is reduced by a cosine factor; second, as the angle of incidence increases, the reflectance of the transparent cover plates increases. For values of incidence angle larger than 50°, the transmittance of the glass plate decreases rapidly, and reflective losses increase correspondingly. The combined impact of the cosine and reflectance effect is clearly shown in Fig. 3.7. For an angle of incidence of 80° the combined effect lowers the collector efficiency by so much that in the range of $T_{coll} - T_g$ greater than 10°F not only will no heat be collected but the collector actually loses heat if the absorber plate temperature is kept at 130°F.

Figure 3.8 shows the variation of collector efficiency as a function of time of day. It can be seen that between 10 A.M. and

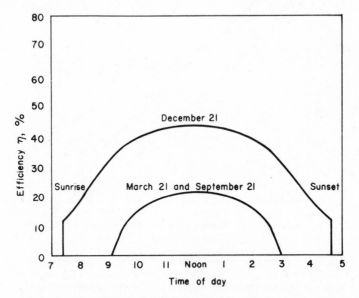

FIG. 3.8 Effect of various hours of the day and days of the year on the efficiency η of a double-plate collector at 40° N latitude under the following conditions: Collector absorber-plate temperature $T_{coll} = 130°$ F, collector outer-plate temperature $T_g = 30°$ F, interplate spacing $D = 3/8$ in., and solar absorptance $\bar{\alpha}_s$ and infrared emittance $\bar{\epsilon}_{ir} = 0.95$. (*Adapted with permission from M. Altman.*[1])

2 P.M. collector efficiency remains relatively constant. The efficiency during winter is higher than in fall and spring because the angle of incidence on a south-facing, vertical collector is smaller in December than in March or September. This seasonal incidence angle effect is the main reason vertical collectors are suitable for winter heating in medium latitudes.

    A performance comparison between a single-plate collector with a selective-black coating and a double-plate collector with either a selective-black or flat-black coating is shown in Fig. 3.9. Since it is not always architecturally feasible to orient a vertical collector exactly to the south, it is important to know the reduction in energy delivery incurred by non-south orientation. From Fig. 3.10, which shows the variation of insolation with collector orientation, it can be observed that a change in orientation of ±20° from the southern direction will not appreciably reduce the amount of solar energy collected.

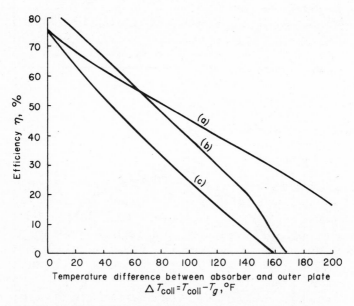

FIG. 3.9 Effect of various temperature differences ΔT_{coll} and various types of absorber coatings on efficiency η under the following conditions: Collector absorber-plate temperature $T_{coll} = 130°F$, interplate spacing $D = 3/8$ in., insolation $I = 150$ Btu/(hr)(ft²), and solar absorptance $\bar{\alpha}_s = 0.95$. (a) Double-plate collector with a selective-black absorber coating having an infrared emittance of 0.10; (b) single-plate collector with a selective-black coating having an infrared emittance of 0.10; (c) double-plate collector with a flat-black absorber coating having an infrared emittance of 0.95. (*Adapted with permission from M. Altman.*[1])

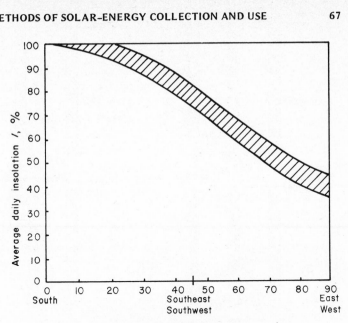

Wall-azimuth angle a_{wa}(deviation from south), deg

FIG. 3.10 Variation of insolation *I* on a vertical surface between November 21st and January 21st at latitudes between 30 and 45°N; the cross-hatched band represents the variation during these two months. (*Adapted with permission from M. Altman.*[1])

PERFORMANCE COMPARISON OF VARIOUS TYPES OF FLAT-PLATE SOLAR COLLECTORS

Figure 3.11 presents various collector configurations tested at the University of Pennsylvania, Philadelphia.[7] Table 3.1 gives high and low estimates for collector component costs for all these collector configurations (in 1975 dollars), and Table 3.2 presents the cost-effectiveness of selected collectors. The discussion here will be restricted to collectors type 1, type 2, type 2A, and type 17B since they are the most efficient, practical designs.

Collector type 1 is the standard type of collector that can be purchased commerically from a number of companies. The performance characteristics of collector types 2, 2A, and 17B were found to be superior to those of a double-plate absorber with flat-black collector surfaces.

Type 1, the reference collector, is the reference standard for subsequent comparisons. It consists of two double-strength (DS) glass panes, enclosing a $\frac{3}{8}$-in. air gap and a flat-black-coated absorber.

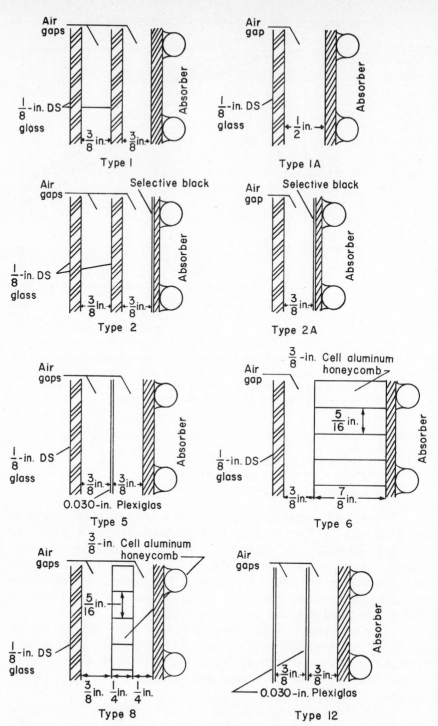

FIG. 3.11 Collector configurations tested at the University of Pennsylvania, Philadelphia; type 1 is the reference collector.[1]

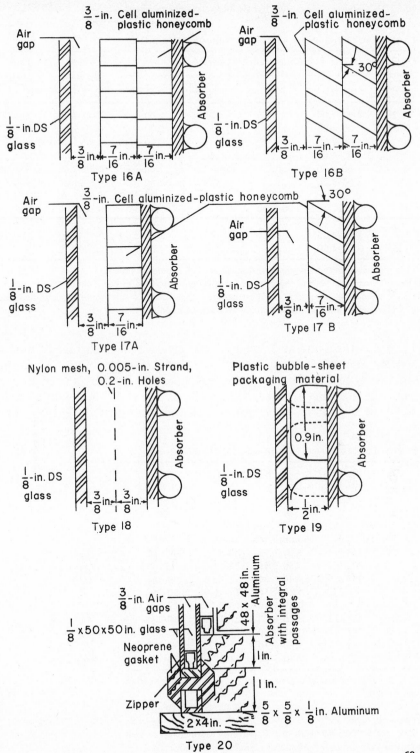

$\frac{3}{8}$-in. Cell aluminized-plastic honeycomb

Air gap

$\frac{1}{8}$-in. DS glass

Absorber

$\frac{3}{8}$in. $\frac{7}{16}$in. $\frac{7}{16}$in.

Type 16 A

$\frac{3}{8}$-in. Cell aluminized-plastic honeycomb

Air gap

$\frac{1}{8}$-in. DS glass

30°

Absorber

$\frac{3}{8}$in. $\frac{7}{16}$in. $\frac{7}{16}$in.

Type 16 B

Air gap

$\frac{3}{8}$-in. Cell aluminized-plastic honeycomb

$\frac{1}{8}$-in. DS glass

Absorber

$\frac{3}{8}$in. $\frac{7}{16}$in.

Type 17A

30°

Air gap

$\frac{1}{8}$-in. DS glass

Absorber

$\frac{3}{8}$in. $\frac{7}{16}$in.

Type 17 B

Nylon mesh, 0.005-in. Strand, 0.2-in. Holes

$\frac{1}{8}$-in. DS glass

Absorber

$\frac{3}{8}$in. $\frac{3}{8}$in.

Type 18

Plastic bubble-sheet packaging material

$\frac{1}{8}$-in. DS glass

0.9in.

Absorber

$\frac{1}{2}$in.

Type 19

$\frac{3}{8}$-in. Air gaps

$\frac{1}{8}$x50x50 in. glass

Neoprene gasket

Zipper

2 x 4 in.

48 x 48 in. Aluminum

Absorber with integral passages

1 in.

1 in.

$\frac{5}{8}$ x $\frac{5}{8}$ x $\frac{1}{8}$in. Aluminum

Type 20

TABLE 3.1　Collector-component Cost Estimates in 1975 Dollars[1]

Fig. 3.11 collector type	Component cost, $/ft² *				Credit for wall, $/ft² †,‡	Total cost, $/ft² †
	Glass†	Plastic†	Honeycomb†	Absorber†		
1, 20	4.00–7.00	...	...	2.00–3.00	0.50–1.00	5.00–9.50
1A	3.00–4.00	...	...	2.00–3.00	0.50–1.00	4.00–6.50
2	4.00–7.00	...	...	2.40–3.40	0.50–1.00	5.40–9.90
2A	3.00–4.00	...	...	2.40–3.40	0.50–1.00	4.40–6.90
6	3.00–4.00	...	0.80–1.00	2.00–3.00	0.50–1.00	4.80–7.50
8	3.00–4.00	...	0.60–1.00	2.00–3.00	0.50–1.00	4.60–7.50
10	...	2.50–4.50	...	2.00–3.00	0.50–1.00	3.50–7.00
11	...	3.00–5.00	...	2.00–3.00	0.50–1.00	4.00–7.50
12	...	2.20–4.20	...	2.00–3.00	0.50–1.00	3.20–6.70
16	3.00–4.00	...	0.90–1.20	2.00–3.00	0.50–1.00	4.90–7.70
17	3.00–4.00	...	0.70–1.20	2.00–3.00	0.50–1.00	4.70–7.70

*A general contractor's overhead and profit of 35% should be added.
†Ranges indicate low and high values for the item described in the column.
‡Since siding and other wall components are not needed if a solar system is installed, the cost of materials and labor associated with these components is saved.

Type 2 is the same as Type 1, but a selective-black surface of nine 4 × 4 in. aluminum foils is epoxied to the absorber. Type 2A is the same as type 2 but has a single glass plate separated by a $\frac{3}{8}$-in. air gap from the absorber surface. Type 17B has one glass plate with a single layer of 7/16-in. honeycomb, inclined upward 30° from the

TABLE 3.2　Cost-effectiveness of Selected Collectors[1]

Fig. 3.11 collector type	Relative efficiency η_r,* %	Estimated cost C_{est},* $/ft²	$\dfrac{\eta_r{}^*}{C_{est}}$ ft²/$	Relative cost-effectiveness,* %
1	100	5.00–9.50	20–10	100†
1A	65 ± 16	4.00–6.50	20–8	80–100
2	113 ± 3	5.40–9.90	22–11	110
2A	119 ± 6	4.40–6.90	28–16	140–160
6	75	4.80–7.50	16–10	80–100
8	68	4.60–7.50	15–9	75–90
12‡	87	3.20–6.70	27–13	130–135
16A	65 ± 10	4.90–7.70	15–7	70–75
16B	84 ± 11	4.90–7.70	19–9	90–95
17B	97 ± 5	4.70–7.70	22–12	110–120

*Ranges indicate low and high values for the item described in the column.
†The reference value to which other values in this column are compared.
‡The plastic material used in this collector was unstable at high temperatures and therefore not suitable for this application.

horizontal, placed against (in contact with) the absorber plate. A
$\frac{3}{8}$-in. air gap exists between the honeycomb and the single plate of
glass. The honeycomb layers are aluminized, phenolic-resin fiber glass.

Experimental measurements of the collector efficiency of a
double-glazed collector of type 1 in Fig. 3.11 were made at the
University of Pennsylvania for various operating temperatures and
insolation levels, and results are shown in Fig. 3.12.[7] The National
Bureau of Standards standard method for testing solar collectors is
presented in App. A.

Collector Performance Summary

The type 1 double-plate, flat-black collector that serves as the
reference proved to be one of the best, having a low heat loss, high
performance, and high cost-effectiveness. The reference-collector
design could be improved in three ways, however:

1. Significant thermal performance improvement
2. Manufacturing processes that would result in cost reduction while
   maintaining performance
3. Significant cost reduction that could justify some decrease in
   performance.

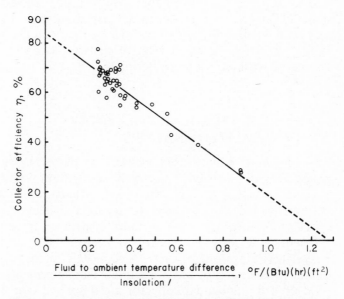

FIG. 3.12 Effect of the ratio of temperature difference to insolation
on efficiency for two double-glazed, flat-plate collectors of type 1
(experimental data). (*Adapted with permission from Lior and
Saunders.*[7])

The most promising collector for improvement method 1 is type 2A, the single-plate collector with a selective-black absorber coating. The thermal performance of type 2A is approximately 20 percent higher and its cost-effectiveness 40 to 60 percent greater than that of the reference collector. While a double-plate collector, black absorber type 2, has a higher thermal performance than its single-plate counterpart (type 2A), particularly under low ambient-temperature conditions, the cost-effectiveness of the type-2 collector is less.

The most significant candidates for improvement method 2 are collectors where one or all of the glass plates could be replaced by plastic. However, no satisfactory plastic has been developed that will have an acceptable lifetime and stability under high-temperature conditions and long-term exposure to ultraviolet radiation.

The best candidate for improvement method 3 is collector Type 17B, a single-plate collector with 1 7/16-in. aluminized, phenolic-resin, fiber glass honeycomb (with cells inclined 30°). Although its thermal performance is slightly degraded, its cost-effectiveness is 10 to 20 percent greater than that of the reference collector. It does not have any of the potential drawbacks of the selective-black absorber coating and represents an excellent back-up configuration to the type 2, which has a selective-black-coated absorber.

Thus, in conclusion, type 2A has a relative cost-effectiveness of 140 to 160 percent greater than that of type 1 and appears to be the most efficient of the four absorbers discussed. Type 17B has a relative cost-effectiveness of 110 to 120 percent greater than that of type 1, and type 2 has a relative cost-effectiveness of 110 percent compared with the standard flat-plate collector.

FLAT–PLATE SOLAR–COLLECTOR: GENERALIZED ANALYSIS

The most widely used measure of performance of a flat-plate collector is the efficiency η, that is, the ratio of delivered heat to the insolation. Although very important, efficiency is not a sufficiently descriptive index by which to select a collector since the more efficient a collector is, the more expensive it usually is. A less efficient, but also less costly, collector might provide the most cost-effective option. In Chap. 4 economic factors are considered in detail. In this section the most important factors affecting performance of a liquid-cooled collector are described, and the method for calculating efficiency is outlined.

The most important properties of a working collector are:

1. Collector temperature
2. Type of collector surface
3. Number and type of covers

These three parameters, along with insolation level, are described analytically in this section so that collector delivery may be calculated.

Covers

The only reliable material currently available for collector covers is glass. However, certain types of glass are more appropriate for collector use than others. The single most important property of glass for solar collectors is the percent of solar radiation it transmits. This transmittance depends directly on the refractive index of the glass. Some typical glass transmittance properties for a normally incident beam of light are shown in Table 3.3. This table shows that the reflection losses are always about 8 percent, but the absorptance losses vary by a factor of 7. The absorptance losses, in turn, depend directly on the iron oxide content of the glass. The higher the iron oxide content, the greater the absorptance losses. The amount of iron in a given glass may be determined from the manufacturer. A simple inspection can indicate high or low iron content: the more iron oxide, the greener a broken edge of the glass will appear.

The reflectance of a plate of glass also varies with the incidence angle i. An expression (the Fresnel equation) to calculate the reflection of beam radiation from one surface of a transparent solid $\bar{\rho}_{b1}$ is given in standard optics texts and in Eq. (3.1):

$$\bar{\rho}_{b1} = \frac{1}{2} \left[\frac{\sin^2 (i - r_a)}{\sin^2 (i + r_a)} + \frac{\tan^2 (i - r_a)}{\tan^2 (i + r_a)} \right] \tag{3.1}$$

TABLE 3.3 Properties of Glass

	Type of glass		
Property	Ordinary float	Sheet lime	Crystal white
Iron oxide content, %	0.10–0.13	0.05	0.01
Refractive index	1.52	1.51	1.50
Light transmittance (normal), %	85–81	87–85	90.5
Glass thickness, in.	0.125–0.1875	0.125–0.1875	0.1875
Reflectance, loss, %	8.2–8.0	8.1–8.0	8.0
Absorptance loss, %	6.8–11.0	4.9–7.0	1.5

Where r_a is the angle of refraction, and

$$\sin r_a = \frac{\sin i}{r_{i,\text{coll}}} \tag{3.2}$$

where $r_{i,\text{coll}}$ is the refractive index of a collector cover. If the incidence angle is exactly $0°$, a special form of the Fresnel equation must be used:

$$\bar{\rho}_{b1} = \frac{(r_{i,\text{coll}} - 1)^2}{(r_{i,\text{coll}} + 1)^2} \tag{3.3}$$

The reflectance from several glass surfaces $\bar{\rho}_{bn}$ is then given by

$$\bar{\rho}_{bn} = 1 - \frac{1 - \bar{\rho}_{b1}}{1 + (2n - 1)\bar{\rho}_{b1}} \tag{3.4}$$

where n is the number of collector covers (polarization ignored).

The reflective loss for diffuse radiation, $\bar{\rho}_d$ which cannot be expressed as a simple equation and must be computed by integrating over the sky, is given in Table 3.4.

Type of Collector Surface

Although much research has been done on selective surfaces, few selective surfaces are available today that have sufficient durability experience and low cost to warrant use on flat-plate collectors. The only commercial selective surface available with lengthy field experience is the proprietary Miromit Selective Black™ * used on Miromit solar water heaters. Nonselective, black

*® Registered trademark, Miromit Ltd., Tel Aviv, Israel.

TABLE 3.4 Diffuse Reflectance from Glass*

No. of surfaces n	Reflectance for diffuse radiation $\bar{\rho}_d$
0	0.0
1	0.16
2	0.24
3	0.29

*Adapted from Löf and Tybout (1972)[9] with permission from the publishers.

TABLE 3.5 Angular Variation
of Absorptance of Lamp-black
Paint*

Incidence angle i	Absorptance $\bar{\alpha}_{s,b}$
0–30	0.96
30–40	0.95
40–50	0.93
50–60	0.91
60–70	0.88
70–80	0.81
80–90	0.66

*Adapted from Löf and Ty-
bout (1972)[9] with permission
from the publishers.

paints with an appropriate primer are recommended for general use at this time. One such paint is Nextel^TM *. The solar absorptance and infrared emittance are 0.96 for this product. The angular variation of the absorptance for a typical nonselective black surface is shown in Table 3.5. The absorptance for diffuse radiation $\bar{\alpha}_{s,d}$ may be taken as

$$\bar{\alpha}_{s,d} = 0.90 \qquad (3.5)$$

Collector Temperature and Thermal Losses

The thermal losses of a flat-plate collector depend on the operating temperature of the collector. The operating temperature is determined by the ultimate use to which the heated fluid is to be put and can range from 90 to 210°F or higher. The fluid flow rate through a collector may be determined from the heat balance:

$$q_a = \dot{m} c_f (T_{coll,in} - T_{coll,out}) \qquad (3.6)$$

where q_a = absorbed energy per unit collector area, Btu/(hr)(ft^2)
$\dot{m}$ = fluid flow rate per unit collector area, lb$_m$/(hr)(ft^2)
c_f = specific heat of fluid, Btu/(lb$_m$)(°F)
$T_{coll,in}$ = collector transport-fluid-inlet temperature, °F
$T_{coll,out}$ = collector transport-fluid-outlet temperature, °F
Equation (3.6) may be solved for the required mass flow rate $\dot{m}$:

$$\dot{m} = \frac{q_a}{c_f (T_{coll,out} - T_{coll,in})} \qquad (3.7)$$

*®Registered trademark, 3M Company, St. Paul, Minn.

For a pumped water system, the inlet-outlet temperature difference is 10 to 20°F; for a thermosiphon, about 40°F. The average collector fluid temperature T_{coll} may be taken as the average of the inlet and outlet temperatures for liquid cooled collectors,

$$T_{coll} = \frac{T_{coll,in} + T_{coll,out}}{2} \tag{3.8}$$

An expression for the thermal loss L_t from such a collector by radiation and convection has been developed[4] and is given here without proof (back and edge losses assumes small).

$$L_t = \frac{T_{coll} - T_{out}}{\dfrac{1}{h_{c1}} + \dfrac{(n/c)}{\sqrt[4]{(T_{coll} - T_{out})/n + f}}} + \frac{\sigma(T_{coll}^4 - T_{out}^4)}{\dfrac{1}{\bar{\varepsilon}_{ir,c}} + \dfrac{2n + f - 1}{\bar{\varepsilon}_{ir,g}} - n} \tag{3.9}$$

where h_{c1} = coefficient of heat transfer from the outermost collector cover, Btu/(hr)(ft²)(°F)

c = factor to account for collector tilt in the expression for coefficient of free convection heat transfer on glass collector covers, Btu/(hr)(ft²)(°F$^{1.25}$)

$\bar{\varepsilon}_{ir,c}$ = nonselective collector surface emittance

$\bar{\varepsilon}_{ir,g}$ = cover emittance

n = number of glass covers

f = outer collector-cover to non-outer cover heat transfer resistance ratio,

and

$$h_{c1} = 1.0 + 0.35v \tag{3.10}$$

$$c = 0.19 - 0.00078\beta \tag{3.11a}$$

$$\beta = \text{surface tilt angle from horizontal} \tag{3.11b}$$

$$\bar{\varepsilon}_{ir,c} = 0.95 \tag{3.12}$$

$$\bar{\varepsilon}_{ir,g} = 0.90 \tag{3.13}$$

$$f = 0.76 \times 10^{-0.037v} \qquad v \leqslant 8 \tag{3.14a}$$

$$f = 0.36 \times 10^{-0.020(v-8)} \qquad 8 < v \leqslant 17 \tag{3.14b}$$

$$f = 0.24 \times 10^{-0.011(v-17)} \qquad v > 17 \qquad\qquad (3.14c)$$

where v is the wind speed in knots.

Insolation

The insolation on a horizontal surface I_h consists of a direct (beam) component and a diffuse (skylight) component. The proportionate contribution of each to the total horizontal flux is difficult to evaluate. Approximately, the following equation may be used to compute the diffuse component $I_{h,d}$[9]

$$I_{h,d} = 0.78 + 1.07\alpha + 6.17CC \qquad\qquad (3.15)$$

where α = solar-altitude angle, deg
CC = cloud cover ($CC = 0$ indicates a clear sky, $CC = 10$ indicates the sky is fully covered with clouds)*
The horizontal beam component $I_{h,b}$ is the difference.

$$I_{h,b} = I_h - I_{h,d} \qquad\qquad (3.16)$$

The incident beam component normal to the collector $I_{b,coll}$ is then

$$I_{b,coll} = \frac{I_{h,b}\ \cos i}{\sin \alpha} \qquad\qquad (3.17)$$

Delivery and Efficiency

The amount of energy absorbed by the absorber plate I_{coll} is the insolation diminished by cover and absorbing-surface reflectance losses (second order ground reflection effects are ignored):

$$I_{coll} = I_{b,coll}(1 - \bar{\rho}_{bn})\bar{\alpha}_{s,b} + I_{h,d}(1 - \bar{\rho}_d)\bar{\alpha}_{s,d} \qquad\qquad (3.18)$$

The value of I_{coll} is a property of the collector independent of the operating temperature of the collector.
The amount of energy delivered to the working fluid q_a is I_{coll} diminished by thermal losses L_t (see Eq. (3.9)),

$$q_a = I_{coll} - L_t \qquad\qquad (3.19)$$

---

*See Fig. C.3(a), "Local Climatological Data," in App. C for examples of various sky-cover conditions.

Finally, the collector efficiency η is the ratio of output, q_a to input, $I_b + I_{h,d}$:

$$\eta = \frac{q_a}{I_b + I_{h,d}} \tag{3.20}$$

Flat-plate, air-cooled solar collectors may use either the space above or below the absorber plate as the air conduit. Heat losses are smaller and the collection efficiency larger when the air duct is placed below the absorber plate. For this design the equations developed above for the liquid-cooled flat-plate collector can be used with only minor modifications for air systems. For the arrangement in which air flows between the innermost collector cover and the absorber plate, no theoretical analysis presently exists and performance data are sparse.

Bliss[3] has refined the thermal model of a liquid-cooled flat-plate collector described above by accounting for the absorber plate temperature gradients between the fluid tubes and along the direction of fluid flow. Two efficiency factors are defined. The first accounts for the temperature gradient between the fluid tubes and is called the *fin efficiency.* The computation of fin efficiency is treated in standard heat transfer textbooks. The second factor that Bliss developed accounts for the absorber-plate temperature rise along the fluid conduit and is called the *plate efficiency.* (Methods for calculation of plate efficiency are described by Bliss.[3]) Plate and fin efficiencies in excess of 97 percent are readily achievable by proper design Klein[5] has considered thermal capacitance effects on the performance of flat-plate collectors and conclude that thermal transients are not important in the design of systems that utilize flat-plate collectors of conventional design.

The steps in efficiency calculation may be tabulated as follows:

1. Calculate collector insolation $I_{b,coll}$ and $I_{h,d}$ from Eqs. (3.15) to (3.17).
   Input: Insolation I_h; cloud cover (CC); solar altitude angle α; collector tilt β; latitude L

2. Calculate absorbed radiation I_{coll} from Eq. (3.18).
   Input: Number of covers (assumed glass) n

3. Calculate delivered energy q_a from Eqs. (3.9) to (3.14) and (3.19).
   Input: Wind speed v, knots; collector temperature T_{coll}, °R; ambient temperature T_{out}, °R; collector physical properties

4. Calculate efficiency η from Eq. (3.20).

EXAMPLE

Using the data listed below, calculate collector efficiency η.

Factor	Specification or description
Location and latitude L	Boulder, Colo.; latitude 40°N
Date and time	January 1; 11:30–12:30 local solar time
Insolation I_h, ly/min	0.5
Cloud cover (CC)	0
Collector tilt β, deg	55°N (latitude ± 15°)
No. of covers n	2
Wind speed v, knots	1
Collector temperature T_{coll}, °F	124 (= 584°R)
Ambient outside temperature T_{out}, °F	22 (= 482°R)

Step 1: Calculate insolation I_{coll} and $I_{h,d}$.
Solar declination $\delta_s = -23°$ (see Fig. 2.15)
Local solar-hour angle $H_s = 0°$
Solar altitude α (from Eq. (2.23))

$$\sin \alpha = \sin \delta_s \sin L + \cos \delta_s \cos L \cos H_s = 0.454$$

$$\alpha = 27°$$

Diffuse component $I_{h,d}$ (from Eq. (3.15))

$$I_{h,d} = 0.78 + 1.07\alpha + 6.17CC = 29.6 \text{ Btu/(hr)(ft}^2)$$

Beam component $I_{h,b}$ (from Eq. (3.16))

$$I_{h,b} = I_h - I_{h,d} = 81 \text{ Btu/(hr)(ft}^2)$$

Incidence angle i (from Eq. (2.29))

$$\cos i = \cos H_s \cos \delta_s \cos (L - \beta)$$
$$+ \sin \delta_s \sin (L - \beta) = 0.990$$
$$i = 8°$$

Beam component incident on collector $I_{b,coll}$ (from Eq. (3.17))

$$I_{b,coll} = \frac{I_{h,b} \cos i}{\sin \alpha} = 179 \text{ Btu/(hr)(ft}^2)(\approx I_b)$$

Step 2: Calculate radiation-absorbed I_{coll}.
Single-surface reflectance $\bar{\rho}_{b1}$ (from Eqs. (3.1) and (3.2))

$$\sin r_a = \frac{\sin i}{r_{i,coll}} = \frac{\sin 8°}{1.52} = 0.091$$

$$r_a = 5.25°$$
$$i - r_a = 2.75°$$
$$i + r_a = 13.25$$

$$\bar{\rho}_{b1} = \frac{1}{2}\left[\frac{\sin^2 (2.75)}{\sin^2 (13.25)} + \frac{\tan^2 (2.75)}{\tan^2 (13.25)}\right] = 0.043$$

Multiple-surface reflectance $\bar{\rho}_{bn}$ (from Eq. (3.4) and Table 3.4)

$$\bar{\rho}_{bn} = 1 - \frac{1 - \bar{\rho}_{b1}}{1 + (2n - 1)\bar{\rho}_{b1}} = 0.15$$

$$\bar{\rho}_d = 0.24$$

Absorber-plate absorptance $\bar{\alpha}_{s,b}$ (from Table 3.5 and Eq. (3.5))

$$\bar{\alpha}_{s,b} = 0.96$$
$$\bar{\alpha}_{s,d} = 0.90$$

Compute I_{coll}

$$\begin{aligned}I_{coll} &= I_{b,coll} (1 - \bar{\rho}_{bn})\bar{\alpha}_{s,b} + I_{h,d} (1 - \bar{\rho}_d)\bar{\alpha}_{s,d}\\ &= (179 \times (1 - 0.15) \times 0.96) + (29.6 \times (1 - 0.24)\\ &\quad \times 0.90)\\ &= 165 \text{ Btu/(hr)(ft}^2)\end{aligned}$$

Step 3: Calculate delivered energy q_a
Compute thermal-loss-equation parameters (from Eqs. (3.9) to (3.14a) and (3.19)).

$$c = 0.19 - (0.00078 \times 55°) = 0.147$$
$$f = 0.76 \times 10^{-1(0.037)} = 0.70$$
$$h_{c1} = 1 + 0.35v = 1.35$$
$$\bar{\varepsilon}_{ir,c} = 0.95$$
$$\bar{\varepsilon}_{ir,g} = 0.90$$
$$L_t = 49.5 \text{ Btu/(hr)(ft}^2)$$

$$q_a = I_{coll} - L_t = 115.5 \text{ Btu/(hr)(ft}^2)$$

Step 4: Calculate efficiency
Compute efficiency from Eq. (3.20) as follows:

$$\eta = \frac{q_a}{I_b + I_{h,d}} = \frac{115.5}{179 + 29.6} \times 100\% = 55\%$$

Collector Temperature Extremes

This subsection describes the thermal impacts on a collector that would occur if fluid flow should cease during a period of high insolation. In that case no useful energy is removed and the thermal balance on the collector consists of the solar input component and radiative and convective losses from the absorber plate via the collector covers, edges, and rear surface. Since efficient solar collectors are designed specifically to prevent heat loss from the absorber plate, the absorber plate can reach a relatively high temperature under no-flow conditions. The absorber of a double-cover collector may reach 400°F at noon on a sunny day if fluid flow should cease. The elevated temperature and consequent thermal stress on the collector could cause a breakage of the glass covers on a flat-plate collector, deterioration of selective surfaces, outgassing of collector sealants and surface paints, or melting of the absorber on a concentrator. Fail-safe measures such as a cooling tower in large systems or a method for automatically defocusing a concentrator must be incorporated into every solar system to deal with possible cessation of coolant flow.

Solar collectors are also subjected to low temperature extremes, but their effects are less dangerous in comparison to high-temperature episodes. As described elsewhere in this book, collector protection against low temperature damage is straightforward and uses well known freeze protection techniques. Snow buildup on the surface of flat-plate collectors is typically not a problem. The transmittance of snow to solar radiation is large enough to allow the collector cover to warm sufficiently to melt a thin layer of snow adjacent to the cover. The entire snow cover usually slides from the collector at one time within an hour or two after solar energy falls on the collector. Consistent snow self-purging of this type has occurred on collectors tilted at angles as low as 27 deg from the horizontal.

CONCENTRATING COLLECTOR SYSTEMS

The potential advantages of solar-energy collection systems that use concentrators have long been recognized. A solar steam engine operating from steam produced by a large, parabolic mirror that tracked the sun and concentrated the solar rays on a small absorber placed at its focal point was exhibited at the World's Fair in Paris in 1878 and is illustrated in Fig. 3.13. A 50-hp solar steam engine using a trough-type parabolic reflector with a tubular absorber placed at the focal point of the mirror was built and operated successfully by the American engineer F. Shuman in Egypt in 1913. The mirror in Shuman's collector had a 4.5:1 concentration ratio (CR), i.e., the ratio of the aperture area intercepting the sun's rays to the absorber area. The system, shown in Fig. 3.14, tracked the sun by moving the entire collector to maintain the sun at all times in the plane perpendicular to the projected area of the mirror. The system was used to pump irrigation water from the Nile.

Although these and many other concentrator systems operated successfully from a technical viewpoint, they were economic failures because until recently all concentrator concepts required that the mirror move to track the sun's motions during the day and during the year. Consequently, flat-plate absorber systems, which do not require tracking of the sun, have over the past 30 yr taken

FIG. 3.13 Pifre sun power plant driving a printing press, circa 1878.

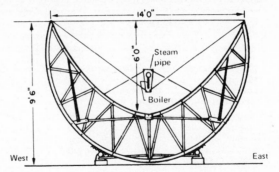

FIG. 3.14 Configuration of the concentrator built by Shuman in Meadi, near Cairo. (*From N. Robinson, "Solar Mechanics," Proceedings of the World Symposium on Applied Solar Energy*, 1955, p. 50. Reprinted by permission.)

predominance because they could be built more economically and installed more cheaply than concentrator systems which require the entire reflecting mirror to track the sun. However, during the past few years at least three new ideas for concentrators that do not require the large reflecting mirror to follow the sun have been proposed. Eliminating the need for the solar reflector to track the sun has improved the economic outlook for concentrator-type, solar-collection systems.

New Concentrator Designs for Heating and Cooling

Three promising systems that recently have been subjected to considerable analysis and development are

1. The SRTA (stationary reflector/tracking absorber) system developed by W. Gene Steward of Environmental Consulting Services at Boulder, Colo. and by J. L. Russell of General Atomics at San Diego, Cal. The Steward SRTA is a compound curvature or dish collector and is described later in this chapter. The Russell SRTA is a single curvature or trough collector. Both are under development for electric power production.
2. The CPC (compound parabolic concentrator) trough system developed by Roland Winston at Argonne National Laboratory, Argonne, Ill.
3. The POC (pyramidal-optical concentrator, solar-energy-collector system) presently under development by the Wormser Scientific Company, at Stamford, Conn.

TABLE 3.6 Advantages and Disadvantages of Stationary-reflector Concentrating Collector Systems (Compared to Flat-plate Collector Systems)

Advantages	Disadvantages
1. Reflecting surfaces require less material and are structurally simpler than flat-plate collectors. For a concentrator system the cost per square foot of solar collecting surface is therefore potentially less than that for flat-plate collectors.	1. Concentrator systems collect less diffuse radiation than flat-plate collectors.
2. The absorber area of a concentrator system is smaller than that of a flat-plate system of the same solar energy collection area, and the insolation intensity is therefore greater.	2. In some stationary reflecting systems it is necessary to have a small absorber track the sun image (SRTA); in others the reflector may have to be adjustable (CPC) to more than one position if year-round operation is desired.
3. Because the area from which heat is lost to the surroundings per square foot of solar-energy collecting area is less than that for a flat-plate collector and because the insolation on the absorber is more concentrated, the working fluid can attain higher temperatures in a concentrator system than in a flat-plate collector of the same solar energy collecting surface.	3. Some systems, e.g., SRTA, require one flexible connection or joint to take the hot working fluid out of the absorber. Such a joint may need maintenance annually.
4. Owing to the small area of absorber per unit of solar-energy collecting area, selective-surface treatment and/or vacuum insulation to reduce heat losses and improve collector efficiency are economically feasible.	4. Reflectance of the solar mirror surface may decrease with time for some surface finishes and may require periodic cleaning and refurbishing.
5. Concentrating systems can be used for electric power generation when not used for heating or cooling. The total useful operating time per year can therefore be larger for a concentrator system than for a flat-plate collector, and the initial installation cost of the system can be regained by savings in energy in a shorter period of time.	5. There is little experience in the operation of stationary reflector-high concentration collector systems, and more research and development work is desirable to obtain experience and confidence in the operation, engineering, and maintenance of such systems.
6. Because the temperature attainable with concentrating systems is higher, the amount of heat which can be stored per unit volume is larger, and consequently the heat storage costs are less for concentrator systems than for flat-plate collectors.	
7. In solar heating and cooling applications the higher temperature of the working fluid attainable with a concentrating system makes it possible to obtain higher efficiencies in the cooling cycle and lower cost for air conditioning with concentrator systems than with flat-plate collectors.	
8. Little or no antifreeze is required to protect the absorber in a concentrator system whereas the entire solar-energy collection surface requires antifreeze protection in a flat-plate collector.	

The main advantages of concentrator systems over flat-plate-type collectors are given in Table 3.6; in brief however, the main advantages are that at a given temperature concentrator systems have a higher collection efficiency, the working fluid can be heated to a much higher temperature than in a flat-type collector, and energy storage per unit volume is greater and therefore less expensive than with a flat plate. Moreover, because the working temperature attainable with a concentrator system is higher, it is possible to generate electricity and utilize the system over a larger fraction of a year. Finally, and probably most importantly, the cost of energy delivered by solar concentrator systems appears to be less than the cost of energy delivered by flat-plate collectors in sunny parts of the world, and concentrator systems also promise considerable economies of scale.

One reason for the reduced cost of energy delivered is clearly illustrated in Fig. 3.15, which shows the cross-sections of a typical flat-plate collector and the reflecting mirror surface for a concentrating absorber-type system. The flat-plate collector requires at least one—but probably two—plates of glass separated by an air space; behind these plates of glass there must be a means of conducting the working fluid and absorbing the solar energy, which can either be a sheet of roll-bond construction, a finned-type tube sheet, or a rectangular-type flow channel for air collectors. Finally, the flat-plate collector requires a layer of insulation to reduce thermal losses from the

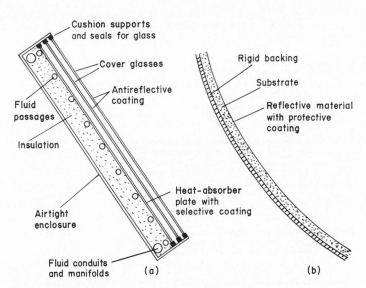

FIG. 3.15 Comparison of construction techniques for (a) flat-plate and (b) concentrating collectors. (*Courtesy of W. Gene Steward.*)

underside. It is estimated that the cost of the materials, including the seals for the glass plates, would amount to at least $3.50/ft² on today's market, and if one estimates that the labor costs will increase at approximately the same rate as the raw materials, it appears unlikely that flat-plate collectors will ever be available, installed, at less than $7.00/ft².

In contrast to the complexity of a flat-plate collector system, the surface of a reflector in a concentrating collector is extremely simple. It consists of a thin layer of reflecting material, which need only be a few molecules thick, and a substrate formed to the appropriate shape that can be made of inexpensive material such as polyurethane, certain plastics, or concrete. Reflectors constructed of urethane or light-weight plastics need a supporting structure to maintain the shape of the reflector. The reflecting material has been estimated to cost about $0.30–0.50/ft², and the supporting structure plus rigid foam about $2/ft² under mass production. Even if it is assumed that these cost figures are optimistic, and even if 50 percent is added to them, it appears entirely possible that a reflecting collector can be manufactured at about $4.50/ft², or approximately two-thirds the cost of flat-type collectors. This estimate includes an allowance for the absorber, the surface area of which is estimated to be about 3 percent of the collection surface area at 10 times the cost. This reduction in cost per square foot of aperture area, coupled with the higher collection efficiency, indicates the cost of thermal energy attainable with a concentrator system will be considerably less than current energy-collection costs with flat-plate collectors.

Operating Trade-offs between Concentrating and Flat-plate Solar Collectors

The main advantages and disadvantages of modern types of concentrating collector systems with stationary reflectors are summarized in Table 3.6. The deterioration of reflecting material mentioned as a possible disadvantage for the concentrating collector systems has been investigated by a research team at the University of Minnesota, Minneapolis,[12] and available data indicate that highly reflective materials that do not deteriorate appreciably with time are already available commercially; their characteristics are shown in Table 3.7. A summary of internal energy storage volumes for systems at different temperatures is shown in Fig. 3.16.

High-temperature storage can reduce the volume and hence the cost of thermal storage since the energy storage per unit volume increases with temperature. Aluminum oxide is an inexpensive and effective solid-phase storage medium since its specific heat is about

TABLE 3.7 Selected Reflectance Data[12]

Type of material	Original reflectance	Exposure time, weeks	Reflectance		Degradation, %
			Uncleaned surface	Cleaned surface	
Aluminized fiber glass (General Dynamics)	0.92	55	0.79	0.82	11
Aluminized acrylic (3M Company)	0.86	50	0.83	0.85	1
Aluminized plexi- glass (Ram Products)	0.80	39	0.71	0.78	3
Anodized aluminum (Alcoa Aluminum)	0.82	34*	. . .	0.79	4

*Accelerated radiation test was used.[12]

twice that of ordinary gravel. Energy storage systems other than thermal have not been developed sufficiently to provide an economical alternative to thermal storage for building heating and cooling.

The main problem with concentrating systems is that they require *direct-beam,* or collimated, *radiation.* The ratio of beam radiation to diffuse radiation for the Four Corners states, the South and southern California and Nevada varies between 3 and 5, but over a large city with smog and fog it may be as low as 1.5. Preliminary estimates indicate, however, that even in a climate such as exists in Iowa the collimated radiation is sufficient to give a distinct advantage to a concentrator over flat-plate-collector system. Since even flat-plate collectors convert only 20 to 30 percent of the diffuse radiation into useful energy, the loss incurred by using a concentrator is not too large. However, in each application the final answer as to whether or not a concentrator or flat-plate collector system is economically most advantageous can only be given after a careful economic analysis.

SRTA System Description

The spherical stationary reflector/tracking absorber (SRTA) concentrator system (patent applied for) discussed here was conceived by W. Gene Steward. The system is based on optical principles which show that, regardless of the sun's location, a fixed spherical mirror can focus most of the incoming solar radiation on a line parallel to the sun's rays. Although this focal line turns about the geometric center of the spherical mirror as the sun moves across the horizon, it is not necessary to move the mirror to maintain the collector in focus; only a small cylindrical absorber, which collects

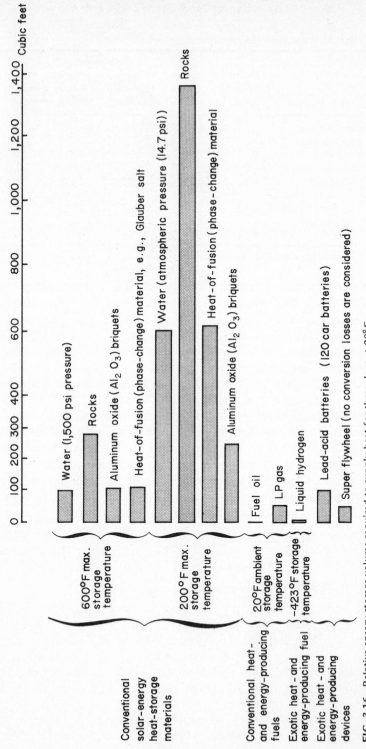

FIG. 3.16 Relative energy-storage volume required to supply heat for three days at 20°F ambient temperature for a 35,000 Btu/degree-day home. (*Courtesy of W. Gene Steward.*)

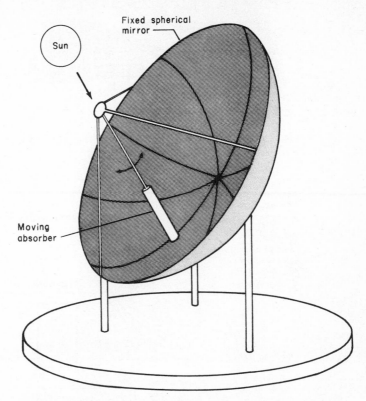

FIG. 3.17 Stationary-reflector/tracking absorber (SRTA) solar collector.

the concentrated energy, need move in a planar motion from morning to night. This absorber movement can be accomplished relatively simply and inexpensively, thereby making the concentrator system advantageous for solar heating and cooling of homes and office buildings as well as for the generation of electrical power. The SRTA collector is shown in Fig. 3.17.

SRTA HOME HEATING SYSTEM

Figure 3.18 is a sketch of the spherical reflector system integrated into the Steward house, located near Boulder, Colo. The collector is capable of supplying nearly all of the heating requirements of this 3,400-ft^2 home located at 8052 ft above sea level. Figure 3.19 is a schematic diagram of the heating and storage system used in the design of this house. The concentrator generates low-pressure steam, which is condensed in the 400 ft^3 storage tank, thereby heating the rocks or Al_2O_3 briquets used to store internal

FIG. 3.18 Architect's drawing of the Steward house, Boulder, Colo. (*Courtesy of Charles Haertling, AIA*.)

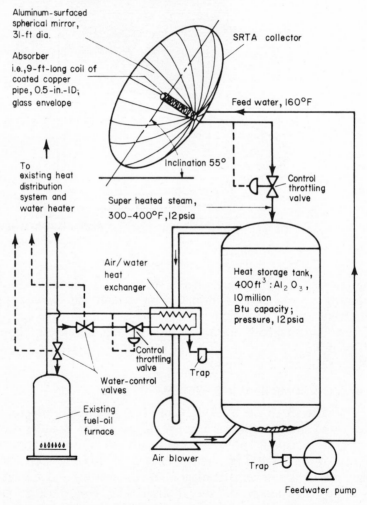

Aluminum-surfaced spherical mirror, 31-ft dia.

Absorber i.e.,9-ft-long coil of coated copper pipe, 0.5-in.-ID; glass envelope

SRTA collector

Feed water, 160°F

Inclination 55°

To existing heat distribution system and water heater

Super heated steam, 300-400°F, 12 psia

Control throttling valve

Air/water heat exchanger

Heat storage tank, $400 \, ft^3$: $Al_2 O_3$, 10 million Btu capacity; pressure, 12 psia

Control throttling valve

Trap

Water-control valves

Existing fuel-oil furnace

Air blower

Trap

Feedwater pump

FIG. 3.19 Home heating system using a stationary-reflector/tracking-absorber (SRTA) solar collector, Steward house, Boulder, Colo.; there is no service-hot-water heating shown.

TABLE 3.8 Cost Estimates for a Custom-built SRTA Collector*

Component or area	Description	Cost† Materials	Labor
Reflector:	35 ft diameter; 120° included angle; 935 ft² frontal area; 1,247 ft² surface area		
Framework	1/2-in. dia. rebar and expanded metal	$ 250	$ 350
Polyurethane foam	2–4 in. thick	416	1,460
Reflective sheeting	Adhesive-backed	515	640
Total		$1,181	$2,450
Absorber:	10-ft long, 6-in. dia. coil		
Copper tubing	1/2-in. inside dia.	54	100
Black coating	Nonselective		20
Glass enclosure	1/4-in thick enclosure	77	190
Fittings and conduit	Copper	95	100
Tracking system	Solid-state photodiode	75	100
Total		$ 301	$ 426
Heat storage:	2-MBtu capacity		
Steel tank	600 gal	160	100
Insulation	12-in. fiber glass	75	125
Heat-storage material	Al_2O_3 solid; 25% void space	330	60
Feedwater pump with controls	1/8 hp	120	30
Total		$ 680	$ 315

*The estimates in this table are for the type of collector used in the house shown in Fig. 3.18.
†No credit for wall replacement is included; costs in 1975 dollars.

energy for use during the night or during cloudy days (see Chap. 4 "Air Cooled Solar Collection Systems" for advantages of gravel storage.) For the installation shown in Fig. 3.19 storage for as long as five days under adverse weather conditions is possible. The system is also capable of generating electricity by driving a small steam turbine from the steam developed in the absorber. Cost estimates for an SRTA solar-energy collection system similar to that in the Steward house, made in detail on a custom-fabrication basis, are shown in Table 3.8.

SRTA OPTICAL ANALYSIS
AND DESIGN

The design and sizing of a spherical SRTA solar-energy conversion system is based on basic optical characteristics outlined below.[6] Consider a spherical mirror, as shown in Fig. 3.20, and divide the frontal surface into n equal areas, i.e.,

$$\pi r_{a1}^2 = \pi(r_{a2}^2 - r_{a1}^2) = \cdots = \pi(r_{an}^2 - r_{an-1}^2)$$

Then select a coordinate system in which x is the distance from the center point of the spherical mirror along the ordinate and y is the distance from the center of the sphere along the abscissa. If the y axis is aligned with the direction of the sun, any ray from the sun at x will be reflected onto the y axis as shown. To make the system dimensionless, divide x and y by r, the sphere radius, or

$\bar{x} = x/r$ = radial distance to any incident ray$/r$
$\bar{y} = y/r$ = distance to y intercept$/r$
It can easily be verified that $\bar{y}$ and $\bar{x}$ are related by

$$\frac{1}{\bar{y}} = 2\sqrt{1 - \bar{x}^2}$$

If the frontal area of the spherical mirror is oriented in such a way that at noon it is perpendicular to the sun's rays, the incident and

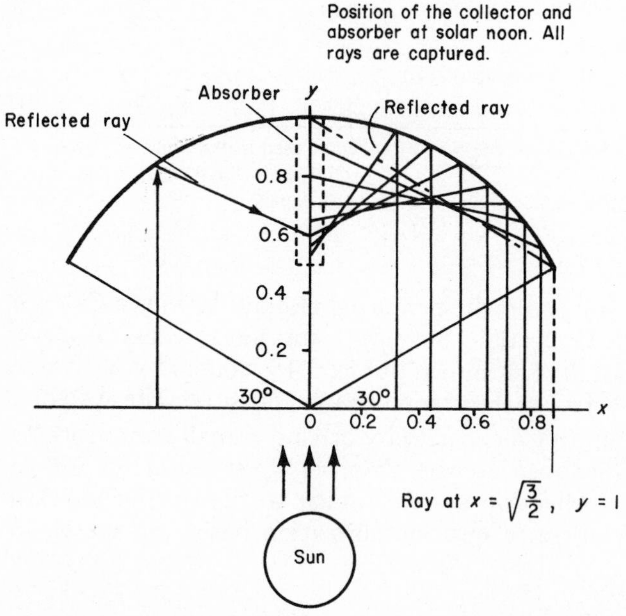

FIG. 3.20 Reflecting characteristics of a perfect spherical mirror of an SRTA collector at solar noon (at this angel of the collector and absorber all rays are captured); rays shown on the right mark equal-area circles and represent equal amounts of energy. (*Courtesy of W. Gene Steward.*)

reflected rays follow the paths shown in Fig. 3.20. It can be seen that the outermost rays that will intercept the $\bar{y}$ axis with one reflection correspond to

$$\bar{y} = 1 \quad \text{and} \quad \bar{x} = \sqrt{\frac{3}{2}} = 0.865$$

In other words, the rays incident within a circle of radius $\bar{x} = 0.865$ are reflected to intercept the $\bar{y}$ axis while rays falling on the mirror surface at $\bar{x} > 0.865$ must be reflected a second time before reaching the $\bar{y}$ axis. If the design is to avoid re-reflection, the value of $\bar{x}$ must be limited to 0.865 and the total included mirror angle limited to 120°. Then, if the diameter of the aperture area, which corresponds to the aperture area of a flat-plate collector, is d_a, the radius of curvature of the mirror is

$$r = \frac{d_a}{2\bar{x}_{max}}$$

For example, for a 35-ft-dia. frontal area, the radius of curvature is (35/2) × 0.865, or 20.2 ft. The size of the frontal area determines the amount of solar energy that becomes available for collection, and the efficiency of the collector system determines the fraction of this energy which can actually be used for heating or cooling a building or for power generation.

The size of the solar-energy absorber in a SRTA system is based on the width of the solar image and the maximum absorber length necessary to capture all the reflected rays. As shown in Fig. 3.20, the largest absorber diameter requirement occurs when $\bar{x}$ is 0 ($\bar{y} = \frac{1}{2}$); L_{coll}, the length of the collector for a 120°-included-angle mirror, should be $L_{coll} = r/2$.

When the sun's position is off the axis of symmetry, i.e., before or after noon, some of the rays will require two reflections before reaching the collector, and a few will not intersect the absorber. But since the maximum averaged area from which rays do not reach the absorber in one reflection is less than 4 percent of the total mirror area, this loss can be neglected in an engineering design. For any reasonably reflecting mirror surface, a second reflection will not reduce the amount of energy available for collection by more than 20 percent. Consequently, the maximum error incurred by neglecting this change in optical pattern with time of day in a thermal analysis will be less than 1 percent that is, 0.2 × 4 percent. It should be emphasized that all rays intercepted by the frontal area at any time during the day, irrespective of the position of the sun relative to this

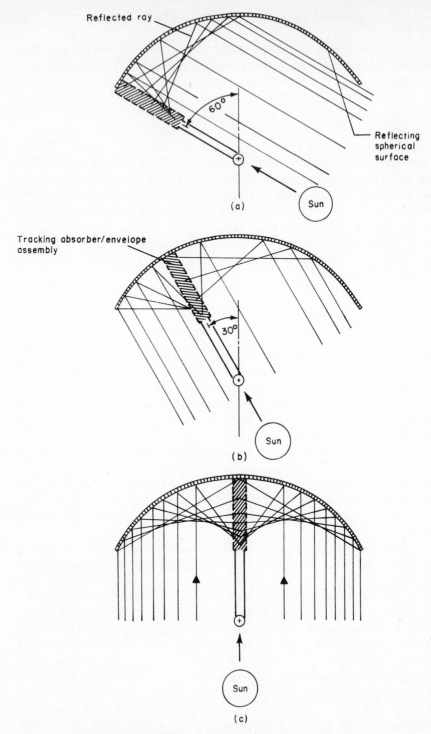

FIG. 3.21 Positions of the absorber during a day's operation. (*a*) 60° incidence angle; (*b*) 30° incidence angle; (*c*) normal incidence. (*Courtesy of W. Gene Steward.*)

mirror, are reflected onto the absorber with the same focal pattern (see Figs. 3.21 and 3.22).

For the design of the absorber, the size of the sun's image reflected on the $\bar{y}$ axis must be known. Since the distance from the mirror surface to the $\bar{y}$ axis is not the same for all reflected rays, the solar image will be larger at the focal line near the mirror surface and smaller near the pivot point end of the focal line. The cross-section diameter of the sun's image d_i at the focal line can be obtained by multiplying the angle subtended by the sun's disk (approximately 30 minutes of arc) and the length of the reflected ray. At $\bar{y} = 1$ for the minimum image diameter,

$$d_{i,\min} = L_{\text{coll}} \frac{\pi}{360}$$

At $\bar{y} = \frac{1}{2}$ for the maximum image diameter,

$$d_{i,\max} = 2L_{\text{coll}} \frac{\pi}{360}$$

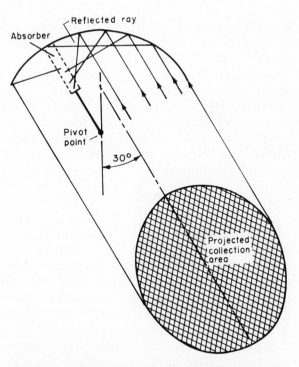

FIG. 3.22 Reflecting characteristics of a perfect spherical mirror at 10 A.M. local solar time showing projected collector aperture area. (*Courtesy of W. Gene Steward.*)

The image of the sun on the focal line is approximately a conical surface having an area A_i given by

$$A_i = \pi \, \frac{(\pi/360) + (\pi/360)2}{2} \, L_{coll}^2 = \frac{\pi^2}{120} L_{coll}^2$$

whereas the intercepting area of the collector A_p at any time of the day is approximately

$$A_p = \pi r^2 \, \bar{x}_{max}^2 \, \cos i$$

where i is the incidence angle.

A theoretical average geometric concentration factor $\bar{F}c$ can be defined as the ratio of the concentrator-projected surface area to the area of the sun's image A_i on the collecting surface. The temporal average for a day is obtained from the relation

$$\bar{F}_c = \frac{1}{\theta_1} \int_0^{\theta_1} F_c \, di \tag{3.21}$$

where θ_1 is the incidence angle i at mirror sunrise and $F_c = A_p/A_i$. For an average December day in Colorado, the average ideal concentrator ratio $\bar{F}_c$ is 187. This value of the concentration factor obtained from Eq. (3.21) does not take into account surface irregularities in the reflecting mirror. In practice, the absorber will be five or six times larger than the theoretical value to compensate for surface irregularities and to absorb some diffuse radiation on hazy or cloudy days. The spherical stationary reflector/tracking absorber (SRTA) concept has been proven by a scale model designed and operated by Steward.

SRTA THERMAL AND ECONOMIC ANALYSIS

Kreider and Steward[6] have simulated the performance of the spherical SRTA collector with a computer. The simulation considered all first-order radiative, convective, and conductive exchange mechanisms through which energy is transferred between the sun and absorber coil, between the coil and its envelope, and between the envelope and the environment. Performance effects of mirror reflectance $\bar{\rho}_m$, insolation level I, heat-transfer-fluid flow rate $\dot{m}$ concentration ratio CR, and surface selectivity $(\bar{\alpha}_s/\bar{\epsilon}_{ir})$ were considered. (A *selective surface* is a surface having a large solar absorptance $\bar{\alpha}_s$ and a small solar emittance $\bar{\epsilon}_{ir}$.) Figures 3.23 and 3.24 show performance maps indicating the effects on efficiency of several of

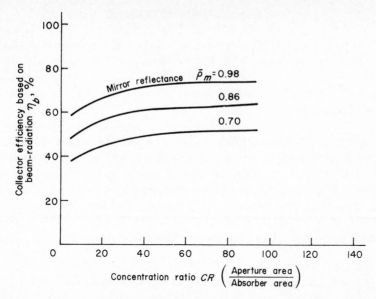

FIG. 3.23 Effect of various mirror reflectance $\bar{\rho}_m$ values and concentration ratios CR on efficiency η_b under the following conditions: Insolation $I = 300$ Btu/(hr)(ft^2), inlet temperature $T_{coll,in} = 160°$ F, heat-transfer-fluid flow rate $\dot{m} = 0.15$ lb$_m$/(hr)(ft^2) and average absorptance $\bar{\alpha}_s$ and infrared emittance $\bar{\epsilon}_{ir} = 0.9$.[6]

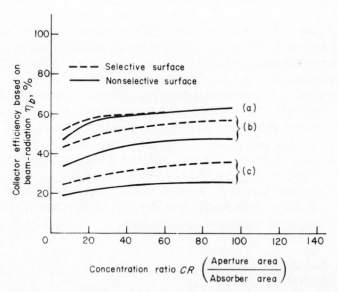

FIG. 3.24 Effect of solar absorptance $\bar{\alpha}_s$ and infrared emittance $\bar{\epsilon}_{ir}$ and fluid flow rate $\dot{m}$ on efficiency η_b under the following conditions: Mirror reflectance $\rho_m = 0.86$, insolation $I = 300$ Btu/(hr)(ft^2), inlet temperature $T_{coll,in} = 160°$ F, average selective-surface absorptance $\bar{\alpha}_s = 0.9$, average selective-surface infrared emittance $\bar{\epsilon}_{ir} = 0.1$, and average non-selective-surface absorptance $\bar{\alpha}_s$ and infrared emittance $\bar{E}_{ir} = 0.9$. Fluid flow rates $\dot{m}$: (a) 0.10 lb$_m$/(hr)(ft^2); (b) 0.05 lb$_m$/(hr)(ft^2); (c) 0.025 lb$_m$/(hr)(ft^2).[6]

these parameters, where efficiency η_b is based on the beam component of radiation. For the performance curves shown, liquid water at an inlet temperature of 160°F was in turn heated, boiled, and superheated within the absorber coil as an example of one mode of SRTA operation.

In the cost comparison between a flat-plate collector and a concentrating collector system, it is imperative that a systems analysis be performed since one does not pay for the surface area but rather the amount of heat or useful energy per unit surface area delivered. If the useful rate of heat delivered by a collector is defined as q_a, the amount of incident radiation as I, and the efficiency of the collector as $\eta = q_a/I$, a preliminary cost comparison between a flat-plate and a SRTA system can be based on[14]

$$\frac{A_{fp}}{\eta_{fp}} \times \text{Cost per square foot of flat-plate collector}$$

where A_{fp} is the surface area of a flat-plate collector and η_{fp} is the efficiency of a flat-plate collector, and

$$\frac{\pi d_a^2}{\eta_{\text{SRTA}}} \times \frac{\text{Cost per square foot of mirror surface}}{\text{Frontal area/mirror area}}$$

where η_{SRTA} is the efficiency of a SRTA collector and the cost of the SRTA abosrber is prorated on a mirror-surface-area basis. For complete cost companions, the cost of the entire solar system for each proposed design must be considered.

Preliminary cost estimates for large and small units indicate that in sunny regions of the world SRTA concentrator systems are less expensive than flat-plate collector systems for heating and air conditioning and are suitable for providing at least part of the electrical power demands of large office buildings, shopping centers, commercial structures, and other building complexes, such as prisons and governmental buildings. Further work is necessary, however, to acquire operating experience, determine maintenance costs, and prove the long-term economic advantages of the collection system.

The Compound Parabolic Concentrator Collector

The compound parabolic concentrator[14] (CPC) is a nontracking solar collector consisting of two sections of a parabola of second degree located symmetrically about the collector midplane. The two sections form a single-curvature or trough-like solar concentrator with an angular acceptance of $2 \times \theta_{max}$ as shown in Fig. 3.25. The acceptance depends on the ratio of aperture area W_e to the absorber area W_a and can be quantified by the relation

$$\theta_{max} = \sin^{-1} \frac{W_a}{W_e}$$

The quantity W_e/W_a is the CPC concentration ratio CR. The collector should be oriented in an east-west direction and tilted toward south at an angle β from the horizontal plane. When the angle γ $(=|\pi/2 - \beta - C|)$ shown in Fig. 3.25 is less than θ_{max}, the CPC accepts both direct and diffuse components of sunlight. When the angle γ is greater than θ_{max}, the CPC accepts only diffuse skylight over a portion of the aperture equal to the absorber area. Beam insolation incident on a CPC collector outside the acceptance angle does not reach the absorber area W_a but is reflected from the sidewalls back through the aperture W_e.

The theoretical depth of the collector d_{coll} depends on the concentration ratio CR according to the relationship

$$\frac{d_{coll}}{W_a} = \frac{CR + 1}{2} \sqrt{CR^2 - 1} \tag{3.22}$$

In practice it has, however, been found advantageous to use a value of d_{coll} one third smaller than that dictated in Eq. (3.22).

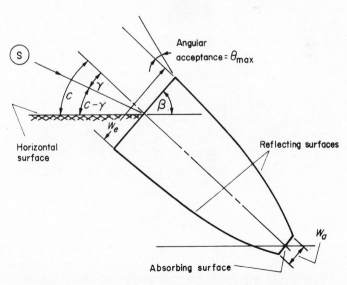

FIG. 3.25 Cross-section of compound parabolic concentrator (CPC).

TABLE 3.9 Typical Collector Dimensions

Nominal concentration ratio CR	Collector dimensions, in.		
	Aperture width W_e	Absorber width W_a	Actual depth d_{coll}
3	27.7	9.44	36
5	18.0	3.56	36
10	12.0	1.20	36

Truncation is advantageous because for any given concentration ratio a significant reduction in mirrored surface changes the angular acceptance only slightly. It is in the truncated region, however, where the greatest number of reflections of sunlight between the aperture and the absorber take place. Removing this high-reflectance-density zone reduces the number of reflections of normally incident incoming light by about 20 percent. Only the less costly truncated collector is considered economically viable. Typical CR values and collector dimensions for a 3-ft-deep CPC are tabulated in Table 3.9.

Figure 3.26 shows how a single CPC can be installed in a gas station, modified to become an office building (the design is proposed by C. Haertling, AIA, Boulder, Colo.) Figure 3.27 shows several CPCs in series as an alternative to an optically selective surface in a high-performance flat-plate collector.

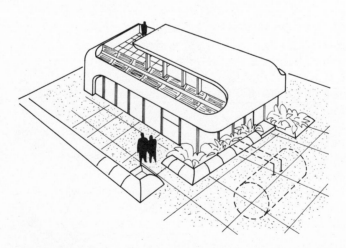

FIG. 3.26 CPC collector incorporated into a previously constructed building. (*Courtesy of Charles Haertling, AIA.*)

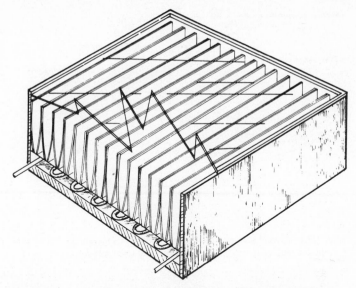

FIG. 3.27 Multiple-unit, compound-parabolic-concentrator (CPC) collector with glass convection shield. (*From R. Winston, "Principles of Solar Concentrators of a Novel Design, Solar Energy, 1974, vol. 16, p. 93. Copyright 1974 by Pergamon Press. Reprinted by permission.*)

Pyramidal-Optical-concentrator (POC) Solar-energy-collection System*,†

The use of the pyramidal-optical focusing system reduces by a factor of 2 to 6 the size, and hence the weight and cost of the flat-plate collector required. Consequently the over-all cost of the solar energy system is reduced by factors of 1.6 to 2. Integration of the system is also greatly simplified with structures of conventional architectural design, and the effect on the esthetic appearance of the building is minimized.

A full-scale experimental prototype of the pyramidal-optics solar-energy collection system has been built, and initial tests have been performed that have proven the feasibility of the system.

Solar-energy systems used for ·heating and cooling usually employ flat-plate collectors. The area of flat-plate collectors required for optimum heating and cooling structures varies with the size of the structure, the heat demand of the structure, and the climate

*E. M. Wormser, Solar House Heating System Using Reflective Pyramidal Optical Condensing System, U.S. Patent 3814302.
†Adapted from a brochure by E. M. Wormser. All claims are by E. M. Wormser.

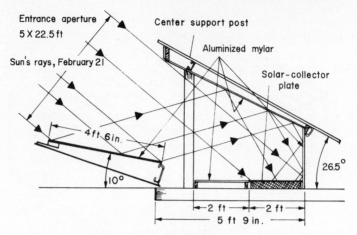

FIG. 3.28 Typical section of a pyramidal-optical solar-energy collection system showing center support and door in an open position.

conditions of the locations. Typical solar-energy heating systems employ collector areas ranging in size from 125 to 600 ft^2.

For optimum collection of solar energy, the collectors are oriented perpendicular to the sun at noon in local winter. This results in an elevation of approximately 60° in latitudes between 30 and 40°N. Depending on the size collector area required, a major portion of the southern elevation of the roof structure has to be devoted to the flat-plate solar collector.

The Falbel pyramidal-optics, solar-energy collector system combines a unique proprietary optical system* with conventional flat-plate collectors. The overall system, shown in Fig. 3.28, can be built into the attic structure of a conventional house. It consists of a movable hinged reflective panel which opens from and closes into the attic structure and which is set at the correct angle to reflect the sun's radiation into the pyramidal reflector as shown in Fig. 3.29. The second reflecting surface, a two-dimensional pyramid, focuses the solar energy onto a flat-plate collector located near the apex of the pyramid. Depending on the detail optical design of the pyramidal reflector, the solar energy is concentrated by a factor of 2 to 6 times. This factor is defined as the *optical gain factor*. The size of the flat-plate collector required is reduced by this factor.

Since the flat-plate collector is located inside the attic of the house and the movable hinged reflector panel is closed when not in use, the collector and the optical reflector are protected, to a degree,

*E. M. Wormser, Solar House Heating System Using Reflective Pyramidal Optical Condensing System, U.S. Patent 3814302.

from rain, ice, snow, and dust. This degree of protection reduces the demands for weather resistance of the critical elements of the solar system. In addition, the installation is greatly simplified by having the flat-plate collector mounted horizontally inside the attic space, permitting its mounting directly on the ceiling joists of the building.

The area and hence the overall size, weight, and cost of the flat-plate collector are reduced in proportion to the optical gain of the pyramidal optical reflector from a minimum of a factor of 2 to a maximum in excess of 6.

The reflection efficiency at each surface can ideally approach 90 percent and since the sun's rays either impinge directly or reflect on average twice before reaching the collector, the overall optical efficiency is estimated at better than 80 percent. The "net optical gain" in solar intensity is the geometric optical gain of the pyramidal reflector, which ranges from 2 to 6 and is modified by the optical efficiency of 80 percent, which results in a "net optical gain" ranging from 1.6 to 4.8.

The movable hinged reflective panel that forms the principal portion of the entrance aperture of the pyramidal optical system can be adjusted to be at the optimum angle for the elevation of the sun, regardless of season and geographical location. While the optical system is adjusted for the best angle to receive direct sunlight, it actually subtends a large solid angle of sky. Hence the system will collect diffuse solar radiation in addition to the direct rays from the sun.

The optical reflector is a configuration requiring relatively low accuracy. It can be constructed using building construction techniques. The shape of the pyramidal optical collector essentially follows the inside contour of the attic roof. The reflector shape can be formed using relatively thin construction material to which

(a)   (b)

FIG. 3.29 Door used in pyramidal-optical solar-energy collector. (a) Door in closed position; (b) door in open position.

reflective aluminized plastic or thin, anodized, reflective aluminum sheet is attached. A similar technique is used to provide a flat, reflective lining to the movable, hinged reflector. The cost of the material and additional labor when the pyramidal optical reflector is added to the attic structure of the house is relatively low and is not affected by variations in collector size.

It is estimated that the optical focusing assembly, i.e., the movable reflective panel and the associated controls, will cost a total of $250 to $350 per module, regardless of size, if custom-made in small quantities. In reasonable quantities the cost per module should be approximately one-half this amount, or $125 to $175 each.

From cost computations it has been shown that for a net optical gain of 3.5 the pyramidal optics system has a cost advantage of a factor of 1.6 for the minimum-size, domestic solar-energy heating system of 125-ft^2 equivalent collector area, increasing to a factor of 2 for the 350-ft^2 equivalent collector-area system.

SOLAR CELLS FOR DIRECT CONVERSION OF SOLAR ENERGY TO ELECTRIC POWER

Direct solar-energy conversion to produce electric power through the photovoltaic effect is one of the most attractive means of using solar energy. Direct conversion of solar energy could solve the energy crisis provided production cost of solar cells could be reduced sufficiently. The main questions to ask in considering direct conversion are

1. What area must be covered by solar cells to generate the desired amount of electric power
2. How much would electric power generated by a photovoltaic solar energy conversion system cost?

Today, only one kind of photovoltaic solar energy is commercially available. This cell is the silicon single-crystal p-n junction cell, which has been used to generate electric power for space vehicles and satellites. Figure 3.30 shows a cross section of such a solar cell. For a single-crystal silicon cell, the dimensions l and b are approximately

$$l = 1 \mu \ (4 \times 10^{-5} \text{ in.})$$

$$b = 800 \mu \ (0.020 \text{ in.})$$

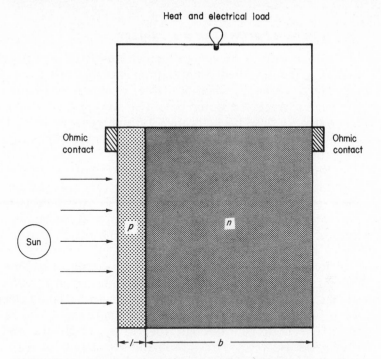

FIG. 3.30 Cross section of a solar cell. For single-crystal silicon cells, l is approximately 1 μ and b is typically 800 μ; for thin-film cadmium sulfide cells, the p region is a copper sulfide approximately 0.2 μ thick and the n region is cadmium sulfide with b approximately 20 μ thick.

On the surface of the earth, with a cell temperature of about 75°F, such a solar cell can operate at a peak efficiency of about 14 percent. At noon on a clear day at sea level, the average insolation is about 1 kW/m², but over the course of the year this level of power input is available only for approximately 6 hr per day.

Because of the intermittency of the availability of solar energy, storage would be required for periods of overcast skies, rain, and snowfall, as well as during the night. Unfortunately, the technology of large-scale energy storage is relatively undeveloped, due at least in part to the fact that the traditional methods of generating electricity utilize fossil or nuclear fuels which do not require this type of storage. As an order of magnitude, a house with a 2,000-ft² roof using 10 percent efficient solar cells could produce in one clear day enough energy to charge about 50 ft³ of commercial lead-acid storage batteries. However, if lead-acid batteries were used to store a substantial fraction of all the electric power needs of homes in the United States, the supply of lead would soon be depleted.

Alternate Forms of Energy Storage

Other storage means are being considered at the present, for example, the so-called "hydropump storage." Hydropump storage uses electric energy, when available, to pump water into reservoirs and then extracts the stored potential energy on demand by allowing the water to flow back through turbine generators; or the electric energy generated by photocells may be used to compress air and run a gas turbine when energy is required.

Storage of energy in a rotating mass is another method currently under investigation. In this type of storage, a motor generator would operate as a motor when excess electrical power is available during sunlight hours and set a flywheel into rotation. Energy would be extracted by using the rotating mass to drive a motor generator as a generator.

Recent studies suggest that the possibility of energy storage in the form of hydrogen gas could provide a third type of energy-storage solution. The hydrogen would be produced by electrolysis of water, an inherently efficient process, and energy could be recovered from the hydrogen through fuel cells, which are also efficient converters. However, safety problems are associated with the use of hydrogen, and considerable research and development would have to be done before energy storage in the form of hydrogen gas becomes operational.

Direct Conversion Economics

The most serious impediment to large-scale photovoltaic solar-energy conversion lies in the cost of currently available, reliable, long-life, efficient solar cells. At present, only two types of semiconductor solar cells are at a level of development to permit a reasonable analysis of their potential cost. One of these is based on the single-crystal silicon solar cell referred to previously. The other is based on the thin-film cadmium sulfide cell. Each falls short of meeting the requirements for economically viable solar-energy conversion systems, though for different reasons.

The silicon cell is reliable, has a long life, and has a conversion efficiency in excess of 10 percent. Unfortunately, the current cost of silicon cell arrays as determined by the use on unmanned space satellites is about $7,000/m^2$. A recent cost comparison with projected nuclear power suggests that the maximum acceptable cost of solar arrays intended for deployment on rooftops of buildings would be no more than $30/m^2$. The principal cause of the high cost of silicon solar-cell conversion systems is the need to make single crystals, and even the most optimistic analyses do not indicate possibility for cost reduction by a factor of 100.

Owing to the high cost of silicon cells, the thin-film cadmium-sulfide cell is very attractive. The active part of this cell, as shown in Fig. 3.30 is a thin (20-μ) polycrystalline film of cadmium sulfide onto which an even thinner (0.2 μ) layer of a copper-sulfur compound film is grown. Recently, the E. I. Dupont Company estimated that large areas of this kind of cell could eventually be produced for approximately \$5/m². However, the current level of understanding of the photovoltaic effect in this system is not sufficient to allow controllable fabrication of reliable, long-lived cells from cadmium sulfide. Moreover, the efficiency of these cells is only in the vicinity of 5 percent, and it is not known whether it can be increased to a level equivalent to that achieved with a silicon cell system.

In summary, at the present time direct conversion of solar energy is too expensive to be economically competitive. Considerably more research and development are required before it can be ascertained whether or not the cost of producing photovoltaic direct solar-energy conversion systems can become economically competitive. On the other hand, the area required for direct power generation is not unreasonable, and methods of energy storage are either available or could be developed within a reasonable period of time. If the cost of direct solar conversion systems can be reduced sufficiently to produce electric power economically, it may be possible to combine direct conversion of solar energy with conventional flat-plate solar energy collectors, as shown in Fig. 3.31.

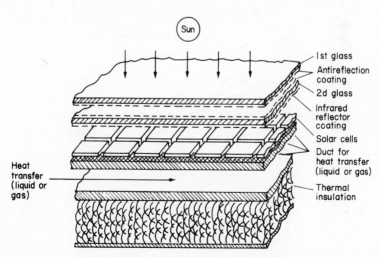

FIG. 3.31 Proposed structure of combination thermal-photovoltaic solar collector.

SUMMARY

In this chapter flat-plate collectors and certain concentrating collectors, which have few or no moving parts and can be used economically in heating and cooling of buildings, have been analyzed. Thermal and optical parameters have been quantified to show their interaction in solar thermal collectors. A parametric sensitivity study of the liquid-cooled flat-plate collector showed that the most important determinants in its performance are the collector operating temperature, the number of glass covers, the absorber surface properties, and the solar incidence angle. A detailed analytical model of this popular collector design is also presented.

A parametric sensitivity study of a stationary-reflector, concentrating collector showed that surface reflectance, concentration ratio, and fluid flow rate (directly related to collector temperature) are the three most important parameters determining its performance. Absorber surface treatment and incidence angle are less important for a SRTA concentrating collector than for a flat-plate collector.

A convenient way to summarize the results of this chapter and to compare several of the collectors described is to plot their efficiency versus the temperature difference-insolation ratio. Figure 3.32 shows performance data for five collectors which are listed below:

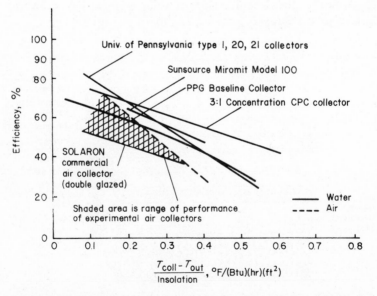

FIG. 3.32 Performance comparison of several commercial and experimental solar collectors of practical design.

1. A commercial air-cooled flat-plate collector with double glazing (SOLARON Corp., Courtesy of G. O. G. Löf)
2. A commercial water-cooled flat-plate collector with double glazing (PPG Industries, Inc. Baseline Collector)
3. A commercial water-cooled flat-plate collector with single glazing, and selective surface (Daylin, Inc., Sunsource Collector)
4. Three experimental flat-plate collectors with double glazing (Types 1, 20, 21, University of Pennsylvania; see Fig. 3.11)
5. An experimental CPC concentrating collector with a concentration ratio of 3 and one cover (courtesy R. Winston, Argonne National Laboratory)

The superiority of the concentrator at higher operating temperatures is evident from the performance curves. For example, at a typical winter operating condition corresponding to a value of $(T_{coll} - T_{out})/I$ of 0.40, the efficiency of the flat-plate collectors is in the range of 25 to 40 percent. At the same operating condition, the efficiency of the CPC concentrator is about 55 percent. Alternatively, to deliver a given quantity of heat with a flat-plate collector could require up to twice the collection surface as the low concentration ratio CPC collector would require.

Preliminary cost estimates for concentrators on a mass production basis indicate that they may be capable of delivering thermal energy at lower cost and greater reliability than flat-plate collectors. Concentrators, by virtue of their high fluid delivery temperature, are also capable of providing steam for the generation of electric power during the periods when thermal building loads are below system capacity thereby increasing the overall use factor of such systems in large buildings such as shopping centers, airports, post offices, and schools. In view of the relative lack of durability experience on the new generation of concentrators, more field tests are necessary to prove the predicted long-term economic advantages of such solar conversion systems.

REFERENCES

1. Altman, M., et al.: Conservation and Better Utilization of Electric Power by Means of Thermal Energy Storage and Solar Heating, NSF/RANN/SE/GI 27976/PR73/5, University of Pennsylvania, Philadelphia, 1973.
2. Bennet, I.: Monthly Maps of Mean Daily Insolation for the United States, Solar Energy, vol. 9, p. 145, 1965.
3. Bliss, R. W.: The Derivation of Several "Plate Efficiency Factors" Useful in the Design of Flat-plate Solar Heat Collectors, Solar Energy, vol. 3, p. 55, 1959.

4. Hottel, H. C., and B. B. Woertz: The Performance of Flat-plate Solar Heat Collectors, *Trans. ASME,* vol. 64, p. 91, 1942.

5. Klein, S. A., J. A. Duffie, and W. A. Beckman: Transient Considerations of Flat Plate Solar Collectors, ASME Paper 73-Wa/Sol-1, 1973.

6. Kreider, J. F.: Thermal Performance Analysis of the Stationary Reflector/ Tracking Absorber (SRTA) Solar Concentrator, *Journal of Heat Transfer,* vol. 97, ser. C, no. 3, 1975; Kreider, J. F., and W. G. Steward: The Stationary Reflector/Tracking Absorber Solar Concentrator, International Solar Energy Society Annual Mtg., Fort Collins, Colo. 1974; Kreith, F., and W. G. Steward: Optical Design Characteristics of a Stationary Concentrating Reflector/Tracking Absorber Solar Energy Collector, *Appl. Optics,* vol. 14, no. 7, pp. 1509 ff., 1975.

7. Lior, N., and A. P. Saunders: *Solar Collector Performance Studies,* NSF/RANN/SE/GI 27976/TR73/1, National Center for Energy Management and Power, University of Pennsylvania, Philadelphia, August 1974.

8. Liu, B. Y. H., and R. C. Jordan: The Interrelationship and Characteristic Distribution of Direct, Diffuse, and Total Solar Radiation, *Solar Energy,* vol. 4, p. 1, 1960.

9. Löf, G. O. G., and R. A. Tybout: A Model for Optimizing Solar Heating Design, ASME Paper 72-Wa/Sol-8, 1972.

10. Norris, D. J.: Correlation of Solar Radiation with Clouds, *Solar Energy,* vol. 12, p. 107, 1968.

11. Norris, D. J.: Solar Radiation on Inclined Surfaces, *Solar Energy,* vol. 10, p. 72, 1966.

12. Sparrow, E. M., *et al.*: *Research Applied to Solar Thermal Power Systems,* NSF/RANN/SE/GI 34871/PR73/4, University of Minnesota, Minneapolis, 1974.

13. Whillier, A.: *Solar Energy Collection and Its Utilization for Home Heating,* doctoral dissertation, MIT, Cambridge, Mass., 1953.

14. Winston, R.: Principles of Solar Concentrators of a Novel Design, *Solar Energy,* vol. 16, p. 93, 1974.

CHAPTER 4

Solar Heating of Buildings

For which of you, intending to build a
tower, sitteth not down and counteth the
cost, whether he have sufficient to finish it?

LUKE 14:28

Solar heating of buildings will be the first significant application of solar energy in the latter twentieth century. The technology is well developed, the systems are known to be viable, and the costs of solar heating are competitive with some conventional fuels in many parts of the United States. In this chapter the methods of sizing a solar system, estimating its performance, and evaluating cost-benefit ratios are described in detail.

HEATING REQUIREMENTS

To determine the heating requirement of a building it is necessary to calculate the rate of heat loss from the building. Assuming that the building is in a quasi-steady thermal state, the heat loss consists of four principal parts: heat transfer through the walls, ceilings, and windows, and infiltration. Any solar heating system, for example the typical solar-heated building shown in Fig. 4.1, consists of five major components:

1. Collector
2. Storage
3. Auxiliary heater
4. Distribution system (e.g., radiators, forced air system, or floor coils)
5. Controls and flow devices (e.g., fans, pumps, and dampers)

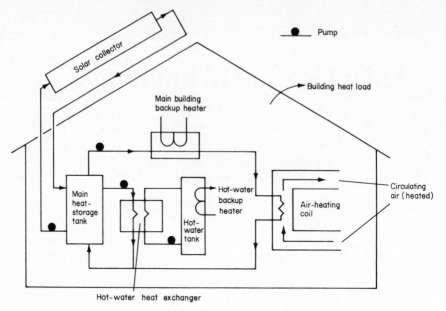

FIG. 4.1 Solar residential heating system schematic diagram—water-cooled collector.

The Degree-day Unit

To relate the heat loss from a building to the design of a solar heating system it is convenient to use the *degree-day* as the unit of heat loss. Long experience in the heating and ventilating industry has shown that in general a building need only be heated when the mean outside temperature during a 24-hr period is below 65°F. When this temperature of 65°F is used as a base, each degree that the mean daily temperature is below 65°F is called a *degree-day unit*. A mean temperature of 50°F then corresponds to 15 heating degree-days. For calculations of monthly and seasonal totals, daily degree-day units are added. The outside temperature to be used in the design of a heating system for most structures can be obtained from tables in the *ASHRAE Guide and Data Book*, for example. Outside temperatures used in the design of solar heating systems are also given in App. C, Tables C.1 and C.2 and Fig. C.3(*a*). Table 4.1 shows average degree-days and design outside temperatures for a number of cities in the United States.

The amount of heat to be supplied by the heating plant, whether it is a solar heater, a conventional heater, or a combined solar–conventional heater, must equal the sum of the heat loss through the walls, the amount of heat required to warm the ventilating air, and the air entering by infiltration. If *Q* cubic feet per hour of air are introduced by ventilation and infiltration, the heat required to bring this air to room temperature is

$$q_v = Qc_a \, \rho(T_{in} - T_{out}) \quad \text{Btu/hr} \tag{4.1}$$

where c_a = specific heat of air, Btu/$(\text{lb}_m)(°F)$
ρ = density of air, lb_m/ft^3
T_{in} = room temperature $°F$,
T_{out} = outside temperature, $°F$

Calculation of hourly heat losses through structures is usually based on steady-state conditions. Unsteady state calculations are

TABLE 4.1 Normal Degree-days and Design Outside Temperatures*, †

State	City	Degree-days, Sept. 1– May 31	Design outside temp., °F	State	City	Degree-days, Sept. 1– May 31	Design outside temp., °F
Alaska	Juneau	8,088	−5	Mont.	Helena	8,250‡	−39
Ala.	Birmingham	2,780	12	Nebr.	Omaha	6,160	−17
Ariz.	Phoenix	1,698	36	Nev.	Reno	6,036‡	3
Ark.	Little Rock	2,982	8	N.H.	Concord	7,612‡	−11
Calif.	San Francisco	3,421‡	37	N.J.	Trenton	5,068§	2
Colo.	Denver	6,132‡	−12	N. Mex.	Albuquerque	4,389	8
Conn.	Hartford	6,139	−2	N.Y.	New York	5,050§	5
D.C.	Washington	4,333	10	N.C.	Raleigh	3,369	14
Fla.	Jacksonville	1,243	28	N. Dak.	Bismarck	9,033‡	−31
Ga.	Atlanta	2,826	11	Ohio	Cleveland	5,950	0
Hawaii	Honolulu			Okla.	Oklahoma City	3,647	−1
Idaho	Boise	5,890‡	−10	Ore.	Portland	4,632‡	10
Ill.	Chicago	6,310	−11	Pa.	Harrisburg	5,258	4
Ind.	Indianapolis	5,611	−8	R.I.	Providence	6,125	1
Iowa	Des Moines	6,446‡	−13	S.C.	Columbia	2,435	19
Kans.	Topeka	5,209	−8	S. Dak.	Rapid City	7,535‡	−22
Ky.	Louisville	4,434	−2	Tenn.	Nashville	3,513	3
La.	New Orleans	1,317	26	Tex.	Fort Worth	2,361	8
Maine	Portland	7,681‡	−9	Utah	Salt Lake City	5,866	−1
Md.	Baltimore	4,787	8	Vt.	Burlington	7,865‡	−17
Mass.	Boston	5,791	0	Va.	Richmond	3,955	11
Mich.	Detroit	6,404‡	−4	Wash.	Spokane	6,852‡	−16
Minn.	Minneapolis	7,853‡	−23	W. Va.	Elkins	5,733	−4
Miss.	Vicksburg	2,000§	15	Wis.	Milwaukee	7,206‡	−15
Mo.	St. Louis	4,699	−5	Wyo.	Cheyenne	7,652‡	−19

*Adapted from "Marks' Standard Handbook for Mechanical Engineers," 7th ed., McGraw-Hill Book Company, New York, 1967, by permission of the publishers.
†All readings except those followed by a section mark (§) were taken at the airport of the city listed.
‡Degree-days for the entire year.
§ Readings were taken within the city listed.

required only for buildings with large masses subject to significant excursions in temperature (a situation not often encountered in homes). As shown in Chap. 2 the rate of heat loss through a wall, a window, or a ceiling in the steady state is given by

$$q_{wa} = UA(T_{in} - T_{out}) \tag{4.2}$$

where U is the over-all heat-transfer coefficient.

EXAMPLE

Calculate the heat load on a house for which the wall area is 2,000 ft^2, the roof area is 3,000 ft^2, and the window area totals 1,000 ft^2. The construction of the wall is shown in Fig. 4.2; the calculation of the U factor for the walls (U_{wa}) follows (heat-loss estimates for these calculations are based on property values presented in "Marks' Standard Handbook for Mechanical Engineers").*

SOLUTION

$$U_{wa} = \frac{1}{(1/6) + (1/1.0) + (0.5/0.25) + (3.5/0.27) + (1/1.1) + (0.5/1.41) + (1/1.65)}$$

| Outside convection | Wood | Drywall | Glass wool | Air space | Gypsum board | Inside convection |

*Theodore Baumeister, Ed., "Marks' Standard Handbook for Mechanical Engineers," 7th ed., McGraw-Hill Book Company, New York, 1967, Ch. 12.

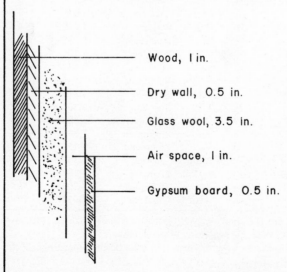

Wood, 1 in.

Dry wall, 0.5 in.

Glass wool, 3.5 in.

Air space, 1 in.

Gypsum board, 0.5 in.

FIG. 4.2 Wall construction for calculation of U factor.

$$U_{wa} = \frac{1}{0.166 + 1.0 + 2.0 + 12.96 + 0.909 + 0.35 + 0.606}$$

$$= 0.0556 \ Btu/(hr)(ft^2)(°F)$$

The heat loss through the windows depends on whether they are single- or double-glazed. In this example, single-glazed windows are installed, and a U factor equal to 1.13 $Btu/(hr)(ft^2)(°F)$ is used. (If double-glazed windows were installed, the U factor would be 0.45 $Btu/(hr)(ft^2)(°F)$.)

The roof is constructed of 1-in.-thick foam and $\frac{1}{2}$-in. siding. The calculation of the U factor for the roof U_{rf} is shown below.

$$U_{rf} = \frac{1}{\underbrace{(1/6) + (1/0.1)}_{\text{Foam}} + \underbrace{(0.5/1.0) + (1/1.65)}_{\text{Siding}}}$$

$$= \frac{1}{0.166 + 10 + 0.5 + 0.606}$$

$$= 0.0887 \ Btu/(hr)(ft^2)(°F)$$

If the respective areas and U factors are known, the rate of heat loss per hour for the walls, windows, and roof q_x can be calculated.

$$q_{wa} = (2,000 \ ft^2) \times 0.0556 \ Btu/(hr)(ft^2)(°F) = \ \ \ 111 \ Btu/(hr)(°F)$$

$$q_{wi} = (1,000 \ ft^2) \times 1.13 \ Btu/(hr)(ft^2)(°F) \ \ \ = 1,130 \ Btu/(hr)(°F)$$

$$q_{rf} = (3,000 \ ft^2) \times 0.0887 \ Btu/(hr)(ft^2)(°F) = \ \ \ 266 \ Btu/(hr)(°F)$$

$$\text{Total } q_x = 1,507 \ Btu/(hr)(°F)$$

If double-glazed windows were used, the heat loss would be reduced to 827 $Btu/(hr)(°F)$.

The infiltration and ventilation rate Q_r for this building is 100 cfm; thus, from Eq. (4.1),

$$q_v = 100 \times (0.070 \ lb/ft^3) \times (0.24 \ Btu/(lb)(°F) \times (60 \ min/hr)$$

$$= 101 \ Btu/(hr)(°F)$$

The total rate of heat loss q_{tot} on a daily basis is the sum of q_x and q_v:

$$q_{tot} = (1,507 + 101) \times 24 = 36,600 \text{ Btu}/(^\circ F) \text{ per day}$$

This calculation is simplified for purposes of illustration. Heat losses through the slab surface and edges have been neglected, for example. The procedure to be used for heat-loss calculation is described in detail in the *ASHRAE Handbook of Fundamentals*.

DESIGN OF A COLLECTOR FOR A BUILDING

Three major parameters must be specified before a collector of a given design is installed on a building: collector tilt β, number of covers n, and collector area A_{coll}.

Collector Tilt

Detailed studies by Löf and Tybout[10] have shown that the optimum, i.e., least cost per Btu delivered, collector tilt *for building heating* is approximately the *latitude plus 15°*. For hot-service-water heating only, the optimum tilt is an angle equal to the latitude.

Number of Glass Covers

Löf and Tybout[10] found that the optimum number of collector covers varies with climate, as shown in Table 4.2.

TABLE 4.2 Optimal Number of Collector Covers*

Climate type	Example location	No. of covers
Tropical and subtropical desert	Phoenix	1
Tropical savannah	Miami	1
Tropical or subtropical steppe	Albuquerque	1 or 2
Mediterranean or dry-summer subtropical	Santa Barbara	1 or 2
Humid subtropical	Charleston	2
Marine West Coast	Seattle	2
Humid continental, warm summer	Omaha	2
Humid continental, cool summer	Boston	2
Cold, northern latitudes	Canada, Alaska	3

*Adapted from Löf and Tybout (1973)[10] by permission of the publishers.

Determination of Collector Area
by a Simplified, Approximate Method

No general rule presently exists to define an optimal collector area. The area must be determined for each building in each location by use of a computerized economic analysis based on hourly heat demand and hourly meteorological data. This method will be described later in this chapter under "Computerized System Optimization."

The performance of a flat-plate collector depends upon several factors that interact in a complex manner. The collector construction method, orientation, and operating temperature, combined with meteorological conditions—solar radiation level, air temperature, and wind speed—and fluid flow rate through the collector are the primary factors affecting performance as described in Chap. 3. Since climatological parameters may vary rapidly and in a random manner over short time periods, an accurate performance prediction for a solar system can be made only if engineering heat transfer and efficiency calculations are performed on a time scale compatible with the time scale in which these meteorological changes occur. It is therefore necessary to calculate the collector output, heat removal from storage, and building heat loss each hour of the year. Computations that provide accurate results can only be performed by computer, as described in "Computerized System Optimization"; an approximate method for which a computer is not required will be presented in this subsection.

THE APPROXIMATE METHOD

The results of this method must be used with care since they are based on long-term averages of weather data and do not reflect the random, rapidly changing nature of geophysical phenomena. The method will be described and an example will be worked in detail.

In summary, the method consists of three steps that, when taken together, will approximately predict the annual delivery of a solar system:

1. Calculate the amount of solar radiation which can be collected each month.
2. Calculate the building heat load and hot-water load for each month.
3. Determine whether or not the amount of solar energy collected is sufficient to provide the monthly energy requirement. If it is not sufficient, determine the deficit to be supplied by auxiliary fuel each month.

To calculate these factors certain input parameters must be specified:

1. Monthly solar radiation
2. Percent of diffuse and direct radiation in total radiation*
3. Collector tilt, size, and efficiency (long term)
4. Heat load and hot water load of building

Monthly solar radiation data are available from the National Weather Service or can be obtained from Figs. C.2(a–i) in App. C. Percent of diffuse and direct radiation is difficult to determine, but can be estimated from data in "Low Temperature Engineering Application of Solar Energy." If the system is designed for heating only, the collector tilt may be determined from the rule that the angle of the collector from the horizontal should be the local latitude plus 15°. The collector area should be selected to provide between 60 and 80 percent of the heat load unless auxiliary fuel costs are unusually high (greater than $8/MBtu); if fuel costs exceed $8/MBtu, a greater percentage of the load may be carried economically by a solar system. The collector area can only be determined after the calculations are completed, so a trial area must be assumed for the first step. Collector efficiency must be a long-term average at the proposed operating temperature and can be taken from manufacturer's specifications. Heat load and hot-water load are determined by the construction used in the building and the expected rate of hot-water use. The architect can usually provide this information. A method for calculating these factors has also been described previously in this chapter. The tilt factor in Table 4.3 is the average value of $(\cos i/\sin \alpha)$, the optical gain factor realized by tilting a collector up from the horizontal.

EXAMPLE

Using the data in the table at the top of page 119, determine the percentage of solar heating that can be provided to the building analyzed in the example earlier in this chapter.

SOLUTION

The method of calculation is summarized in Tables 4.3 and 4.4.

---

*Richard Jordan, Ed., "Low Temperature Engineering Application of Solar Energy," ASHRAE, New York, 1967, Ch. 1.

Factor	Specification or description
Location and latitude	Denver Colo.; latitude 40.0°N
Collector tilt, deg	Latitude + 15° = 55°
Collector efficiency, %	32
Hot-water load, Btu/hr	1,200*
Heating load, Btu/degree-day	36,600
Ratio of diffuse-to-total insolation	0.25
Collector area, ft²	1,000
Meteorological data	†

*Average for a family of three.
†See Table 4.3.

TABLE 4.3 Calculation of Expected Solar Delivery to Collector

Month	Horizontal insolation Btu/(d)(ft²)	×	Ratio of diffuse-to-total = insolation	Diffuse component + Btu/(d)(ft²)	Direct component × Btu/(d)(ft²)	Tilt factor	× Efficiency =	Avg. daily collection, Btu/(d)(ft²)
Jan	960		0.25	240	720	2.36	0.32	621
Feb	1,270		0.25	318	952	1.82	0.32	656
Mar	1,630		0.25	408	1,222	1.33	0.32	651
Apr	1,950		0.25	488	1,462	0.96	0.32	605
May	2,180		0.25	545	1,635	0.73	0.32	556
Jun	2,440		0.25	610	1,830	0.64	0.32	570
Jul	2,360		0.25	590	1,770	0.69	0.32	580
Aug	2,200		0.25	550	1,650	0.88	0.32	641
Sep	1,840		0.25	460	1,380	1.20	0.32	677
Oct	1,410		0.25	353	1,057	1.65	0.32	671
Nov	1,025		0.25	256	769	2.20	0.32	623
Dec	840		0.25	210	630	2.56	0.32	583

TABLE 4.4 Expected Solar Heating Delivered to Building

		Heating requirements and availability, MBtu/mo			
Month	No. of heating degree-days	[(Hot-water demand	+ Heating demand)	− Solar-energy available] =	Deficit if any
Jan	1,132	0.893	41.43	19.25	23.07
Feb	938	0.806	34.33	18.37	16.77
Mar	887	0.893	32.46	20.18	13.17
Apr	558	0.864	20.42	18.15	3.10
May	288	0.893	10.54	17.24	...
Jun	66	0.864	2.42	17.10	...
Jul	6	0.893	0.22	17.98	...
Aug	9	0.893	0.33	19.87	...
Sep	117	0.864	4.28	20.31	...
Oct	428	0.893	15.66	20.80	...
Nov	819	0.864	29.98	18.69	12.15
Dec	1,035	0.893	37.88	18.07	20.70
Total	6,283	10.510	229.95	226.01	88.96

The percent of heating demand supplied by the solar source is then

$$\frac{229.95 - 88.96}{229.95} \times 100\% = 62\%$$

It should be noted that this analysis assumes storage capacity sufficient to use all collected energy. Such a large storage is probably not the most economical amount of storage since it would be sufficient for several cloudy days and therefore would be fully used only a few times during a normal heating season.

This method which uses monthly data can also be used with daily solar data. If this were to be done, the number of calculations would indicate the need for a programmable desk calculator.

STORAGE OF THERMAL ENERGY

Energy may be stored in insulated tanks ($U \cong .05$ Btu/(hr)(ft^2) ($^\circ$F)) containing a thermal storage medium:

1. Water (most popular with water-cooled collectors)
2. Rocks or gravel (used with hot-air collector systems)
3. Fusible salts (not yet perfected for reliable, low-temperature storage)

A comparison of typical rock and water storage costs is given in Table 4.5. Of the two materials now available for practical storage at low temperature, water is superior because of its lower material cost and lower volume required per Btu stored. Rock does have two advantages, however:

1. Rock is more easily contained than water.
2. Rock acts as its own heat exchanger, which reduces total system cost.

TABLE 4.5   Comparison of Rock- and Water-Storage Costs

Storage material	Qty. energy delivered by storage material, Btu/$	Tank volume required per unit of heat delivered, Btu/ft^3
Water	2.3×10^6	9,350
Rocks	1.6×10^4	3,000*

*20% void volume.

Storage at high temperature, where water cannot be used in liquid form without an expensive, pressurized storage tank, requires a solid storage medium. The final choice of storage medium must be made in the context of the complete system type and cost.

Water may be stored in galvanized metal tanks, in concrete tanks, or in fiber glass tanks. The availability of inexpensive steel tanks has diminished rapidly recently, and this limited availability of steel tanks along with their corrosion problems, indicates that a concrete or fiber glass tank is preferred. Concrete tanks are difficult to seal reliably, so some expense penalty may be encountered with their use. Whatever storage vessel is selected, it should be insulated and provided with internal access for maintenance.

Because water is quite heavy, it can require special precautions in designing the tank support pads in the building floor. A 600 ft^2 collector system should have about 9,000 lb of water for storage. A thicker floor slab is obviously required to support such a large weight.

Optimal Storage Volume

There is an optimal amount of water storage per square foot of collector. In a small storage volume the storage temperature will be high; thermal losses from storage will be large, and stored energy can supply heat for only a short time. The cost of large storage volumes is prohibitive since the storage capacity of the system will rarely be fully used, resulting in a high marginal cost of storage associated with the small load factor; in addition storage temperature will frequently be too low to be useful. The optimal storage volume per square foot of collector is independent of building location, of heat load, and of insolation. The optimal value ranges from *10 to 15 lb of water per square foot of* collector for nearly all locations in the conterminous United States.

Other Practical Design Determinants

Since solar energy heating systems operate at temperatures relatively close to the temperatures of the spaces to be heated, storage must be capable of delivering and receiving thermal energy at relatively small temperature differences. The designer must consider the magnitude of these driving forces in sizing heat exchangers, pumps, and air blowers. The designer must also consider the nonrecoverable heat losses from storage—even though storage temperatures are relatively low, surface areas of storage units may be large and heat losses therefore appreciable.

Some investigators have proposed heating the storage medium with conventional fuels to maintain its temperature at useful levels during sunless periods. This approach has two major flaws:

1. If storage is heated with conventional fuels, it cannot be heated with solar energy when available; therefore, some collected solar energy cannot be used.
2. If storage is partially heated with conventional fuels, the collector inlet temperatures will be higher and efficiency lower than it would be if storage were not boosted. Therefore, the useful return on the solar system investment would be diminished.

In conclusion it should be emphasized that storage heating with conventional fuels is uneconomical in any practical solar thermal system designed to date.

ECONOMIC ANALYSIS

Life-cycle Costing

In this discussion the cost of additional money invested in a solar system is determined. It is assumed the homeowner will finance the added solar cost along with the home mortgage. The added cost of the solar-system investment each year is then compared to the cost of fuel saved each year; from this comparison it can be determined whether or not solar heating is economically viable for a given building in a given location throughout the predicted life of the system. This approach is called *life-cycle costing.* Life-cycle costing, rather than initial-investment costing, is the appropriate way to determine the cost-benefit ratio for a solar system because initial-investment costing does not take into account the cost of fuel saved during the life of the solar system.

As shown in basic engineering economics texts, initial, additional capital cost is converted to an annual basis by use of the equation

$$C_h = C_{h,\text{tot}} \times \text{CRF} \tag{4.3}$$

where C_h = annual additional cost of the solar system (neglecting power and maintenance costs, which can be made quite small in a well-designed system), \$/yr

$C_{h,\text{tot}}$ = total additional initial investment in solar hardware, \$

CRF = capital recovery factor, \$/\$/yr

$$\text{CRF} = \frac{i_d(1 + i_d)^t}{(1 + i_d)^t - 1} \qquad (4.4)$$

where i_d = annual discount (or interest) rate, \$/\$/yr

t = expected lifetime of the solar system, yr

Values of 1/CRF are given in App. C, Table C.6, and the CRF is described in detail after the following example.

EXAMPLE

Compute the annual cost of a solar-energy system with the characteristics tabulated below.

Factor	Specification
Expected system lifetime t, yr	20
Discount rate, %	8
Collector area A_c, ft²	200
Collector cost, \$/ft²	4
Storage cost, \$/lb stored	0.05
Storage amount, lb/ft²	12.5
Control-system cost, \$	100
Miscellaneous costs (e.g., pipes, pumps, motors, heat exchangers), \$	200 + (0.50/ft²)

SOLUTION

To obtain the total cost of the solar energy system $C_{h,\text{tot}}$ described above, add the costs for the controls, the collector, the storage, and miscellaneous items, or,

$$C_{h,\text{tot}} = \$100 + (\$4 \times 200) + (\$0.05 \times 12.5 \times 200)$$
$$+ [\$200 + (0.50 \times 200)]$$

$$= \$1,325$$

From App. C, Table C.6, for a capital recovery factor CRF of 20 yr at 8 percent

$$\text{CRF}_{20 \text{ yr, } 8\%} = \frac{1}{9.818} = 0.102$$

The annual cost C_h is then

$$C_h = \$1,325 \times 0.102 = \$135/\text{yr}$$

COST OF DELIVERED SOLAR ENERGY

The cost of delivered solar energy C_{se} is defined as

$$C_{se} = \frac{\text{additional annual cost of solar hardware } C_h}{\text{total annual solar-energy delivery to building}} \qquad (4.5)$$

The value of C_{se}, expressed in dollars per million Btu, is the index that must be used to select the optimal solar system. Properties of any one component, say, collector efficiency or cost, are not appropriate criteria for selecting a solar-system.

Cost-effectiveness of Solar Heating Systems with Increasing Fuel Cost

To determine the economically justifiable investment in a solar-heating system based on expected savings in heating operating costs over the life of the system, conventional compound-interest calculations can be used. The future value X of a sum of money whose present worth is P invested at an annual interest rate i_{ann} over a period of t years is

$$X = P(1 + i_{ann})^t \qquad (4.6)$$

Consequently, the present worth of a sum X payable t years from now is

$$P = \frac{X}{(1 + i_{ann})^t}$$

In the "self-amortizing" mortgage commonly used for single-family residences, a *constant* monthly payment is used both to repay the interest on the unpaid balance B and to repay part of the principal P. Since the unpaid balance diminishes each month, the apportionment between interest and amortization of principal changes each month over the life of the mortgage: the interest portion decreases and the amortization portion increases. The compounded interest value of such a mortgage with constant annual payments of P_{ann} is

$$X = P_{ann} \frac{(1 + i_{ann})^t - 1}{i_{ann}} \qquad (4.7)$$

Combining Eqs. (4.6) and (4.7) and solving for P_{ann} gives the constant annual payment

$$P_{ann} = P \frac{i_{ann}(1 + i_{ann})^t}{(1 + i_{ann})^t - 1} = P \times CRF \qquad (4.8)$$

The present worth P is then

$$P = P_{ann} \frac{(1 + i_{ann})^t - 1}{i_{ann}(1 + i_{ann})^t} \qquad (4.9)$$

If the annual payment P_{ann} (say for heating fuel for a building) is not constant, but increases at an annual rate j, in dollars per dollars per year, owing to price escalation (in constant dollars), the analogous expression is

$$P = P_0 \frac{(1 + i_{eff})^t - 1}{i_{eff}(1 + i_{eff})^t} \qquad (4.10)$$

where P_0 is the initial annual payment, and the effective interest rate i_{eff} is

$$i_{eff} = \frac{1 + i_{ann}}{1 + j} - 1 \cong i_{ann} - j$$

That is, the effective interest rate i_{eff} is approximately the difference between the actual interest rate i_{ann} and the rate of price escalation j.

Equation (4.10) provides the answer to the question: What is the economically justifiable principal $C_{h,tot}$ a building owner can invest in a solar heating system if the present annual savings in heating costs is P_0, a cost that is increasing at an annual rate of j and for which the interest rate for borrowing is i_{ann}?

EXAMPLE

The owner of a four-unit apartment building of masonry construction is considering the retrofit installation of a solar heater. Under the existing climatic conditions it is expected that the solar unit can supply 87.5×10^6 Btu/yr and will thus save 1,460 gal of fuel oil (with an effective heating value of 60,000 Btu/gal and an efficiency of 43 percent in the existing heating plant). At 1975 prices, this savings due to solar represents a saving of $510 ($0.35/gal of fuel oil saved). If the owner can borrow money for the solar retrofit at 10 percent per year and wants to repay the loan in 7.8 yr, how much should the owner invest in the solar heating system?

SOLUTION

If

P_0 = \$510 (savings in 1975 resulting from substitution of solar energy for fuel oil)
j = 0.12 (expected increase in fuel cost)
i_{ann} = 0.10 (interest rate for loan)
t = 7.8 (time to repay loan, in years)

then

$$i_{eff} = \frac{1 + 0.10}{1 + 0.12} - 1 = -0.0179 \cong i_{ann} - j \text{ (effective interest rate)}$$

and

$$P = 510 \times \frac{0.982^{7.8} - 1}{-0.0179\,(0.982)^{7.8}} = \$4,300 \text{ (present worth)}$$

The owner can pay \$4,300 or less for the retrofit installations.*

*Repeat if owner can extend repayment time to 15 yr and government-backed, FHA solar-improvement loans at 5 percent interest are available. Observe the large effect of interest rate on the calculations.

Fuel-oil costs have more than doubled recently in many locations of the country, particularly in the New England States; coal and natural-gas costs are expected to increase sharply. Based on the fossil-fuel situation nationwide and worldwide, a continuing cost increase is almost a certainty. Many experts predict long-term increases of 10 percent to 15 percent per year for many years to come.

Steadily increasing fuel costs—and fuel costs are the largest part of the operating costs saved through solar heating—have a significant influence on the economically justifiable capital investment in a solar heating system. This influence is shown in Table 4.6 for a number of different lifetimes of the system, interest rates, and annual operating cost increases. For example, assume a 15-yr life and an annual interest rate of 10 percent: an investment of \$95.38 is justified for each dollar per month in present operating costs saved. If operating costs increase 5 percent per year, the proper investment is \$128.00 (an increase of 35 percent); if operating costs increase 10 percent per year, the correct investment is \$178.00 (an increase of almost 90 percent). This increase in justifiable investment demonstrates the

TABLE 4.6 Economically Justifiable Investment in Solar-heating Systems for Each Dollar per Month ($12/yr) in Present Savings

Projected system lifetime, yr	Projected fuel cost increase, %/yr	Interest rate, %/yr		
		7	10	12
5	0	50.76	47.54	45.59
	5	56.98	53.21	50.94
	10	63.91	59.53	56.88
	15	71.62	66.54	63.47
	20	80.18	74.30	70.77
10	0	86.95	77.06	71.46
	5	108.83	95.38	87.82
	10	137.30	119.05	108.86
	15	174.33	149.63	135.91
	20	222.42	189.11	170.70
15	0	112.76	95.38	86.13
	5	156.01	128.80	114.54
	10	221.56	178.58	156.36
	15	321.62	253.41	218.59
	20	474.79	366.49	311.79
20	0	131.16	106.77	94.46
	5	198.95	155.28	133.89
	10	318.33	238.10	199.77
	15	532.84	383.02	312.95
	20	922.51	640.55	510.99

significant effect of rising fuel costs on investment in solar-heating systems. Figure 4.3 shows this effect, which increases with expected annual increases in fuel costs but decreases with increasing interest rates.

COMPUTERIZED SYSTEM OPTIMIZATION

From the analysis discussed in "Flat-plate Solar-Collector Performance—Generalized Analysis" in Chap. 3 or "Design of a Collector for a Building," earlier in this chapter, it is possible to determine the amount of solar energy available for building and domestic hot-water heating for any location in the United States on an annual basis. There is clearly an optimal-collector size for a given building, a given cost of money, a given fuel inflation rate, and a given collector-system component cost. If a collector is small, energy cannot be collected economically because of the overwhelming effect of fixed system costs (controls, pumps, heat exchangers, etc.) that remain unchanged no matter how small the collector becomes. If a collector is large,

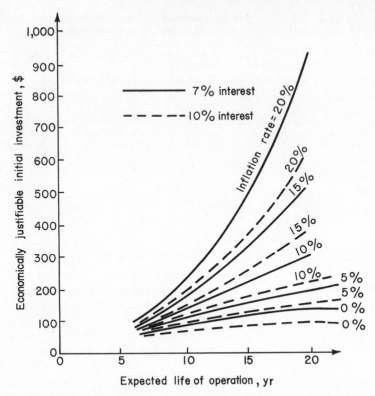

FIG. 4.3 Economically justifiable investment in a solar-energy system for each dollar per month at current prices of fuel saved for different system operating lifetimes. Curves represent different rates of fuel-cost increase and different interest rates.

say, large enough to provide 95 percent of the heating demand, it will be fully used only a small part of the time; i.e., its *load factor* is too small to be economical.

Figure 4.4 shows qualitatively the typical annual cost C_h and delivery-to-load of a solar system. The average cost of solar energy ($/MBtu) C_{se} is simply the quotient of the C_h cost curve and delivery curve. At very small collector areas A_{coll}, the delivery is vanishingly small and C_{se} is large. At large values of A_{coll}, the delivery curve is flat but the cost curve continues to rise. As a result C_{se} is also large for large A_{coll}. Between the large and small collectors lies the optimal collector size, where the cost increase in collector area results in the optimal increase in benefit, i.e., energy delivery.

It has been shown that solar collector optimization calculations must be conducted on an hourly basis for an entire heating season for an optimum size to be found.[3] Calculations on a monthly, weekly, or even daily basis can either result in no optimum size being found or a highly erroneous optimum size being found. The massive computational effort required for such an optimization can only be handled by a digital computer.

The major steps of the computer model are displayed in Fig. 4.5. The symbols used in the figure are defined in the nomenclature.

The computer optimization need only be performed to determine the optimal collector area since optimal collector tilt, number of covers, and system storage can be determined from the results described previously in this chapter. The method is simply to select a collector area and run the computer model through a typical year's hourly meteorological data. The results of the computer run show the average cost of solar energy C_{se} for the selected collector area. Several additional collector areas are selected and the value of the average cost C_{se} computed. The results of a set of such runs may be plotted as shown in Fig. 4.6. Each point on the curve represents a computer run for a year's meteorological data. Many more runs were made than are shown on the curve. A collector area of 120 ft² results in the lowest value of C_{se}.

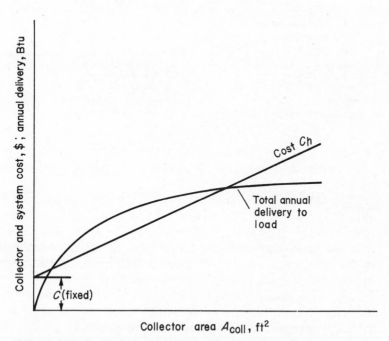

FIG. 4.4 Typical annual energy delivery and amortized annual cost for a solar system; C_{fixed} is the portion of annual solar system cost independent of collector area.

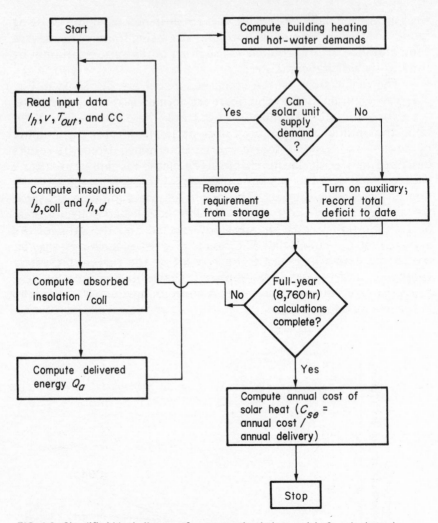

FIG. 4.5 Simplified block diagram of computer-simulation model of a solar-heated building.

The specification of the average cost-curve minimum (C_{se} = \$3.75/MBtu, A_{coll} = 120 ft²) does not determine the most economic mix of solar and auxiliary fuel, however. The average cost of electricity is also shown in Fig. 4.6. It is easily seen that at *any operating point* between *A* and *B,* solar energy costs less than electrical energy. To determine the *best operating point,* the total cost of energy—solar plus auxiliary—must be determined. The minimum point of the total cost curve is the best mix of solar and auxiliary energy. Total cost is given by

$$C_{tot} = Q_{tot}f_sC_{se} + Q_{tot}(1 - f_s)C_{aux}$$

or,

$$C_{tot} = Q_{tot}[f_sC_{se} + (1 - f_s)C_{aux}] \tag{4.11}$$

where C_{tot} = total annual cost of building and water heating provided by solar energy and auxiliary energy, \$

f_s = fraction of energy provided by the solar system (abscissa of Fig. 4.6)

Q_{tot} = total heat load for one year, MBtu

C_{aux} = average cost of auxiliary energy delivered, \$/MBtu

Figure 4.7 shows the total cost curve corresponding to the solar performance shown in Fig. 4.6. The auxiliary (back up) heating system is electrical with a delivered energy cost of \$4.50/MBtu. The minimum total cost is about \$144/yr corresponding to about 70 percent solar heating. Values read from Fig. 4.6 were used to complete Tables 4.7 and 4.8 from which Fig. 4.7 was plotted.

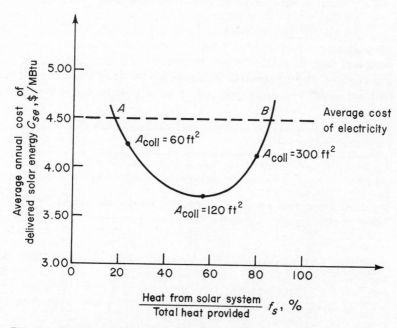

FIG. 4.6 Cost comparison of solar and back-up (electrical) heating for a house with a 36-million Btu load per year; the effect of collector-area size A_{coll} on the average annual cost of solar energy C_{se} is also shown. (Cost and performance data are based on computer projections using actual climatic data.)

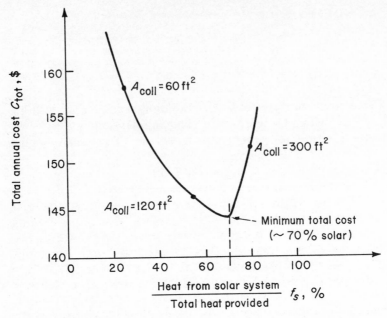

FIG. 4.7 Total cost curve for example solar-heated house modeled in Fig. 4.6.

An alternative method of finding the best solar/auxiliary mix is by equating the marginal costs of solar and auxiliary energy. The marginal cost of solar energy (measured in dollars per million Btu) is the cost of adding additional capacity to deliver another 1 million Btu per year. The marginal cost of solar energy can be thought of as

TABLE 4.7 Calculation of Annual Operating Cost for a Combined Solar/Auxiliary Heating System

Proportion of heating from solar system, %	Proportion of heating from aux. system, %	Qty. of heat delivered by solar system, MBtu/yr	Qty. of heat delivered by aux. system, MBtu/yr	Ann. cost of heat delivered by solar system, $*	Ann. cost of heat delivered by aux. system, $†	Tot. cost for combined system, $
20	80	7.2	28.8	31.60	129.60	161.20
30	70	10.8	25.2	43.70	113.40	157.10
40	60	14.4	21.6	55.00	97.20	152.20
50	50	18.0	18.0	67.00	81.00	148.00
60	40	21.6	14.4	80.30	64.80	145.10
70	30	25.2	10.8	95.70	48.60	144.30
75	25	27.0	9.0	106.00	40.50	146.50
80	20	28.8	7.2	121.00	32.40	153.40

*Average annual cost per MBtu, which varies with quantity used, is read from Fig. 4.6 and is given in Table 4.8.
†Average annual cost per MBtu is $4.50.

TABLE 4.8 Average Annual Cost per MBtu of Delivered Solar Energy C_{se} in a Combined Solar/Auxiliary Heating System

Proportion of heating delivered by solar system, %	Average annual cost of delivered solar energy C_{se}, $/MBtu
20	4.40
30	4.05
40	3.82
50	3.72
60	3.72
70	3.80
75	3.92
80	4.20

the slope of the solar system cost curve, i.e., the functional relationship between the annual cost of delivered solar energy C_h and the annual delivery. The marginal costs of solar and auxiliary energy are shown in Fig. 4.8. The intersection of the two curves is approximately at the 70 percent solar point, the same result as the total cost-curve minimum provided. By interpolation, the collector area required to deliver the optimal proportion of solar energy (70 percent) is about 180 ft^2.

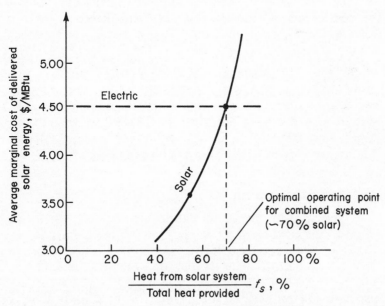

FIG. 4.8 Marginal cost curve for example solar-heated house modeled in Fig. 4.6.

All the operations just described can be routinely programmed on a computer, which can provide the user with the optimum mix of solar and auxiliary directly by numerical means. Any auxiliary energy supply may be analyzed in the same manner as that used for electric energy.

Although the technology of solar systems is well-established for first-generation (flat-plate) systems, a readily usable system design procedure does not yet exist. The only method presently available for accurate design is a computer simulation, as described above, but it is expected that in the not-too-distant future more expeditious design procedures will be developed. A situation analogous to that in estimating building heat loads exists in the design of solar collectors. A computer simulation of each building can be done either by means of detailed models or by use of the degree-day approach described earlier in this chapter. The degree-day method is accurate to ±10 percent and is universally used. The solar-design engineer should soon have a method analogous to the rapid degree-day method in which he uses tables or nomograms available for his climatic zone. These tables can be entered with values of collector cost, building heat load, auxiliary fuel cost, etc., to determine the optimal collector area directly. Two architectural firms in the United States are now preparing such design guides. Large, zonally heated buildings will continue to require the simulation approach, however.

SOLAR SERVICE HOT–WATER HEATING

Solar water heating is the direct use of solar energy that has been practiced most extensively in the last two decades. It is the most viable of all low-temperature solar energy applications; it will probably be the first wide use of solar energy in the last quarter of the 20th century because the initial investment is small and the system is used throughout the year. This high use factor results in a larger load factor than in a solar heating system. Solar hot-water heating components are now available commercially.

Approximate Sizing

The annual average insolation on a surface tilted at an angle equal to the latitude in the midlatitudes of the United States is about 2,000 Btu/ft^2 per day. If the average collection efficiency is 30 percent, 600 Btu/ft^2 per day are delivered by the collector system. The service hot-water demand can be taken as 12 gal per day per person or 36 gal per day for a family of three, on the average. If the tap water is 50°F and the hot water 160°F, about 900 Btu/gal are required for water heating. For a delivery of 36 gal per day, 33,000 Btu per day are needed. If the collector delivers 600 Btu/ft², approximately a 55 ft² collector is needed. This collector is about one-tenth the size of the collector required for heating a typical building in the midlatitudes. A rule of thumb is that 1.5 ft² of collector are required for each gallon of hot water to be delivered. The storage tank should be large enough to hold about a 2-day supply of hot water, or 75 gal for a family of three.

System Design

Since the hot-water system loads vary little during a year, the best angle of tilt is that equal to the latitude, that is, $\beta = L$. The density difference between the tap water and the collector outlet water permits the use of density-driven, or *thermosiphon,* circulation. The capital and operating cost of a pump can thereby be avoided. Such a system is shown in Fig. 4.9. After sunset, a thermosiphon system can reverse its flow direction and lose heat to the environment during the night. To avoid reverse flow, the top header of the absorber should be at least 1 ft below the cold leg fitting on the storage tank, as shown. To provide heat during long, cloudy periods, an electrical immersion heater can be used as a backup for the solar system. A nonfreezing fluid should be used in

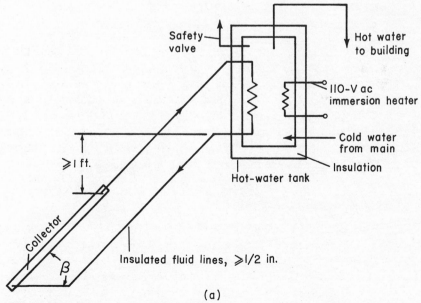

(a)

FIG. 4.9 Thermosiphon loop used in a natural-circulation, service hot-water heating system. (a) Simplified view.

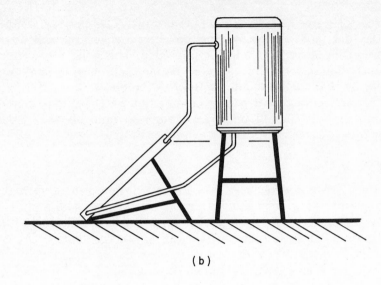

(b)

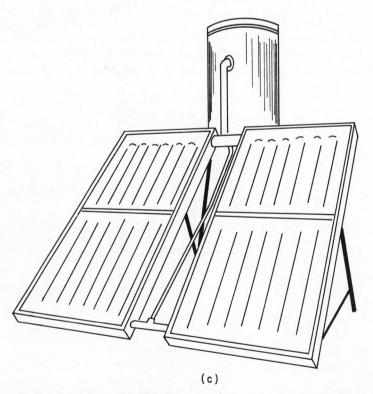

(c)

FIG. 4.9 (*continued*)   Thermosiphon loop used in a natural-circulation, service hot-water heating system. (*b*) Side view; (*c*) front view.

the collector circuit. The thermosiphon system is one of the least expensive solar hot-water systems and should be used whenever possible.

Since the driving force in a thermosiphon system is only a small density difference and not a pump, larger-than-normal plumbing fixtures must be used to reduce pipe friction losses. In general, one pipe size larger than would be used with a pump is satisfactory. Under no conditions should piping smaller than 1/2 in. National Pipe Thread be used. The flow rate through a thermosiphon system is about 1 gal/(ft^2)(hr) in bright sun.

If a thermosiphon, which requires that the storage tank be above the collector, cannot be accommodated in a structure, a forced-circulation system is necessary. The additional components would include a pump, motor, and a pump controller (a differential thermostat between tank and collector). A forced circulation system is shown in Fig. 4.10. To prevent reverse flow through this type of system, a check valve is placed in the heat-transfer-fluid loop as shown in Fig. 4.10. The same flow rate as that used in a thermosiphon system, 1 gal/(ft^2)(hr), should be used.

In both systems a nonfreezing heat transfer fluid is used for two reasons:

1. Collectors cannot withstand the pressures in most city water mains:
2. Collectors in many locations would freeze in winter if water were used in the collector.

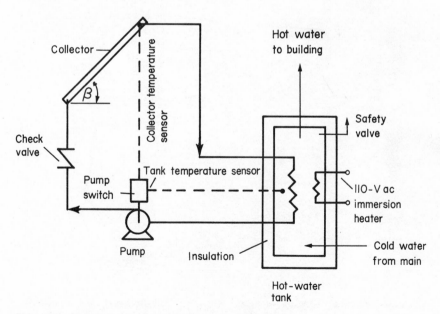

FIG. 4.10 Forced-circulation, service hot-water heating system.

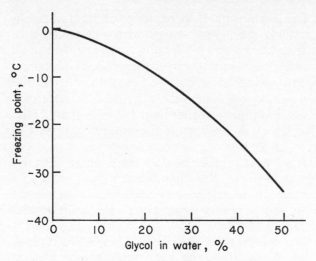

FIG. 4.11 Effect of glycol concentration on the freezing point
of water.

Sunworks, Inc. suggests using propylene glycol as the antifreeze since
it is non-toxic. Figure 4.11 shows the freezing-point depression of
water with ethylene or propylene glycol.

A compact, solar hot-water heater is shown in Fig. 4.12. The
tank is divided into an upper and lower half and is integrally

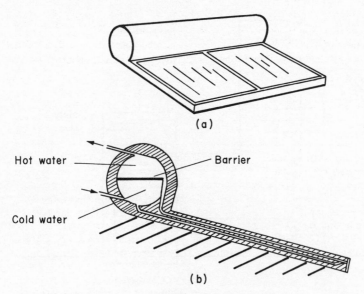

FIG. 4.12 Compact solar hot-water heater. (*a*) Outside view; (*b*) sectional
view.

connected with the collector. Because the thermosiphon principle is used, this design is not suited for freezing climates.

AIR-COOLED, SOLAR COLLECTION SYSTEMS

Advantages and Disadvantages

The air-cooled solar collector, an alternative to the more common water-cooled collector, has a number of attractive features that merit consideration for some solar heating applications. If only space heating is required, an air-cooled collector may have sufficient advantages, when used with rock-bed storage, to make it an economical system. The advantages of a hot-air system are given in Table 4.9.

The first three disadvantages of an air-cooled system all follow from the poor heat-transfer properties of air compared to water. To transfer the same amount of heat between the same temperatures, an air system requires 50 to 100 times the heat-exchanger area, depending on flow rates, because heat-transfer coefficients associated with air are so small.

Thermal Storage and Operation

The primary advantage of a hot-air–rock-bed system is that the storage bed acts as the solar-heated air-to-storage heat exchanger, the storage, and the storage-to-room-air heat exchanger. Because the rock bed accomplishes so many functions, the initial capital investment, and consequently, the cost of solar energy, can be reduced compared to some liquid systems.

TABLE 4.9 Advantages and Disadvantages of an Air-cooled Collector System

Advantages	Disadvantages
1. Freezing is not a problem.	1. Service water is heated slowly.
2. Leaks are not a problem.	2. Solar-assisted air conditioning is not feasible (see Chap. 5).
3. Corrosion is not a problem.	3. The load factor is low since air conditioning and water heating are not economically feasible.
4. Rock storage acts as its own heat exchanger.	
5. Can be one of the least-expensive types of systems for the do-it-yourself homeowner.	4. Air ducts are much larger.
	5. Fan horsepower requirements are larger than pump requirements in the liquid system.
	6. Rock storage can become loaded with dust unless special precautions are taken.

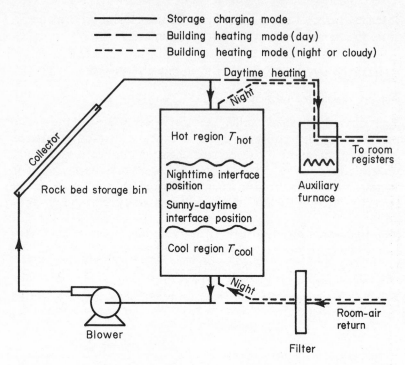

FIG. 4.13   Air-cooled collection system schematic diagram showing typical hot-cold zone interface locations.

Figure 4.13 shows a hot-air system. During the storage-charging mode, all heated air flows through the rock bed at a low flow rate determined by the desired temperature rise in the collector, usually 2 to 3 cfm/(ft²). As progressively more heat is stored, the interface between the hot and cold regions of storage moves downward in the storage bin. The air returning to the collector is at the temperature of the cool region of storage, about 70°F.

During daytime heating on a sunny day, air from the collector is diverted directly to the building instead of to storage. During sunny periods when no heating demand exists, storage is charged by warm air from the collector. During the nighttime or cloudy daytime heating modes, heat is removed from storage by a counterflow of air through the rock bed, as shown in Fig. 4.13. The outlet temperature from storage is close to the daytime collection (inlet) temperature since the air being heated passes through the hottest zone of storage last. As progressively more heat is removed from storage, the interface between the hot and cold regions of storage moves upward, as shown in Fig. 4.13.

The storage medium for air systems has typically been 1- to 2-in.-dia. granite or river-bed rocks, which cost approximately $5.00/ton. A more cost-effective storage medium is crushed bauxite, which has a specific heat twice that of gravel and costs approximately $6.00/ton. An air filter is required between the heated space and storage to eliminate dust buildup in the gravel bed (dust would reduce the heat-transfer coefficient to the rock pieces and increase the bed pressure drop). The recommended amount of storage to be used is roughly the same as for a liquid system—about 60 lb of rock per square foot of collector. Design information on the expected pressure drop through pebble-bed storage is contained in Chap. 5, section on "Non-mechanical Air Conditioning Systems."

Collector Designs

Several air-cooled collector designs are shown in Figs. 4.14 and 4.15. Relatively little performance data for such collectors has been published to provide designers with the information necessary to predict system performance and determine optimal system configurations reliably.

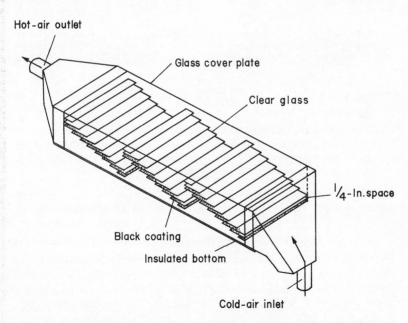

FIG. 4.14 Löf solar collector. (*Adapted, with permission, from Proc. World Symp. Appl. Solar Energy, Phoenix, Ariz., 1955, published by Stanford Research Inst., Menlo Park, Calif., 1956.*)

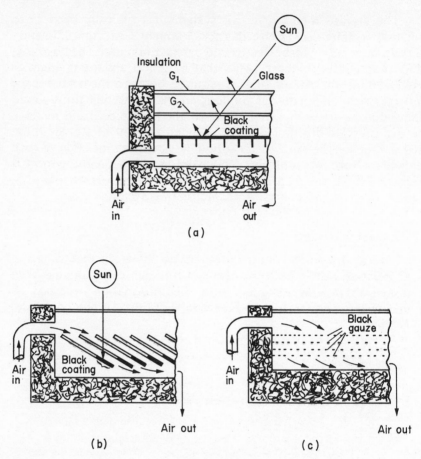

FIG. 4.15 Air-cooled solar collectors. (*a*) Double-glazed, finned absorber plate; (*b*) single-glazed, overlapping glass plate; (*c*) single-glazed, porous absorber. (*Adapted, with permission, from Proc. World Symp. on Appl. Solar Energy, Phoenix, Ariz., 1955, published by Stanford Research Inst., Menlo Park, Calif., 1956.*)

SUMMARY

In this chapter the methods of determining heating requirements of buildings and selecting an optimal solar system for meeting these requirements are presented. Rules, methods, and criteria are given for specifying components of a solar system including collector area, collector tilt, number of covers, and storage volume. The importance of accurate calculation of the size of the most expensive component—the collector—is emphasized. Both approximate and precise methods of collector sizing are given; the former requires only a hand calculator, the latter, a computer.

Design procedures for solar service hot-water heaters are given along with mechanical equipment diagrams and design specifications. A comparison of the advantages and disadvantages of hot-air solar systems indicates that for heating only they are competitive with liquid-cooled collector systems. The limited usage factor of a heating only system diminishes the economic viability of air systems, however.

The economics of solar heating are presented, including the effects of interest rates, inflation, and auxiliary fuel prices. The concepts of life-cycle costing and solar system payout period are developed in tabular and analytical form. Marginal, average, and total costs of solar systems are described and the method of use of each is illustrated by example calculations.

REFERENCES

1. Beckman, W. A.: Radiation and Convection Heat Transfer in a Porous Bed, *Journal of Engineering for Power,* vol. 90A, p. 51, 1968.
2. Bridgers, F., et al.: Performance of a Solar Heated Office Building, *Heating, Piping and Air Conditioning,* p. 165, November, 1957.
3. Buchberg, H., and J. R. Roulet: Simulation and Optimization of Solar Collection and Storage for House Heating, *Solar Energy,* vol. 12, p. 31, 1968.
4. Butz, L.: *Use of Solar Energy for Residential Heating and Cooling,* Master's thesis, University of Wisconsin, Madison, 1973.
5. Close, D. J.: Solar Air Heaters for Low and Moderate Temperature Applications, *Solar Energy,* vol. 7, p. 117, 1963.
6. Close, D. J.: Rock Pile Thermal Storage for Comfort Air Conditioning, *Mech. Chem. Engrg. Trans. Inst. Eng. (Australia),* vol. MC-1, p. 11, 1965.
7. Czarnecki, J. T.: Performance of Experimental Solar Water Heaters in Australia, *Solar Energy,* vol. 2, p. 2, 1958.
8. Davey, E. T.: Solar Water Heating, *Building Materials,* vol. 8, p. 57, 1966.
9. Löf, G. O. G., and R. W. Hawley: Unsteady State Heat Transfer Between Air and Loose Solids, *Industr. Engrg. Chem.,* vol. 40, p. 1061, 1948.
10. Löf, G. O. G., and R. A. Tybout: Cost of House Heating with Solar Energy, *Solar Energy,* vol. 14, p. 253, 1973.
11. Speyer, E.: Solar Energy Collection with Evacuated Tubes, *J. Engrg. for Power,* vol. 87, p. 270, July, 1965.
12. Selcuk, K.: Thermal and Economic Analysis of the Overlapped-Glass Plate Solar-Air Collector, *Solar Energy,* vol. 13, p. 165, 1971.

CHAPTER 5

Solar Cooling of Buildings

*The real cycle you're working on is a
cycle called yourself.*

ROBERT PIRSIG

Solar cooling of buildings represents a potentially significant application of solar energy for building air conditioning in most sunny regions of the United States. Solar-cooling technology is presently not as advanced as solar-heating technology, but research work is expected to close the gap between the two by 1980. Several viable, solar-air-conditioning schemes are described in this chapter and methods for tentative system design are presented in detail.

COOLING REQUIREMENTS

The cooling load of a building is the rate at which heat must be removed to maintain the air in a building at a given temperature. It is usually calculated on the basis of the peak load expected during the cooling season. For a given building the cooling load primarily depends on

1. Design inside and outside dry-bulb temperatures
2. Design inside and outside relative humidities
3. Solar radiation heat load
4. Wind speed

A method of cooling-load calculation is presented in detail in the *ASHRAE Handbook of Fundamentals.*

The steps in calculating the cooling loads of a building are

1. Specify the building characteristics:
   Wall area, type of construction, and surface characteristics
   Roof area, type of construction, and surface characteristics
   Window area, setback, and glass type
   Building location and orientation
2. Specify the outside and inside wet- and dry-bulb temperatures.
3. Specify the solar heat load and wind speed

4. Calculate building cooling loads due to
    Heat transfer through windows
    Heat transfer through walls
    Heat transfer through roof
    Sensible heat gains due to infiltration and exfiltration
    Latent heat gains (water vapor)
    Internal heat sources, such as people, lights, etc.

Equations (5.1) through (5.7) may be used to calculate the various cooling loads for a building; cooling loads due to lights, building occupants, etc. may be estimated from the *ASHRAE Handbook of Fundamentals*. For unshaded or partially shaded windows, the load is

$$Q_{wi} = A_{wi}\left[F_{sh}\bar{\tau}_{b,wi}I_{h,b}\frac{\cos i}{\sin \alpha} + \bar{\tau}_{d,wi}I_{h,d}\right.$$
$$\left. + \bar{\tau}_{r,wi}I_r + U_{wi}(T_{out} - T_{in})\right] \quad (5.1)$$

For shaded windows the load (neglecting diffuse sky radiation) is

$$Q_{wi,sh} = A_{wi,sh}U_{wi}(T_{out} - T_{in}) \quad (5.2)$$

For unshaded walls the load is

$$Q_{wa} = A_{wa}\left[\bar{\alpha}_{s,wa}\left(I_r + I_{h,d} + I_{h,b}\frac{\cos i}{\sin \alpha}\right)\right.$$
$$\left. + U_{wa}(T_{out} - T_{in})\right] \quad (5.3)$$

For shaded walls the load (neglecting diffuse sky radiation) is

$$Q_{wa,sh} = A_{wa,sh}[U_{wa}(T_{out} - T_{in})] \quad (5.4)$$

For the roof the load is

$$Q_{rf} = A_{rf}\left[\bar{\alpha}_{s,rf}\left(I_{h,d} + I_{h,b}\frac{\cos i}{\sin \alpha}\right) + U_{rf}(T_{out} - T_{in})\right] \quad (5.5)$$

For sensible heat infiltration and exfiltration the load is

$$Q_i = \dot{m}_a(h_{c,out} - h_{c,in}) \quad (5.6)$$

For moisture infiltration and exfiltration the load is

$$Q_w = \dot{m}_a(W_{out} - W_{in})\lambda_w \quad (5.7)$$

where $\quad Q_{wi}$ = heat flow through unshaded windows of area A_{wi}, Btu/hr

$Q_{wi,sh}$ = heat flow through shaded windows of area $A_{wi,sh}$, Btu/hr

Q_{wa} = heat flow through unshaded walls of area A_{wa}, Btu/hr

$Q_{wa,sh}$ = heat flow through shaded walls of area $A_{wa,sh}$, Btu/hr

Q_{rf} = heat flow through roof of area A_{rf}, Btu/hr

Q_i = heat load due to infiltration/exfiltration, Btu/hr

Q_w = latent heat load, Btu/hr

$I_{h,b}$ = beam component of insolation on horizontal surface, Btu/(hr)(ft^2)

$I_{h,d}$ = diffuse component of insolation on horizontal surface, Btu/(hr)(ft^2)

I_r = ground-reflected component of insolation, Btu/(hr)(ft^2)

W_{out}, W_{in} = outside and inside humidity ratios, lb$_m$ H$_2$O/ (lb$_m$ dry air)

U_{wi}, U_{wa}, U_{rf} = over-all heat-transfer coefficients for windows, walls, and roof, including radiation, Btu/(hr)(ft^2)($^\circ$F)

$\dot{m}_a$ = net infiltration and exfiltration of dry air, lb$_m$/hr

T_{out} = outside dry-bulb temperature, $^\circ$F

T_{in} = indoor dry-bulb temperature, $^\circ$F

F_{sh} = shading factor (1.0 = unshaded, 0.0 = fully shaded)

$\bar{\alpha}_{s,wa}$ = wall solar absorptance

$\bar{\alpha}_{s,rf}$ = roof solar absorptance

i = solar-incidence angle on walls, windows, and roof, deg

$h_{c,out}, h_{c,in}$ = outside and inside air enthalpy, Btu/lb$_m$

α = solar altitude angle, deg

λ_w = latent heat of water vapor, Btu/lb$_m$

$\bar{\tau}_{b,wi}$ = window transmittance for beam (direct) insolation

$\bar{\tau}_{d,wi}$ = window transmittance for diffuse insolation

$\bar{\tau}_{r,wi}$ = window transmittance for ground-reflected insolation

EXAMPLE

Determine the cooling load for a building in Phoenix, Arizona with the specifications tabulated below.

Factor	Description or specification
Building characteristics:	
Roof:	
Type of roof	Flat, shaded
Area $A_{rf,sh}$, ft²	1,700
Walls (painted white):	
Size, north and south, ft	8 x 60 (two)
Size, east and west, ft	8 x 40 (two)
Area A_{wa}, north and south walls, ft²	$480 - A_{wi} = 480 - 40 = 440$ (two)
Area A_{wa}, east and west walls, ft²	$320 - A_{wi} = 320 - 40 = 280$ (two)
Absorptance $\bar{\alpha}_{s,wa}$ of white paint	0.12
Windows:	
Size, north and south, ft	4 x 5 (two)
Size, east and west, ft	4 x 5 (two)
Shading factor F_{sh}	0.20
Insolation transmittance	$\bar{\tau}_{b,wi} = 0.60; \bar{\tau}_{d,wi} = 0.81; \bar{\tau}_{r,wi} = 0.60$
Location and latitude	Phoenix, Ariz.; 33°N
Date	August 1
Time and local-solar-hour angle H_s	Noon; $H_s = 0$
Solar declination δ_s, deg	18°14'
Wall surface tilt from horizontal β	90°
Temperature, outside and inside, °F	$T_{out} = 100; T_{in} = 75$
Insolation I, Btu/(hr)(ft²)	$I_{h,b} = 185; I_{h,d} = 80; I_r = 70$
U factor for walls, windows, and roof	$U_{wa} = 0.19; U_{wi} = 1.09; U_{rf} = 0.061$
Infiltration, lb$_m$ dry air/hr	Neglect
Exfiltration, lb$_m$ dry air/hr	Neglect
Internal loads	Neglect
Latent heat load Q_w, %	30% of wall sensible heat load*

*Approximate rule of thumb for Phoenix.

SOLUTION

To determine the cooling load for the building just described, calculate the following factors in the order listed.

1. Incidence angle for the south wall i

$$\cos i = \cos \delta_s \cos (L - \beta) + \sin \delta_s \sin (L - \beta) = 0.257$$

2. Solar altitude α

$$\sin \alpha = \sin \delta_s \sin L + \cos \delta_s \cos L \cos H_s$$
$$= \cos (L - \delta_s) = \cos 15° = 0.96$$

3. South-window load (from Eq. (5.1))

$$Q_{wi} = 40 \left\{ (0.2 \times 0.6)\left(185 \frac{0.257}{0.96}\right) + (0.81 \times 80) \right.$$
$$\left. + (0.60 \times 70) + [1.09(100° - 75°)] \right\}$$
$$= 5,600 \text{ Btu/hr}$$

4. Shaded-window load (from Eq. (5.2))

$$Q_{wi,sh} = (3 \times 40)[1.09(100° - 75°)] = 3,270 \text{ Btu/hr}$$

5. South-wall load (from Eq. (5.3))

$$Q_{wa} = (480 - 40) \left\{ .12 \left[70 + 80 + \left(185 \frac{0.257}{0.96}\right) \right] \right.$$
$$\left. + 0.19(100° - 75°) \right\} = 12,610 \text{ Btu/hr}$$

6. Shaded-wall load (from Eq. 5.4))

$$Q_{wa,sh} = [(480 + 320 + 320)$$
$$-(3 \times 40)][0.19(100° - 75°)] = 4,750 \text{ Btu/hr}$$

7. Roof load (from Eq. 5.5))

$$Q_{rf} = 1,700[\bar{\alpha}_{s,rf} \times 0 + 0.061(100° - 75°)] = 2,600 \text{ Btu/hr}$$

8. Latent-heat load (30 percent of sensible wall load)

$$Q_{wa} = 0.3[(480 + 480 + 320 + 320)$$
$$- (4 \times 40)][0.19(100° - 75°)] = 2,050 \text{ Btu/hr}$$

9. Infiltration/exfiltration load

$$Q_i = 0$$

10. Total cooling load for the building described in the example.

$$Q_{tot} = Q_{wi} + Q_{wi,sh} + Q_{wa} + Q_{wa,sh} + Q_{rf} + Q_w + Q_i$$

$$Q_{tot} = 30,880 \text{ Btu/hr} \sim 2.5 \text{ tons air conditioning}$$

> This example is simplified for illustrative purposes. Heat loads must be calculated each hour of a design day to determine the maximum load. The maximum cooling load occurs usually between 3 and 4 P.M.

DESIGN OF A COLLECTOR FOR A SOLAR–COOLED BUILDING

Solar energy can be an economical source of energy for building temperature control in areas where the local climate requires both air conditioning and heating, because use of solar energy for heating and air conditioning increases the system usage factor. The combined system shown in Fig. 5.1 is presently installed in the Solar Energy Applications Laboratory, Colorado State University, Fort Collins, Colo. The same solar collector and storage system can be used for combined, year-round heating and cooling operation. Löf and Tybout[6] performed an optimization study of combined systems for the same geographical locations as their solar-heating study referenced earlier.* The general conclusions pertinent to collector design for combined systems used in climates with hot summers are

1. Optimal collector tilt β is equal to latitude L (except in the South where $\beta = L - 10°$).
2. Optimal collector area is always greater than that needed for heating only.
3. Three glass covers are desirable due to higher temperature needs of air conditioning systems; in subtropical steppe climate, e.g., Albuquerque, two are optimal.

The Löf-Tybout study modeled a flat-plate collector and lithium-bromide (LiBr) absorption air conditioner with price based on assumed mass production costs for both. The results are the only guide to solar air-conditioning design presently available. More information on the economics of solar air conditioning is given later in this chapter in "The Economics of Solar Air Conditioning."

The results of work and experience on solar air conditioning to date have not yet culminated in a reliable, inexpensive solar air-conditioning unit, and no system can be recommended for wide use at this time.

---

*G. O. G. Löf and R. A. Tybout, Cost of House Heating with Solar Energy, *Solar Energy*, vol. 14, p. 23, 1973.

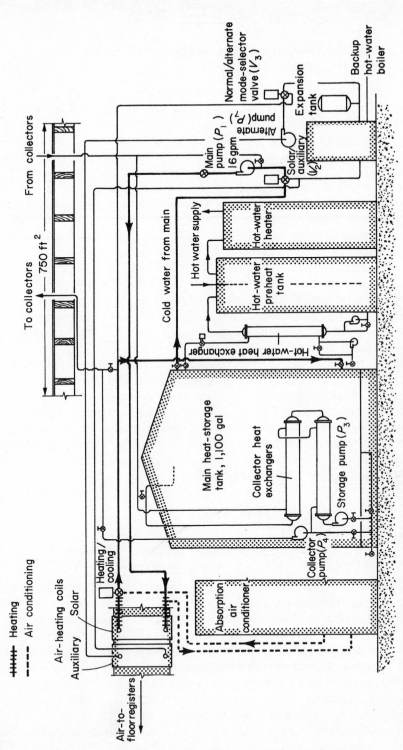

FIG. 5.1 Combined solar heating and cooling system at the Solar Energy Applications Laboratory, Colorado State University, Fort Collins. (*Courtesy of G. O. G. Löf.*[7])

STORAGE OF ENERGY AT HIGH
AND LOW TEMPERATURES

For a detailed description of high-temperature storage of energy the reader may refer to Chap. 4, "Storage of Energy." In an air-conditioning system it is also possible to cool a tank of water or other storage medium if excess cooling capacity is available during operational hours. One of the attractions of this approach is that the temperature difference between cold storage (say, 45°F at the coldest) and building temperature (~70°F) is less than the temperature difference between hot-storage and room temperature. As a result, less cooling effect is lost from cold storage than from hot storage. Moreover, cold storage is preferable to hot storage because losses from hot storage in a building add to the summer air-conditioning load.

In addition, 1 Btu of cold storage is 1 Btu of cooling; 1 Btu of hot storage is equivalent to less than 1 Btu of cooling since the air conditioner is not 100 percent efficient. These advantages of cold storage are offset by the additional cost and upkeep required to install and maintain a second storage unit and the associated plumbing and ducting.

The decision to use hot, cold, or combined storage depends on the building location. In Phoenix one should design for cold storage since the investment in hot storage is uneconomical in a climate where heating loads are small (only 1,500 degree-days per year). In Bemidji, Minn. on the other hand one should design for hot storage since the cooling load is small. It should be noted that the hot-storage tank used for winter heating can be used for summer cold storage by a proper valve arrangement.

VAPOR-COMPRESSION SYSTEMS
AND HEAT PUMPS

Introduction

As mentioned previously, the periodicity and intermittence of solar energy incident on a collector require the use of a conventional backup (auxiliary) system. The size and cost of the total system depend not only on the Btu's collected but also on storage facilities; any concept that can reduce the collection and storage requirements can improve the economics of solar-energy use. One attractive method to achieve a fuller use factor is to utilize a solar-assisted heat pump–air conditioner for all-year climate control. From an energy standpoint the efficiency of a conventional heat-pump system operating in a northern climate is reduced as a result of low ambient temperatures; owing to the reduced efficiency of the heat pump, reliance of the system of electric-resistance heating is necessary. However, solar energy can be used at low ambient temperatures to reduce or often eliminate the need for supplemental resistance heating in such systems.

A *heat pump* may be defined as a system that absorbs energy at low temperature and delivers energy at higher temperature through a vapor-compression or absorption cycle. However, no absorption-cycle heat pumps are presently available. Available heat-pumps use a standard vapor-compression refrigeration cycle operated in reverse: the evaporator is usually placed outdoors and the condenser indoors. The mechanical energy input to the compressor during vapor compression raises the internal energy absorbed at low ambient temperatures to a temperature level useful for space heating. The quantity of heat extracted from such a system can be several times larger than the energy required by the compressor. The basic advantage of a heat pump is this "heating multiplication."

In a solar-assisted heat-pump system designed both to heat and to cool a building, two general modes of operation are possible. As shown in Fig. 5.2(a), the solar heating system and the heat-pump system may be separate. In Fig. 5.2(b) the two cycles are shown connected, and a solar collector assists in reducing the amount of heat that must be supplied by the compressor when heating the building. Obviously, the second arrangement is more efficient.

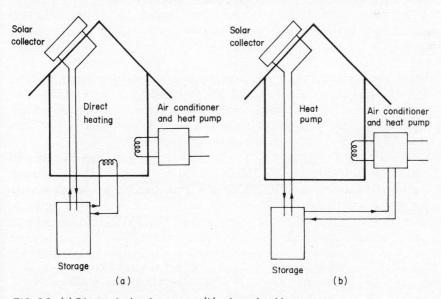

FIG. 5.2 (a) Direct solar heating system; (b) solar-assisted heat pump.

Solar-Powered Heat-pump Cycle

The heat pump can theoretically be used with any thermo-dynamic cooling cycle, such as an absorption, a jet compression, and a mechanical vapor compression cycle. Figure 5.3 illustrates the basic heat flows in a conceptual solar-powered heat pump system in which the sun's energy directly provides shaft work instead of acting as a booster as in Fig. 5.2. Q_4 is the amount of heat removed from the environment during the heating mode. A heat balance of the cycle in Fig. 5.3 gives

$$Q_1 + Q_4 = Q_2 + Q_3 \tag{5.8}$$

where Q_1 = solar heat input
Q_2 = heat rejected in the power cycle
Q_3 = heat rejected in the cooling cycle

The efficiency $\eta_{C,hp}$ of the system when the system acts as a heat pump is given by

$$\eta_{C,hp} = \frac{Q_2 + Q_3}{Q_1} \tag{5.9}$$

The efficiency in cooling, i.e., the heat transferred from a building divided by the energy required to drive the system, is called the *coefficient of performance* (COP) and is given as

$$COP = \frac{\text{cooling effect}}{\text{heat input}} = \frac{Q_4}{Q_1} \tag{5.10}$$

When the system is acting as a heat pump to heat a building, Q_2 and Q_3, the so-called heat rejected are rejected into the building and serve to heat the interior. When the system is acting as an air conditioner, the refrigerant flow is reversed, and heat is removed from the building and rejected to the environment.

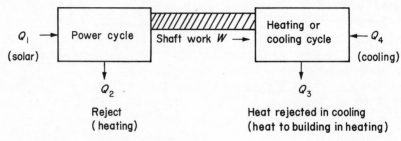

FIG. 5.3 Basic solar-powered heat-pump system.

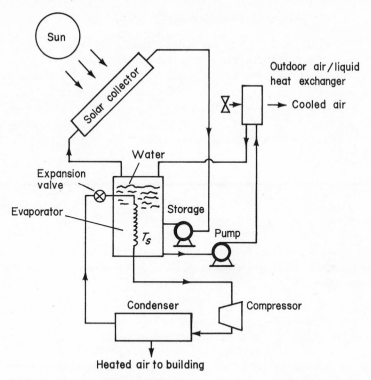

FIG. 5.4 Solar-assisted heat pump system concept using a liquid-to-air heat pump cycle showing the heating mode.

Solar-assisted, Conventional Heat Pump

A possible system in which solar energy augments a heat pump is shown in Fig. 5.4. In this system, thermal energy is stored in the form of sensible heat and is used as the energy source for a liquid-to-air heat pump. During the heating cycle, solar energy is used to increase the temperature of the storage, and therefore the amount of compression work necessary to increase the temperature of the working fluid is reduced. The energy that could be saved in a typical, single-family residence in the northeastern portion of the United States is shown in Fig. 5.5. From September to November as well as from March through April, solar energy is sufficient to supply the entire heat demand. Between November and March, the work input to the compressor is reduced by approximately 50 percent. The coefficient of performance, i.e., the energy delivered to the house divided by the electrical energy to drive the system, shows an appreciable reduction in energy power consumption for the solar-assisted system. Less power is consumed because the average yearly

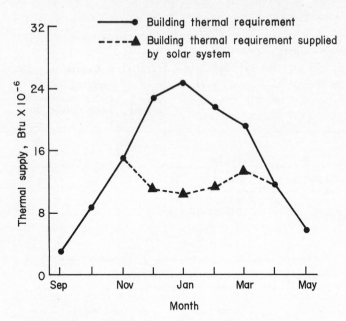

FIG. 5.5 Thermal supply by heating months for solar assisted heat pump concept.

coefficient of performance for a solar-activated system is on the order of 3 in contrast to the yearly coefficient of performance for a conventional heat pump, which is approximately 2. Greater improvements could be expected if concentrator solar collectors were used.

A solar-assisted heat pump (without the fan loop) of the type shown in Fig. 5.4 has been operating successfully for a year in a house constructed in 1974 in Colorado Springs, Colo. The North Campus of the Community College of Denver, the largest solar heated building in the world—300,000-ft² heated floor area, 36,000-ft² of collector—will use this same concept when the building is completed in 1976.

The Rankine Power Cycle

Many power cycles could be considered for a solar-assisted heat pump. One practical, recommended system is based on the Rankine cycle, illustrated in Fig. 5.6. The components are shown schematically on the left-hand side, and on the right-hand side a pressure-enthalpy diagram for the working fluid is presented.

The working fluid starts at state point 1 (P_1) and is pumped to the pressure in the boiler, state point 2 (P_2), where heat is added until the working fluid arrives at state point 3 (P_3) in a superheated form. The working fluid is then expanded at constant entropy to state point 4 (P_4), and heat is rejected at constant pressure in a condenser between points 4 and 1.

The magnitude of the efficiency of a Rankine cycle may be estimated by the so-called *Carnot efficiency* of the cycle, which is the highest efficiency that can be attained between the maximum and minimum temperatures available. The Carnot efficiency η_c, is equal to

$$\eta_C = \frac{T_3 - T_1}{T_3} = 1 - \frac{T_1}{T_3} \tag{5.11}$$

where T_3 is the maximum temperature attainable, and T_1 is the lowest, or sink, temperature. Equation (5.11) clearly shows the advantages of a concentrating collector that could increase the temperature T_3. T_1 is not readily maneuverable since it is, in general, dictated by the outside temperature for a power cycle. Figure 5.7 shows the effect of sink temperature and maximum temperature on the Carnot efficiency.

Any real power cycle can achieve only a fraction of the Carnot efficiency owing to the thermodynamic irreversibility of heat additions and other cycle losses. The efficiency for a real cycle, which can be obtained by examination of Fig. 5.6, is given by

$$\eta = \frac{h_3 - h_4}{h_3 - h_1} \tag{5.12}$$

where h is the enthalpy at the condition indicated by the subscript. The work required by the feed pump is usually small and can be neglected.

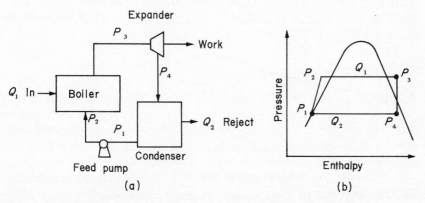

FIG. 5.6 Simple Rankine cycle. (*a*) Components; (*b*) pressure-enthalpy diagram.

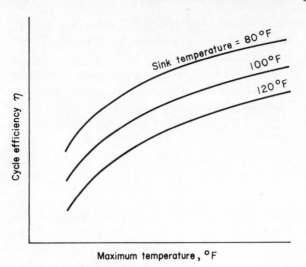

FIG. 5.7 Ideal Carnot-cycle efficiencies as a function of sink temperature and maximum cycle temperature.

The Vapor-compression Cooling Cycle

Figure 5.9 shows a vapor-compression cooling system that is most commonly used for cooling and air conditioning. The only major alternative at the present time is an absorption-cycle cooling system, which will be discussed subsequently.

The Carnot COP for the vapor-compression cycle is given by

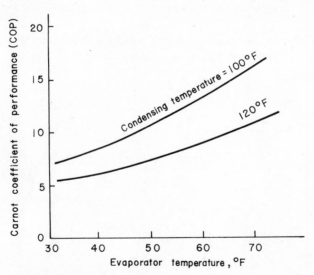

FIG. 5.8 Ideal Carnot refrigeration-cycle coefficient of performance as a function of evaporator and condenser temperatures.

$$COP_C = \frac{T_2}{T_1 - T_2} \tag{5.13}$$

where T_2 = lowest temperature level of the working fluid at which heat from the interior of the house is absorbed

T_1 = upper temperature of the working fluid in the condenser, at which heat is rejected to the atmosphere.

Figure 5.8 shows some typical numbers for Carnot cooling cycles. It can be seen that the COP is greater than unity; i.e., more cooling is accomplished than the shaft work required to drive the system.

Figure 5.9 shows the hardware for the vapor-compression cycle schematically as well as a pressure-enthalpy diagram for the working fluid. It can be seen that the heat absorbed (equal to Q_4 in Fig. 5.3)

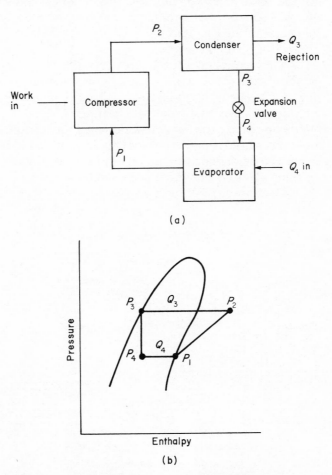

(a)

(b)

FIG. 5.9 Vapor-compression cooling cycle. (a) Components; (b) pressure enthalpy diagram.

is given by $h_1 - h_4$, and the heat rejected (Q_3 in Fig. 5.3) is given by $h_2 - h_3$. The compression work necessary to increase the temperature to the level required by h_2 is given by $h_2 - h_1$. Thus the COP of the vapor-compression cycle is given by

$$COP = \frac{\text{cooling effect}}{\text{work input}} = \frac{h_1 - h_4}{h_2 - h_1} \qquad (5.14)$$

In summary, the heat-pump–air conditioner is a climate control system that employs a vapor-compression cycle both to heat and cool a building. Typically, the heat sink as well as the heat source for such a system is ambient air, although the ground or ground water can be used. The heat pump is simply a heat transformer, which, by the introduction of mechanical work into the cycle, increases the temperature of the working fluid to a level where it can be used as a source for heating the home.

The heat pump is composed of the same components that are used in standard vapor-compression air-conditioning systems, as shown in Fig. 5.10. The only additions in a heat pump are a four-way valve to permit cycle reversal when the heating mode changes to the cooling mode and an expansion device designed specifically for the particular operational (heating or cooling) refrigerant-flow condition. The basic cycle can be designed to operate in many configurations, as discussed in standard texts on air conditioning.

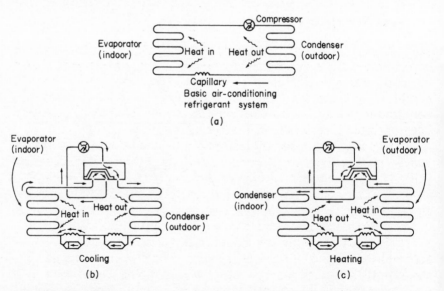

FIG. 5.10 Vapor-compression cycle in (a) basic air-conditioning mode, (b) heat pump cooling mode, and (c) heat pump heating mode.

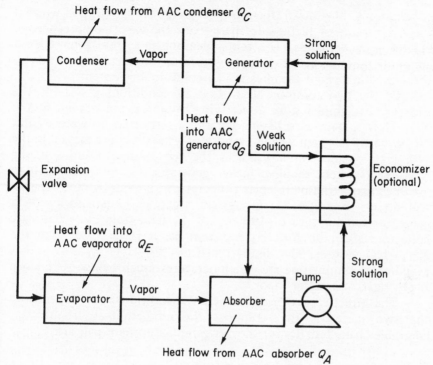

Heat flow from AAC condenser Q_C

Strong solution

Condenser — Vapor → Generator

Heat flow into AAC generator Q_G

Weak solution

Expansion valve

Economizer (optional)

Heat flow into AAC evaporator Q_E

Strong solution

Pump

Evaporator — Vapor → Absorber

Heat flow from AAC absorber Q_A

FIG. 5.11 Absorption-air-conditioner heat and fluid flow diagram with economizer.

ABSORPTION AIR CONDITIONING

Absorption air conditioning is the only air conditioning system compatible with the upper collection-temperature limits imposed by currently available flat-plate collectors. Home-sized absorption air conditioning units are more expensive than vapor-compression air conditioning units, but to date only absorption air conditioning has been operated successfully in a full-scale installation.

Presently, two types of absorption air conditioning systems are widely marketed in the United States: the lithium-bromide–water ($LiBr-H_2O$) system and the ammonia–water (NH_3-H_2O) system. The absorption air conditioning system is shown in Fig. 5.11. Absorption air conditioning differs from vapor-compression air conditioning only in the positive pressure gradient stage (right of the dashed line in Fig. 5.11). In absorption air conditioning systems, the pressurization is accomplished by first dissolving the refrigerant in a liquid (the absorbent) in the *absorber section*, then pumping the solution to a high pressure with an ordinary liquid pump. The low-boiling refrigerant is then driven from solution by the addition of heat in the *generator.* By this means the refrigerant vapor is compressed without the large input of high-grade shaft work the vapor-compression air conditioning demands. The remainder of the system consists of a condenser, expansion valve, and evaporator, identical in function to those used in a vapor-compression air conditioning system.

Of the two common absorption air conditioning systems, the $LiBr-H_2O$ is simpler since a rectifying column is not needed: in the NH_3-H_2O system a rectifying column assures that no water vapor, mixed with NH_3, enters the evaporator where it could freeze. In the $LiBr-H_2O$ system water vapor is the refrigerant. In addition, the NH_3-H_2O system requires higher generator temperatures (250 to 300°F) than a flat-plate solar collector can provide without special techniques. The $LiBr-H_2O$ system operates satisfactorily at a generator temperature of 190 to 200°F, achievable by a flat-plate collector; also, the $LiBr-H_2O$ system has a larger COP than the NH_3-H_2O system. The disadvantage of $LiBr-H_2O$ systems is that evaporators cannot operate at temperatures much below 40°F since the refrigerant is water vapor.

The effective performance of an absorption cycle depends on the two materials that comprise the refrigerant–absorbent pair. Desirable characteristics for the refrigerant–absorbent pair are

1. The absence of a solid-phase absorbent
2. A refrigerant more volatile than the absorbent in order to be separated from the absorbent easily in the generator
3. An absorbent that has a small affinity for the refrigerant
4. A high degree of stability for long-term operations
5. A refrigerant that has a large latent heat so that the circulation rate can be kept at the minimum
6. A low corrosion rate and nontoxicity for safety reasons

The only disadvantage of the $LiBr-H_2O$ pair is the possible problem with crystallization in the generator.

Absorption air conditioners are manufactured by many of the large air conditioning manufacturers in the United States—Carrier, Trane, York, Singer, Arkla-Servel, etc. No manufacturer currently makes a residential-sized (3 to 5 ton) $LiBr-H_2O$ unit. These units are NH_3-H_2O systems. In the past, Arkla-Servel manufactured a smaller residential $LiBr-H_2O$ unit. The line was discontinued some years ago but is being revived because of the recent interest in solar-assisted air conditioning systems. The present trade name of the Arkla-Servel system* is Sol-air.™

---

*®Registered by Arkla Industries, Evansville, Ind.

Performance

The COP of an absorption air conditioner can be calculated from Fig. 5.11.

$$COP = \frac{\text{cooling effect}}{\text{heat input}}$$

$$COP = \frac{Q_E}{Q_G} \tag{5.15}$$

The pump work has been neglected since it is quite small; in some 3- and 5-ton units the pump can even be eliminated entirely and a percolation principle employed instead.

The COP values for absorption air conditioning range from 0.5 for a small, single-stage unit to 0.85 for a double-stage, steam-fired unit. These values are about 15 percent of the COP values that can be achieved by a vapor-compression air conditioner. It is incorrect to compare the COP of an absorption air conditioner with that of a vapor-compression air conditioner, however, because the efficiency of electric power generation or transmission is not included in the vapor-compression air conditioning COP. If the COP of the mechanical system is multiplied by the thermal efficiency of the power plant and the efficiency of the transmission network, it can be shown that the vapor-compression air conditioner has little or no thermal performance advantage over the absorption air conditioning system.

EXAMPLE

A water–lithium-bromide, absorption–refrigeration system such as that shown in Fig. 5.12 is to be analyzed for the following requirements:

1. The machine is to provide 100 tons of refrigeration with an evaporator temperature of 40°F, an absorber outlet temperature of 90°F, and a condenser temperature of 110°F.
2. The approach at the low-temperature end of the liquid heat exchanger is to be 10°F.
3. The generator is heated by a flat-plate solar collector capable of providing a temperature level of 192°F for the evaporation of the refrigerant.

Determine the COP, absorbent and refrigerant flow rates, and heat input required for a 100-ton unit.

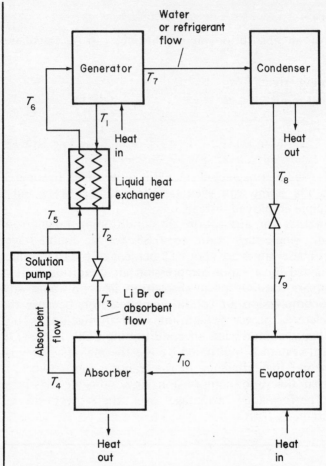

FIG. 5.12 Lithium-bromide-water, absorption–refrigeration cycle (see Table 5.1).

SOLUTION

Analytical evaluation of the LiBr-H_2O cycle requires that several simplifying factors be assumed, for example,

1. At those points in the cycle for which temperatures are specified, the refrigerant and absorbent phases are in equilibrium
2. With the exception of pressure reductions across the expansion device between points 8 and 9 in Fig. 5.12, pressure reductions in the lines and heat exchangers are neglected.
3. The temperature difference at the inlet to the liquid heat exchanger is 10°F.

4. Pressures at the evaporator and condenser are equal to the vapor
   pressure of the refrigerant, i.e., water, as found in steam tables.
5. Enthalpies for LiBr–H_2O mixtures are given in Fig. 5.13.

As the first step in solving the problem, set up a table similar
to Table 5.1; enter values of pressure in the appropriate table
columns, enthalpy, and weight fraction for which sufficient
information is available. For example, at point 8 the temperature
is 110°F, and the vapor pressure of steam corresponding to this
pressure in the condenser is 1.28 psia, or 66 mm of Hg.

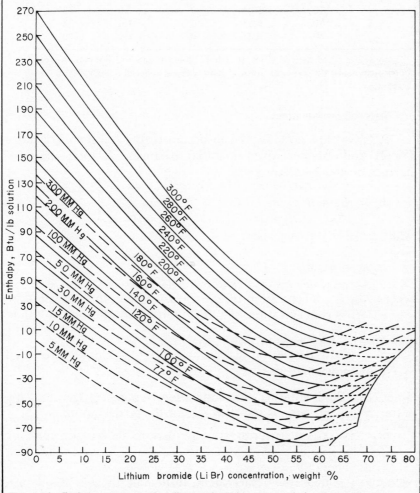

FIG. 5.13 Enthalpy-concentration diagram for lithium-bromide–water combination.

TABLE 5.1 Thermodynamic Properties of Refrigerant and Absorbent for Fig. 5.12

Condition no. in fig. 5.12	Temperature, °F	Pressure, mm Hg	LiBr weight fraction, %	Flow, lb/lb H_2O	Enthalpy, Btu/lb
T_1	192	66.0	0.61	11.2	−30
T_2	100	66.0	0.61	11.2	−70
T_3	100	6.3	0.61	11.2	−70
T_4	90	6.3	0.56	12.2	−75
T_5	90	66.0	0.56	12.2	−75
T_6	163	66.0	0.56	12.2	−38.8
T_7	192	66.0	0	1.0	1147
T_8	110	66.0*	0	1.0	78
T_9	40	6.3*	0	1.0	78
T_{10}	40	6.3	0	1.0	1079

*These values were derived from Joseph H. Keenan and Frederick G. Keyes, "Thermodynamic Properties of Steam," John Wiley & Sons, New York, 1936.

Mass balance equations

Relative flow rates for the absorbent (LiBr) and the refrigerant (H_2O) are obtained from material balances. A total material balance on the generator gives

$$\dot{m}_6 = \dot{m}_1 + \dot{m}_7$$

while a LiBr balance gives

$$\dot{m}_6 X_r = \dot{m}_1 X_{ab}$$

where X_{ab} = concentration of LiBr in absorbent, lb/lb of solution
X_r = concentration of LiBr in refrigerant, lb/lb of solution

Substituting $(\dot{m}_1 + \dot{m}_7)$ for $\dot{m}_6$ gives

$$\dot{m}_1 X_r + \dot{m}_7 X_r = \dot{m}_1 X_{ab}$$

Since the fluid entering the condenser is pure refrigerant, i.e., water, $\dot{m}_7$ is the same as the flow rate of the refrigerant $\dot{m}_r$:

$$\frac{\dot{m}_1}{\dot{m}_7} = \frac{X_r}{X_{ab} - X_r} = \frac{\dot{m}_{ab}}{\dot{m}_r}$$

where $\dot{m}_{ab}$ = flow rate of absorbent, lb/hr
$\dot{m}_r$ = flow rate of refrigerant, lb/hr

Substituting for X_r and X_{ab} from the table gives the ratio of absorbent-to-refrigerant flow rate

$$\frac{\dot{m}_{ab}}{\dot{m}_r} = \frac{0.56}{0.61 - 0.56} = 11.2$$

The ratio of the refrigerant-absorbent solution flow rate $\dot{m}_s$ to the refrigerant-solution flow rate $\dot{m}_r$ is

$$\frac{\dot{m}_s}{\dot{m}_r} = \frac{\dot{m}_{ab} + \dot{m}_r}{\dot{m}_r} = 11.2 + 1 = 12.2$$

Energy balance equations

The enthalpy of the refrigerant–absorbent solution leaving the liquid heat exchanger at point 6 is obtained from an over-all energy balance on the unit, or

$$\dot{m}_s h_5 + \dot{m}_{ab} h_1 = \dot{m}_{ab} h_2 + \dot{m}_s h_6$$

Hence, $h_6 = h_5 + \left[\dfrac{\dot{m}_{ab}}{\dot{m}_s} (h_1 - h_2) \right]$

$$= -75 + \frac{11.2}{12.2} [-30 - (-70)] = -38.2 \text{ Btu/lb of solution}$$

The temperature corresponding to this value of enthalpy and a pressure of 66 mm Hg is found from Fig. 5.13 to be 163° F.

The flow rate of refrigerant required to produce the desired 100 tons of refrigeration (equivalent to 1,200,000 Btu/hr) is obtained from an energy balance about the evaporator,

$$q_{refrig} = \dot{m}_r (h_9 - h_{10})$$

where q_{refrig} is the cooling effect supplied the refrigeration unit, and

$$\dot{m}_r = \frac{1,200,000}{1,079 - 78} = 1,200 \text{ lb/hr}$$

The flow rate of the absorbent is

$$\dot{m}_{ab} = \frac{\dot{m}_{ab}}{\dot{m}_r} \dot{m}_r = 11.2 \times 1,200 = 13,400 \text{ lb/hr}$$

while the flow rate of the solution is

$$\dot{m}_s = \dot{m}_{ab} + \dot{m}_r = 13,400 + 1,200 = 14,600 \text{ lb/hr}$$

The rate at which heat must be supplied to the generator q_{sup} is obtained from the heat balance

$$
\begin{aligned}
q_{sup} &= \dot{m}_r h_7 + \dot{m}_{ab} h_1 - \dot{m}_{ab} h_6 \\
&= [(1,200 \times 1,147) + (13,400 - 30)] - (14,600 \times -38.2) \\
&= 1,540,000 \text{ Btu/hr}
\end{aligned}
$$

This requirement, which determines the size of the solar collector, likely represents the maximum heat load the refrigeration unit must supply during the hottest part of the day.

The coefficient of performance, COP, is

$$\text{COP} = \frac{q_{refrig}}{q_{sup}} = \frac{1,200,000}{1,540,000} = 0.78$$

The rate of heat transfer in the other three heat-exchanger units—the liquid heat exchanger, the water condenser, and the generator—is obtained from heat balances. For the liquid heat exchanger this gives

$$
\begin{aligned}
q_{1-2} &= \dot{m}_{ab}(h_1 - h_2) \\
&= 13,400[(-30) - (-70)] = 540,000 \text{ Btu/hr}
\end{aligned}
$$

where q_{1-2} is heat transferred from the absorbent stream to the refrigerant/absorbent stream. For the water condenser the rate of heat transfer q_{7-8} rejected to the environment is

$$
\begin{aligned}
q_{7-8} &= \dot{m}_r(h_7 - h_8) \\
&= 1,200(1,147 - 78) = 1,280,000 \text{ Btu/hr}
\end{aligned}
$$

The rate of heat removal from the absorber can be calculated from an over-all heat balance on this system:

$$
\begin{aligned}
Q_A &= q_{7-8} - q_{sup} - q_{refrig} \\
&= 1,280,000 - 1,540,000 - 1,200,000 \\
&= -1,460,000 \text{ Btu/hr}
\end{aligned}
$$

Explicit procedures for the mechanical and thermal design as well as the sizing of the heat exchangers are presented in standard heat transfer texts. In large commercial units it may be possible to use higher concentrations of LiBr, operate at a higher absorber tempera-ture, and thus save on heat-exchanger cost. In a solar-driven unit this approach would require a concentrator-type absorber because flat-plate solar collectors cannot achieve a sufficiently high temperature to raise the temperature level in the absorber of an absorption air conditioner much above 190°F.

COMPARISON OF MECHANICAL AND ABSORPTION REFRIGERATION SYSTEMS

Absorption–refrigeration systems operate on cycles in which the primary fluid, a gaseous refrigerant, which has been vaporized in an evaporator, is absorbed by a secondary fluid, called the absorbent. Absorption–refrigeration cycles can be viewed thermodynamically as a combination of a heat-engine cycle and a vapor-compression refrigeration cycle, which are also the two components of a mechanical-refrigeration system. Simplified diagrams for the two methods of providing refrigeration are shown in Figs. 5.14 and 5.15,

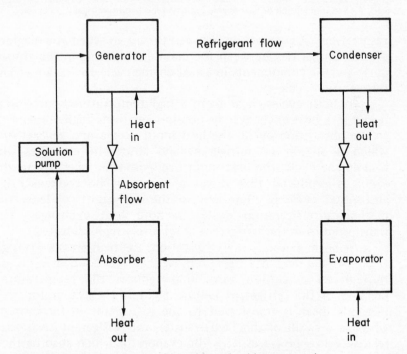

FIG. 5.14 Basic absorption–refrigeration cycle without economizer.

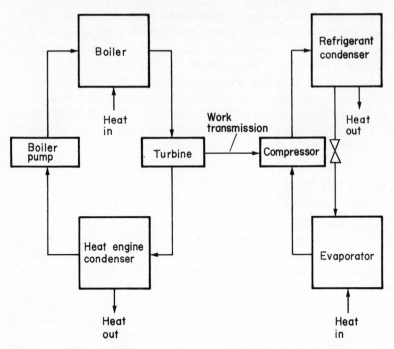

FIG. 5.15 Combination of heat-engine cycle and mechanical-refrigeration cycle.

respectively. A comparison between these two schematic diagrams indicates similarities between the main components in the absorption cycle and the components in a heat-engine cycle driving a mechanical-refrigeration cycle.

In both cycles, heat from a high-temperature source is transferred in a heat exchanger to obtain a relatively high-pressure vapor. In the absorption cycle, the heat input occurs in a generator, from which a stream of refrigerant and absorbent emanates. In the heat-engine cycle, the heat input occurs at the boiler, from which a vapor is produced that drives a turbine. The condenser in the absorption cycle is equivalent to the refrigerant condenser in the mechanical-refrigeration cycle. In both heat exchanges, heat is transferred from the refrigerant at relatively high pressures.

In both methods of refrigeration, the high-pressure refrigerant (from which heat has been removed in the condenser) is passed through an expansion valve that reduces the temperature and pressure of the refrigerant before it enters the evaporator. In both methods, heat is transferred to the refrigerant in the evaporator, where as a result of this heat transfer, the refrigerant is vaporized at relatively low pressures. It is the evaporator which absorbs the heat and provides the refrigeration effect in both methods.

The absorber in the absorption–refrigeration cycle corresponds to the heat-engine condenser in the mechanical-refrigeration cycle. Heat is transferred out of the absorber in the absorption cycle, and out of the heat-engine condenser in the combination method, to an intermediate temperature sink to facilitate the conversion of relatively low-pressure vapor to the liquid state. In the absorption method, the absorbent is mixed with the refrigerant in the absorber. The final similarity between the two systems is the solution pump and the boiler pump. In both systems a small amount of work is necessary to increase the pressure of the liquid before entering the boiler or generator of the cycle.

The turbine, which extracts heat energy from the high temperature vapor from the boiler in the combination cycle thereby transforming heat into work to drive a compressor, does not have a counterpart in the absorption method. In the absorption method, the energy input occurs in the form of heat into the generator, and hence the generator can operate at a temperature of less than 200°F. In some absorption cycles, the heat supply can be provided by flat-plate solar collectors. On the other hand, in the combination heat-engine and mechanical-refrigeration cycle, the turbine drive requires a relatively high-temperature vapor for efficient operation and it is difficult to obtain good performance with a flat-plate solar collector and a concentrator type is required.

The relationship between work and heat for an ideal heat-engine operating on a Carnot cycle is

$$W = q_g \frac{T_h - T_{hs}}{T_h} \tag{5.16}$$

where W = work output rate,
q_g = heat input rate,
T_h = temperature of heat source, and
T_{hs} = temperature of heat sink

The relationship between the work required and the refrigeration load for an ideal mechanical-refrigeration machine operating on the reverse Carnot cycle is

$$-W = q_e \frac{T_{hs} - T_1}{T_1} \tag{5.17}$$

where $-W$ = work input rate,
q_e = rate of refrigeration
T_1 = temperature of the refrigeration load
T_{hs} = temperature of the heat sink

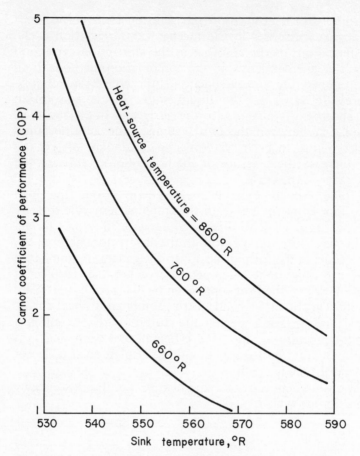

FIG. 5.16 Ideal coefficient of performance for absorption–refrigeration cycles with evaporator (load) temperature T_l of 500° R.

The coefficient of performance for the combination of this engine cycle and the mechanical-refrigeration machine is given by

$$\text{COP} = \frac{q_e}{q_g} = \frac{T_1(T_h - T_{hs})}{T_h(T_{hs} - T_1)} \tag{5.18}$$

Equation (5.18) applies to the ideal absorption–refrigeration process as shown earlier as well as to the combination heat–engine and mechanical-refrigeration cycle. The coefficient of performance is plotted as a function of sink temperature for various temperatures of the heat source for a Carnot cycle with a refrigeration load at 40°F in Fig. 5.16.

NONMECHANICAL SYSTEMS

Australian Rock System

A nocturnal cooling-storage system first tested in 1955 in a desert in the southwest United States has been more recently tested and developed in Australia. The system consists of a large bed of rocks cooled by drawing cool night air across them and exposing them to night sky radiation. During the day, warm inside air may be cooled by circulating it through the rock bed. Augmented cooling can be achieved by drawing the night air through a porous surface having a high emittance at low temperature and facing the night sky. Such a system operates best in a desert climate, where the night skies are clear and humidity is low. In a desert climate, the diurnal temperature variation may be 45°F. Although this system is not an active solar cooling system, it uses the same pebble-bed storage that an air-cooled solar heating system uses for heat storage.

The use of a pebble bed to provide energy storage for both heating and cooling cycles has been described by Close[2] and Dunkle.[3] Figure 4.13 shows how a solar air heater can be combined with pebble-bed storage to provide a heat source for a building. The cycles in Fig. 5.17 illustrate the operation of the same pebble bed when it is used for cooling. During the night, cool air from the outside is brought through an evaporative cooler into the pebble bed

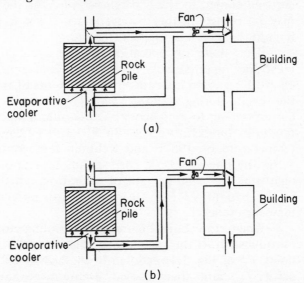

FIG. 5.17 Operation of a pebble bed thermal storage as a source of air conditioning.

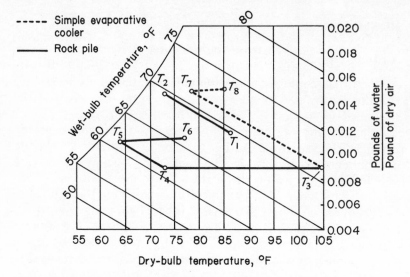

FIG. 5.18 Cooling cycles for pebble bed with and without evaporative cooler.

and is cooled at approximately wet-bulb temperature to a condition approaching saturation. The entire bed is eventually brought to this temperature by passing the cooled air through it for several hours, as shown in Fig. 5.17(a). On the next day, when cooling is required in the building, outside air is drawn vertically downward through the bed and cooled to the bed temperature. As the air leaves the bed it may be cooled further by evaporation and is then passed into the building as shown in Fig. 5.17(b).

The mode of operation of the cycle just described can be illustrated quantitatively by means of an example. The psychometric chart in Fig. 5.18 indicates the state points of the air passing through the cycle during a 24-hr period with environmental conditions corresponding to a night with a wet-bulb temperature of 70°F and a dry-bulb temperature of 86°F, and a day with a dry bulb temperature of 105°F and wet-bulb temperature of 71°F. In Fig. 5.18, the line $T_1 - T_2$ corresponds to evaporative cooling of the nighttime air to 80 percent of saturation, which will cool the rocks in the bed to 73°F if steady state can be achieved at the lowest temperature level.

When the building requires cooling the next day, air is introduced into the pebble bed under conditions corresponding to point T_3 in Fig. 5.18, cooled in the bed to 73°F, corresponding to point T_4, and then cooled evaporatively at constant wet-bulb temperature to a dry-bulb temperature of 63°F, corresponding again to 80 percent saturation. This air is then passed into the building to maintain conditions of 77°F and about 57 percent relative humidity. The increase in internal energy of the air, corresponding to the amount of heat transferred from the hot interior of the building to the coolant, is 3.5 Btu/lb. Thus, for a cooling load of 30,000 Btu/hr or 2.5 tons, the required air circulation rate is about

$$\frac{(30{,}000 \text{ Btu/hr}) \times (14 \text{ cu ft/lb of air})}{(3.5 \text{ Btu/lb of air}) \times (60 \text{ min/hr})} = 2{,}000 \text{ cfm}$$

For comparison, the dotted line in Fig. 5.18 corresponds to a simple evaporative cooler. To maintain a temperature of 85°F in the building, the required air-flow rate is about 30 percent larger than for a rock bed that maintains the building at 77°F.

To determine the size of the pebble bed necessary for a cooling period of 12 hr, a heat balance must be made on the rocks, or

$$V_r \rho_r c_r (T_{a,\text{in}} - T_{a,\text{out}}) = \dot{m}_a c_a (T_{a,\text{in}} - T_{a,\text{out}})\theta \tag{5.19}$$

where
V_r = volume of the rock in the bed, ft^3
ρ_r = density of the rock, lb/ft^3
c_r = specific heat of the rock, Btu/(lb)(°F)
$T_{a,\text{in}}$ = temperature of air entering bed, °F
$T_{a,\text{out}}$ = temperature of air leaving bed, °F
$\dot{m}_a$ = air-flow rate, lb/hr
c_a = specific heat of air, Btu/(lb)(°F)
θ = cooling period, hr

Equation (5.19) assumes that all the thermal energy stored can be extracted at the maximum temperature potential. For less efficient operation the size of the rock pile must be increased.

The size of the pile is approximately

$$V_r = \frac{(2{,}000 \times 0.24)(T_{a,\text{in}} - T_{a,\text{out}})(12 \times 60)}{(85 \times 0.21)(T_{a,\text{in}} - T_{a,\text{out}}) \, 14} = 1{,}400 \text{ ft}^3 \tag{5.20}$$

However, since the pile has empty spaces, the actual volume will be larger by the inverse of the empty-space fraction, defined as

$$\frac{\text{Volume of rock pile} - \text{volume of rocks}}{\text{Volume of rock pile}}$$

To design a rock-pile storage system completely the friction factor for packed beds and the heat-transfer coefficient for air passing through the pile must be known. In, Fig. 5.19 the friction factor for packed beds, f, (defined as 2 times the pressure drop Δp

divided by $\rho_f v_f^2$ times the ratio d_r/l, where l is the packed bed length) is plotted as a function of the *Particle Reynolds number*, Re (defined as $v_f \rho_b d_f/\mu_b$). In Fig. 5.19 v_f is the superficial air velocity (air-flow rate per pile cross-sectional area), d_r is the equivalent spherical diameter of the rocks $(6 V_p/\pi)^{1/3}$, V_p is the rock-particle volume, μ_b is the viscosity of the air in the pile, ρ_b is the density of the air and Δp is the pressure drop across the pile. These parameters must be in a consistent set of units to make the abscissa and the ordinate dimensionless. This correlation, proposed by Dunkle,[3] agrees well with other correlations but is simpler to use. Experimentally measured heat-transfer coefficients from Close[2] are shown in Fig.

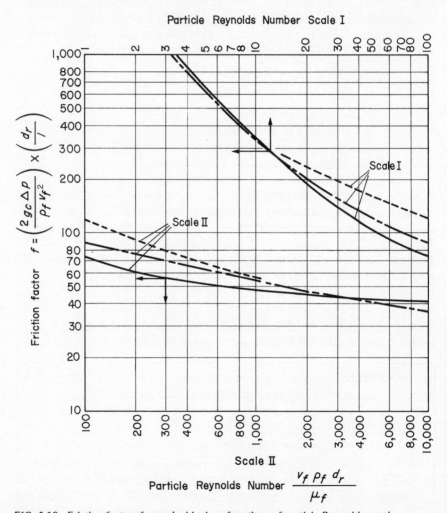

FIG. 5.19 Friction factors for packed beds as functions of particle Reynolds number.

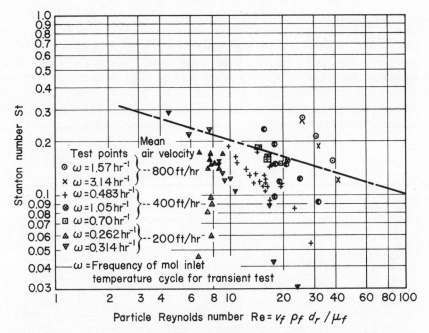

FIG. 5.20 Comparison between measured heat-transfer coefficients for packed beds as a function of particle Reynolds number. (*Adapted from Close.*[2])

5.20 as a function of particle Reynolds number. In Fig. 5.20 the *Stanton number*, St, defined as a heat-transfer coefficient divided by the product of the air density, the velocity, and the specific heat, is plotted again as a function of the particle Reynolds number. Figure 5.21 shows the pressure decrease in inches of water per foot of the depth of the pile and the heat-transfer coefficient as a function of the superficial air velocity in feet per minute for various rock sizes. The following example illustrates the calculations for a rock-pile storage system. For a more detailed analysis the variation in temperature during a 24-hr period must be taken into account.

EXAMPLE

A rock pile is required to store 1 million Btu. Charging and extraction are at constant rates, and each lasts approximately 10 hr. The temperature difference is to be 60° (70°F minimum and 130°F maximum). The maximum allowable pressure decrease is 0.1 in. of water. Approximately 80 percent of the energy stored in the pile can be extracted. Determine the fluid flow rate and ratio of energy storage to energy required to move the air through the bed.

SOLUTION

Assume the rock has a density of 85 lb/ft³ and a specific heat of 0.21 Btu/(lb)(°F); the total volume of rock required for the storage would then be 934 ft³. If this storage is to be charged in 10 hr, the mass flow rate of air, $\dot{m}_a$, may be obtained from a heat balance. If it can be assumed that all the heat is extracted at the stored temperature

$$\dot{m}_a = \frac{10^6 \text{ Btu}}{10 \text{ hr} \times [(0.24 \text{ Btu})/(\text{lb})(°F)] \times 60°F} = 7,000 \text{ lb/hr}$$

The volumetric flow rate for air at 70°F is thus about 1,600 ft³/min. If the maximum allowable pressure decrease is 0.1 in. of water, the ratio of heat stored to energy required to operate the pile is approximately 400:1.

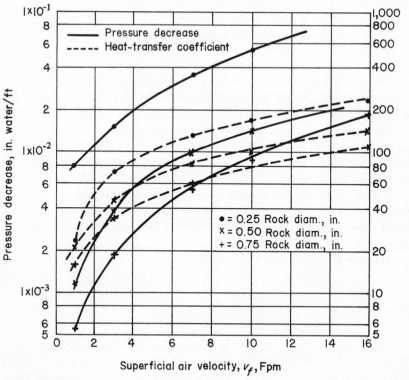

FIG. 5.21 Heat-transfer coefficients and pressure drops for various air flows and rock sizes in a pebble bed.

TABLE 5.2 Dimensions and Air Velocity for Various Rock Beds

Bed dimensions and air velocity	Rock diameter, in.			Design basis
	0.25	0.5	0.75	
Bed height, ft	3.4	5.9	7.4	Maximum allowable
Bed area, ft²	275.5	159	126	pressure drop
Air velocity, fpm	6	10.5	13.1	(maximum height)
Bed height, ft	2.75	4.0	5.0	Intermediate height
Bed area, ft²	340	233.5	186.8	
Air velocity, fpm	4.85	7.07	8.8	
Bed height, ft	2	2.0	2.0	Minimum height
Bed area, ft²	467	467	467	
Air velocity, fpm	3.53	3.53	3.53	

Table 5.2 summarizes the dimensions and air velocities for various pebble beds as compiled by Close[2] for engineering design. It should be noted that best performance is obtained when the rocks are as small as possible and the bed sized for the maximum allowable pressure decrease. The smaller the rocks, however, the more the pile becomes subject to blockage by dust, and the installation of a dust-removal screen may be necessary.

Sky-Therm™ System*

H. R. Hay has built a home in Atascadero, Calif., which uses his patented, passive, solar heating and cooling arrangement. It consists, in summary, of a combined collector-radiator-storage system mounted on the horizontal roof of a one-story building. The combined energy medium consists of 8-in.-deep pools of water enclosed in plastic bags located between beams of the house and on top of a black, plastic liner used to further waterproof the roof. Solar energy heats the water to 85°F, a purposely maintained low temperature. Insulated shutters are then placed over the panels during winter nights to reduce heat loss.[4,5]

During the summer the shutters cover the panels to prevent heating during the day. However, the collection area used for service water heating remains uncovered during daytime. At night the primary ponds are uncovered to permit radiation of the heat collected from the building during the day. The system is currently under test at the Atascadero test site. A smaller building was maintained at a temperature of 70°F ± 2°F for a year in Phoenix, Ariz., by this method.

*®Registered trademark, Skytherm Processes and Engineering, Los Angeles, Calif.

The summer function of such a system relies on nocturnal radiation loss. Such losses are usually in the range of 10 to 35 $Btu/(hr)(ft^2)$. To provide 1 ton-hr of heat removal, 500 ft^2 of radiator surface is required. If the entire roof of a 2,000 ft^2 house were used as a radiator, only 4 tons of heat would be removed, assuming the water ponds do not reach an equilibrium temperature and night heat loss averages 25 $Btu(hr)(ft^2)$. For an 8-hr night this cooling effect amounts to 32 ton-hr. In the Southwest, where the system is proposed for application, additional daytime cooling might be necessary to maintain comfort. In the 2,000 ft^2 house, the water ponds would create a total physical load of 42 tons.

THE ECONOMICS OF SOLAR AIR CONDITIONING

A detailed economic analysis of solar air conditioning today suffers from the lack of knowledge of costs and performance of the $LiBr-H_2O$ system when the system is modified for solar use. Several investigators have studied the system in detail, and a performance picture is beginning to emerge. As described under "Design of a Collector for a Building," earlier in this chapter, Löf and Tybout have modeled a combined solar-heated and solar-cooled residence by computer using current estimates on costs and performance of

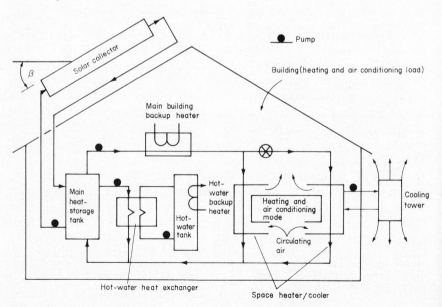

FIG. 5.22 Combined solar heating and cooling system schematic diagram.

TABLE 5.3 Costs Assumed in Löf and Tybout Solar System*

Item	Cost, $	Comments
Storage costs	0.05/lb water	Per lb water stored per ft² collector area (including tank, etc.)
Controls	150	Fixed cost
Pipes, fittings	$100 + 0.10A_{coll}$	Fixed plus variable cost
Motors, pumps	$50 + 0.20A_{coll}$	Fixed plus variable cost
Heat exchangers	$75 + 0.15A_{coll}$	Fixed plus variable cost
Collector	$2.00\ A_{coll}$	Projection based on economics of scale
LiBr-H₂O air conditioning	1,000	Optional

*Equipment useful life is assumed to be 20 yr; discount rate is assumed to be 8% (CRF = 0.102).

a small (3 to 5 ton) $LiBr–H_2O$ absorption air conditioner. Their analysis is much like the approach described in Chap. 4, "Computerized System Optimization," for optimizing solar heating systems. Figure 5.22 shows a detailed schematic diagram, including all components of the combined system.

Table 5.3 shows the additional costs in the solar system used by Löf and Tybout. The model optimization runs showed that

1. In climates where both air conditioning and heating are required, a *combined solar-heating–air-conditioning system is more economical* than either a solar-heating or solar-cooling system by itself. Figure 5.23 shows one of the plots presented in the Löf and Tybout study.
2. *Optimal storage was found to be about 10 to 15 lbₘ/ft²*, the same value determined for the optimal heating-only system described in Chap. 4.
3. *Between 30 and 70 percent of the combined loads could be met by the solar-optimal system*, modeled in the eight climatic zones studied, at a cost of $1.75 to $3.00 per million delivered Btu.
4. Solar energy costs less in nearly all locations than electrical energy used for heating and cooling. The costs were computed in 1970 dollars.

In this study the optimal mix of solar and auxiliary energy was not determined. The optimal mix may be obtained by determining the mix at the point where marginal costs of solar and auxiliary energy are equal, as described in Chap. 4, "Computerized System Optimization."

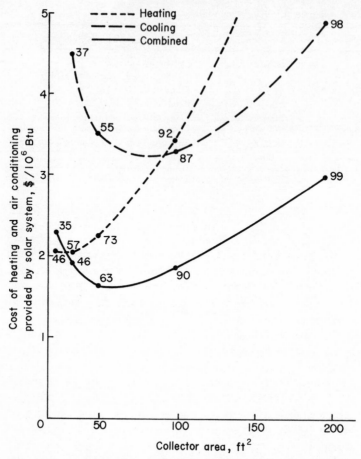

FIG. 5.23 Annual cost of solar heating, solar cooling, and combined solar heating and cooling for a model home in Albuquerque, N. Mex. Numbers on the curves are percentages of load carried by solar under the following conditions: collector tilt β = latitude L = 35.05° N, heating demand = 25,000 Btu/°F per day, system cost amortized for 20 yr at 8%, collector cost with two glass covers = \$2/ft, absorption air conditioner \$1,000 more costly than standard mechanical system. (*Adapted from Löf and Tybout.*[6])

SUMMARY

This chapter deals with various methods by which solar energy can be used to provide building cooling. The first section presents the procedure for calculating the cooling load on a building during the cooling season. A design study of a flat-plate collector–lithium-bromide air conditioner showed that the three most important design parameters for the solar collection system are the tilt angle, number of covers and collector area.

Several thermodynamic cooling cycles are considered in detail in the remainder of the chapter. Performance criteria for the Rankine cycle–powered vapor-compression cycle are described and the use of vapor-compression heat pumps with solar assist is explained. The absorption cycle is also discussed and the method of analyzing the cycle thermodynamically is shown in detail for a lithium-bromide system. A nonmechanical cooling system that uses pebble bed storage as a means of summer cooling in climates where cool nights are frequent is described. Design correlations for pressure drop and heat transfer coefficient for this sytem are given. The last section of the chapter summarizes the economic considerations which determine the viability of solar cooling systems.

REFERENCES

1. Beckman, W. A., and J. A. Duffie: "Modeling of Solar Heating and Air Conditioning," Nat'l Tech. Information Serv., Springfield, Va., 1973.
2. Close, D. J.: Rock Pile Thermal Storage for Comfort Air Conditioning, *Mech. Chem. Engrg. Trans. Inst. Engr. (Australia),* vol. MC-1, p. 11, 1965.
3. Dunkle, R. V.: A Method of Solar Air Conditioning, *Mech. Chem. Engrg. Trans. Inst. Engrs. (Australia),* vol. MC-1, p. 73, 1965.
4. Hay, H. R., and J. I. Yellott: A Naturally Air Conditioned Building, *Mech. Engrg.,* vol. 92, p. 19, 1970.
5. Hay, H. R.: Energy Technology and Solarchitecture, *Mech. Engrg.,* vol. 94, p. 18, 1973.
6. Löf, G. O. G., and R. A. Tybout: Design and Cost of Optimal Systems for Residential Heating and Cooling by Solar Energy, *Solar Energy,* vol. 16, p. 9, 1974.
7. Löf, G. O. G.: "Design and Construction of a Residential Solar Heating and Cooling System," Nat'l Tech. Information Serv., Springfield, Va., 1974.

Postscript

We are all working together to one end,
some with knowledge and design,
and others without knowing what they do.

MARCUS AURELIUS

The field of solar energy utilization is undergoing rapid and accelerating change. Many new ideas and concepts are under active and vigorous development. Physical properties of both conventional and new materials for solar systems are being measured by new and more accurate techniques. Improved data-gathering methods are being instituted by the National Weather Service to provide accurate weather and insolation data for the solar designer.

Although the authors have tried to update this book until a few weeks prior to production, a book in a newly developing technical field is past history. It is necessary for architects, engineers, planners, and installers of solar systems to remain current with developments in the field by keeping updated with both trade publications and the technical literature. The information contained in the appendixes has been found useful to designers in the past, but as more reliable meteorological data and material properties are acquired, it may be replaced by new information.

The Solar Energy Research Institute established by the Congress in 1974 should soon be able to provide current information on all aspects of solar system design, fabrication, and installation as well as economic and performance criteria for rating solar energy conversion systems.

Method of Testing for Rating Solar Collectors Based on Thermal Performance*

by James E. Hill and Tamami Kusuda**

ABSTRACT

The National Bureau of Standards has made a study of the different techniques that could be used for testing solar collectors and rating them on the basis of thermal performance. This document outlines a proposed standard test procedure based on that study. It is written in the format of a standard of the American Society of Heating, Refrigerating, and Air Conditioning Engineers and specifies the recommended apparatus, instrumentation, and test procedure.

Key Words: Solar collector; standard test; thermal performance; solar energy; standard; solar radiation

SECTION 1. PURPOSE

1.1 The purpose of this standard is to provide test methods for determining the thermal performance of solar collectors which heat fluids and are used in systems to provide the thermal requirements for heating, cooling, and the generation of domestic hot water in buildings.

SECTION 2. SCOPE

2.1 This standard applies to solar collectors in which a fluid enters the device through a single inlet and leaves the device through a single outlet. The collector containing more than

*Prepared for the National Science Foundation, NBS, U.S. Dept. of Commerce, Washington, D.C., Interim Rept. NBSIR 74-635, December, 1974.
**The authors are with the Thermal Engineering Systems Section, Center for Building Technology, National Bureau of Standards, Washington, D.C.

one inlet and/or outlet can be tested according to this standard provided that the external piping can be connected in such a way as to effectively provide a single inlet and/or outlet for the determination of the bulk properties of the fluid entering and leaving the collector. The fluid can be either a gas or liquid but not a mixture of the two. The collector can be a concentrating collector provided that the aperture or interception area for the device can be determined. The collector may have the capability of rotating so as to track the sun.

2.2   This standard is not applicable to those configurations in which the flow into the collector and out of the collector cannot be reduced effectively to one inlet and one outlet. This standard is not applicable to those collectors in which the thermal storage unit is an integral part of the collector such that the collection process and storage process cannot be separated for the purpose of making measurements.

2.3   This standard does not address factors relating to cost or consideration of requirements for interfacing with a specific heating and cooling system.

2.4   The present version of the standard provides test methods for determining the steady-state efficiency of solar collectors. The transient response of solar collectors cannot be determined with the test methods outlined herein.

SECTION 3. DEFINITIONS

3.1   Ambient Air

Ambient air is the outdoor air in the vicinity of the solar collector being tested.

3.2   Absorber

The absorber is that part of the solar collector that receives the incident solar radiation and transforms it into thermal energy. It is usually a solid surface through which energy is transferred to the transfer fluid; however, the transfer fluid itself could be the absorber in the case of a "black liquid".

3.3   **Aperture**

The aperture is the opening or projected area of a solar collector through which the unconcentrated solar energy is admitted and directed to the absorber.

3.4   **Concentrating Collector**

A concentrating collector is a solar collector that contains reflectors, lenses, or other optical elements to concentrate the energy falling on the aperture onto a heat exchanger of surface area smaller than the aperture.

3.5   **Concentrator**

The concentrator is that part of a concentrating collector which directs the incident solar radiation onto the absorber.

3.6   **Cover Plate**

The cover plate designates the diathermanous material or materials covering the aperture and most directly exposed to the solar radiation. These materials are generally used to reduce the heat loss from the absorber to the surroundings and to protect the absorber.

3.7   **Flat-plate Collector**

A flat-plate collector is a solar collector in which the solid surface absorbing the incident solar radiation is essentially flat and employs no concentration.

3.8   **Gross Cross-sectional Area**

Gross cross-sectional area is the overall or outside area of a flat-plate collector. It is usually slightly larger than the absorber area since it includes the framework required to hold the absorber.

3.9   **Incident Angle**

The incident angle is the angle between the sun's rays and the outward drawn normal from the solar collector.

3.10 Insolation

Insolation is the rate of solar radiation received by a unit surface area in unit time (W/m^2, $Btu/(h \cdot ft^2)$).

3.11 Instantaneous Efficiency

The instantaneous efficiency of a solar collector is defined as the amount of energy removed by the transfer fluid per unit of transparent frontal area over a given 15 minute period divided by the total incident solar radiation onto the collector per unit area for the 15 minute period.

3.12 Integrated Average Insolation

The integrated average insolation is the total energy per unit area received by a surface for a specified time period divided by the time period (W/m^2, $Btu/(h \cdot ft^2)$).

3.13 Pyranometer

A pyranometer is a radiometer used to measure the total incident solar energy per unit time per unit area upon a surface which includes the beam radiation from the sun, the diffuse radiation from the sky, and the shortwave radiation reflected from the foreground.

3.14 Pyrheliometer

A pyrheliometer is a radiometer used to measure the direct or beam radiation on a surface normal to the sun's rays.

3.15 Quasisteady

Quasisteady is the term used in this document to describe the state of the solar collector test when the flow rate and temperature of the fluid entering the collector is constant but the exit fluid temperature changes "gradually" due to the normal change in insolation that occurs with time for clear sky conditions.

3.16 Solar Collector

A solar collector is a device designed to absorb incident solar radiation and to transfer the energy to a fluid passing in contact with it.

3.17 Total Incident Insolation

Total incident insolation is the total energy received by a unit surface area for a specified time period (J/m^2).

3.18 Transfer Fluid

The transfer fluid is the medium such as air, water, or other fluid which passes through or in contact with the solar collector and carries the thermal energy away from the collector.

3.19 Transparent Frontal Area

The transparent frontal area is the area of the transparent frontal surface for flat-plate collectors.

3.20 Standard Air

Standard air is air weighing 1.2 kg/m^3 (0.075 lb/ft^3), and is equivalent in density to dry air at a temperature of 21.1°C (70°F) and a barometric pressure of 1.01 x 10^5 N/m^2 (29.92 in. of Hg)

3.21 Standard Barometric Pressure

1.01 x 10^5 N/m^2 (29.92 in. of Hg)

SECTION 4. CLASSIFICATIONS

4.1   Solar collectors may be classified according to their collecting characteristics, the way in which they are mounted, and the type of transfer fluid they employ.

  4.1.1 *Collecting characteristics.* A non-concentrating or "flat-plate" collector is one in which the absorbing surface for solar radiation is essentially flat with no means for concentrating the incoming solar radiation. A concentrating or "focusing" collector is one which usually contains reflectors or employs other optical means to concentrate the energy falling on the aperture onto a heat exchanger of surface area smaller than the aperture.

  4.1.2 *Mounting.* A collector can be mounted to remain stationary, be adjustable as to tilt angle (measured from the horizontal) to follow the change in solar declination, or be designed to track the sun. Tracking is done by employing either an equatorial mount or an altazimuth mounting, for the purpose of increasing the absorption of the daily solar irradiation.

4.1.3 *Type of fluid.* A collector will usually use either a liquid or a gas as the transfer fluid. The most common liquids are water or a water-ethylene glycol solution. The most common gas is air.

SECTION 5. REQUIREMENTS

5.1 Solar collectors shall be tested for rating in accordance with the provisions set forth below and in Section 8.

5.1.1 The size of collector tested shall be large enough so that the performance characteristics determined will be indicative of those that would occur when the collector is part of an installed system. If the collector is modular and the test is being done on one module, it should be mounted and insulated in such a way that the back and edge losses will be characteristic of those that will occur during operation on a structure.

5.1.2 The collector shall be mounted in a location such that there will be no significant energy reflected or reradiated onto the collector from surrounding buildings or any other surfaces in the vicinity of the test stand for the duration of the test(s). This will be satisfied if the ground and immediately adjacent surfaces are diffuse with a reflectance of less than 0.20. If significant reflection will occur, provision shall be made to shield the collector by the use of a non-reflective shield. In addition, the test stand shall be located so that a shadow will not be cast onto the collector at any time during the test period.

5.1.3 The test(s) shall be conducted on days having weather conditions such that the 15 minute integrated average insolation measured in the plane of the collector or aperture, reported, and used for the computation of instantaneous efficiency values shall be a minimum of

630 W/m² (199.8 Btu/(h · ft²)). Specific values that can be expected for clear sky conditions are shown in Tables A.1 through A.6, taken from reference [1]. More accurate estimates can be made using the tables in conjunction with clearness numbers.*

5.1.4 The orientation of the collector shall be such that the incident angle (measured from the normal to the collector surface or aperture) is less than 45° during the period in which test data is being taken. Angles of incidence can be estimated from Tables A.7 through A.12 taken from reference [2]. More accurate estimates can be made using the procedures outlined in references [3], p. 393 or [4], pp. 283-292.

5.1.5 The air velocity across the collector surface of a flat-plate collector or aperture of a concentrating collector during the test(s) shall be measured. The measurement shall be made at a distance of approximately 1 m (3.3 ft) from the collector along the direction it faces and at a height corresponding to the center of the collector panel.

5.1.6 The *range* of ambient temperatures for all reported test points comprising the "efficiency curve" shall be less than 30°C (54°F).

5.1.7 The transfer fluid used in the solar collector shall have a known specific heat which varies by less than 0.5% over the temperature range of the fluid during a particular 15 minute test period.

SECTION 6. INSTRUMENTATION

6.1 Solar Radiation Measurement

6.1.1 A pyranometer shall be used to measure the total short-wave radiation from both the sun and the sky. The instrument shall have the following characteristics [5]:

6.1.1.1 *Change of response due to variation in ambient temperature.* The instrument shall

*Reference [3], p. 394, Figure 4.

TABLE A.1 Solar Position and Insolation Values for 24 Degrees North Latitude

Date	Solar time AM	PM	Solar position Alt	Azm	BTUH/sq. ft. total insolation on surfaces Normal	Horiz.	South facing surface angle with horiz. 14	24	34	44	90
Jan 21	7	5	4.8	65.6	71	10	17	21	25	28	31
	8	4	16.9	58.3	239	83	110	126	137	145	127
	9	3	27.9	48.8	288	151	188	207	221	228	176
	10	2	37.2	36.1	308	204	246	268	282	287	207
	11	1	43.6	19.6	317	237	283	306	319	324	226
	12		46.0	0.0	320	249	296	319	332	336	232
	Surface daily totals				2766	1622	1984	2174	2300	2360	1766
Feb 21	7	5	9.3	74.6	158	35	44	49	53	56	46
	8	4	22.3	67.2	263	116	135	145	150	151	102
	9	3	34.4	57.6	298	187	213	225	230	228	141
	10	2	45.1	44.2	314	241	273	286	291	287	168
	11	1	53.0	25.0	321	276	310	324	328	323	185
	12		56.0	0.0	324	288	323	337	341	335	191
	Surface daily totals				3036	1998	2276	2396	2436	2424	1476
Mar 21	7	5	13.7	83.3	194	60	63	64	62	59	27
	8	4	27.2	76.8	267	141	150	152	149	142	64
	9	3	40.2	67.9	295	212	226	229	225	214	95
	10	2	52.3	54.8	309	266	285	288	283	270	120
	11	1	61.9	33.4	315	300	322	326	320	305	135
	12		66.0	0.0	317	312	334	339	333	317	140
	Surface daily totals				3078	2270	2428	2456	2412	2298	1022
Apr 21	6	6	4.7	100.6	40	7	5	4	4	3	2
	7	5	18.3	94.9	203	83	77	70	62	51	10
	8	4	32.0	89.0	256	160	157	149	137	122	16
	9	3	45.6	81.9	280	227	227	220	206	186	46
	10	2	59.0	71.8	292	278	282	275	259	237	61
	11	1	71.1	51.6	298	310	316	309	293	269	74
	12		77.6	0.0	299	321	328	321	305	280	79
	Surface daily totals				3036	2454	2458	2374	2228	2016	488
May 21	6	6	8.0	108.4	86	22	15	10	9	9	5
	7	5	21.2	103.2	203	98	85	73	59	44	12
	8	4	34.6	98.5	248	171	159	145	127	106	15
	9	3	48.3	93.6	269	233	224	210	190	165	16
	10	2	62.0	87.7	280	281	275	261	239	211	22
	11	1	75.5	76.9	286	311	307	293	270	240	34
	12		86.0	0.0	288	322	317	304	281	250	37
	Surface daily totals				3032	2556	2447	2286	2072	1800	246
Jun 21	6	6	9.3	111.6	97	29	20	12	12	11	7
	7	5	22.3	106.8	201	103	87	73	58	41	13
	8	4	35.5	102.6	242	173	158	142	122	99	16
	9	3	49.0	98.7	263	234	221	204	182	155	18
	10	2	62.6	95.0	274	280	269	253	229	199	18
	11	1	76.3	90.8	279	309	300	283	259	227	19
	12		89.4	0.0	281	319	310	294	269	236	22
	Surface daily totals				2994	2574	2422	2230	1992	1700	204

Date	Solar time		Solar position		BTUH/sq. ft. total insolation on surfaces						
	AM	PM	Alt	Azm			South facing surface angle with horiz.				
					Normal	Horiz.	14	24	34	44	90
Jul 21	6	6	8.2	109.0	81	23	16	11	10	9	6
	7	5	21.4	103.8	195	98	85	73	59	44	13
	8	4	34.8	99.2	239	169	157	143	125	104	16
	9	3	48.4	94.5	261	231	221	207	187	161	18
	10	2	62.1	89.0	272	278	270	256	235	206	21
	11	1	75.7	79.2	278	307	302	287	265	235	32
	12		86.6	0.0	280	317	312	298	275	245	36
	Surface daily totals				2932	2526	2412	2250	2036	1766	246
Aug 21	6	6	5.0	101.3	35	7	5	4	4	4	2
	7	5	18.5	95.6	186	82	76	69	60	50	11
	8	4	32.2	89.7	241	158	154	146	134	118	16
	9	3	45.9	82.9	265	223	222	214	200	181	39
	10	2	59.3	73.0	278	273	275	268	252	230	58
	11	1	71.6	53.2	284	304	309	301	285	261	71
	12		78.3	0.0	286	315	320	313	296	272	75
	Surface daily totals				2864	2408	2402	2316	2168	1958	470
Sep 21	7	5	13.7	83.8	173	57	60	60	59	56	26
	8	4	27.2	76.8	248	136	144	146	143	136	62
	9	3	40.2	67.9	278	205	218	221	217	206	93
	10	2	52.3	54.8	292	258	275	278	273	261	116
	11	1	61.9	33.4	299	291	311	315	309	295	131
	12		66.0	0.0	301	302	323	327	321	306	136
	Surface daily totals				2878	2194	2342	2366	2322	2212	992
Oct 21	7	5	9.1	74.1	138	32	40	45	48	50	42
	8	4	22.0	66.7	247	111	129	139	144	145	99
	9	3	34.1	57.1	284	180	206	217	223	221	138
	10	2	44.7	43.8	301	234	265	277	282	279	165
	11	1	52.5	24.7	309	268	301	315	319	314	182
	12		55.5	0.0	311	279	314	328	332	327	188
	Surface daily totals				2868	1928	2198	2314	2364	2346	1442
Nov 21	7	5	4.9	65.8	67	10	16	20	24	27	29
	8	4	17.0	58.4	232	82	108	123	135	142	124
	9	3	28.0	48.9	282	150	186	205	217	224	172
	10	2	37.3	36.3	303	203	244	265	278	283	204
	11	1	43.8	19.7	312	236	280	302	316	320	222
	12		46.2	0.0	315	247	293	315	328	332	228
	Surface daily totals				2706	1610	1962	2146	2268	2324	1730
Dec 21	7	5	3.2	62.6	30	3	7	9	11	12	14
	8	4	14.9	55.3	225	71	99	116	129	139	130
	9	3	25.5	46.0	281	137	176	198	214	223	184
	10	2	34.3	33.7	304	189	234	258	275	283	217
	11	1	40.4	18.2	314	221	270	295	312	320	236
	12		42.6	0.0	317	232	282	308	325	332	243
	Surface daily totals				2624	1474	1852	2058	2204	2286	1808

1 BTUH/SQ. FT. = 3.152 W/m²
NOTE: 1) Based on data in Table 1, p. 387 in ref. [3]; 0% ground reflectance; 1.0 clearness factor.
   2) See Fig. 4, p. 394 in [3] for typical regional clearness factors.
   3) Ground reflection not included on normal or horizontal surfaces.

TABLE A.2 Solar Position and Insolation Values for 32 Degrees North Latitude

Date	Solar time AM	PM	Solar position Alt	Azm	BTUH/sq. ft. total insolation on surfaces Normal	Horiz.	South facing surface angle with horiz. 22	32	42	52	90
Jan 21	7	5	1.4	65.2	1	0	0	0	0	1	1
	8	4	12.5	56.5	203	56	93	106	116	123	115
	9	3	22.5	46.0	269	118	175	193	206	212	181
	10	2	30.6	33.1	295	167	235	256	269	274	221
	11	1	36.1	17.5	306	198	273	295	308	312	245
	12		38.0	0.0	310	209	285	308	321	324	253
	Surface daily totals				2458	1288	1839	2008	2118	2166	1779
Feb 21	7	5	7.1	73.5	121	22	34	37	40	42	38
	8	4	19.0	64.4	247	95	127	136	140	141	108
	9	3	29.9	53.4	288	161	206	217	222	220	158
	10	2	39.1	39.4	306	212	266	278	283	279	193
	11	1	45.6	21.4	315	244	304	317	321	315	214
	12		48.0	0.0	317	255	316	330	334	328	222
	Surface daily totals				2872	1724	2188	2300	2345	2322	1644
Mar 21	7	5	12.7	81.9	185	54	60	60	59	56	32
	8	4	25.1	73.0	260	129	146	147	144	137	78
	9	3	36.8	62.1	290	194	222	224	220	209	119
	10	2	47.3	47.5	304	245	280	283	278	265	150
	11	1	55.0	26.8	311	277	317	321	315	300	170
	12		58.0	0.0	313	287	329	333	327	312	177
	Surface daily totals				3012	2084	2378	2403	2358	2246	1276
Apr 21	6	6	6.1	99.9	66	14	9	6	6	5	3
	7	5	18.8	92.2	206	86	78	71	62	51	10
	8	4	31.5	84.0	255	158	156	148	136	120	35
	9	3	43.9	74.2	278	220	225	217	203	183	68
	10	2	55.7	60.3	290	267	279	272	256	234	95
	11	1	65.4	37.5	295	297	313	306	290	265	112
	12		69.6	0.0	297	307	325	318	301	276	118
	Surface daily totals				3076	2390	2444	2356	2206	1994	764
May 21	6	6	10.4	107.2	119	36	21	13	13	12	7
	7	5	22.8	100.1	211	107	88	75	60	44	13
	8	4	35.4	92.9	250	175	159	145	127	105	15
	9	3	48.1	84.7	269	233	223	209	188	163	33
	10	2	60.6	73.3	280	277	273	259	237	208	56
	11	1	72.0	51.9	285	305	305	290	268	237	72
	12		78.0	0.0	286	315	315	301	278	247	77
	Surface daily totals				3112	2582	2454	2284	2064	1788	469
Jun 21	6	6	12.2	110.2	131	45	26	16	15	14	9
	7	5	24.3	103.4	210	115	91	76	59	41	14
	8	4	36.9	96.8	245	180	159	143	122	99	16
	9	3	49.6	89.4	264	236	221	204	181	153	19
	10	2	62.2	79.7	274	279	268	251	227	197	41
	11	1	74.2	60.9	279	306	299	282	257	224	56
	12		81.5	0.0	280	315	309	292	267	234	60
	Surface daily totals				3084	2634	2436	2234	1990	1690	370

Date	Solar time		Solar position		BTUH/sq. ft. total insolation on surfaces						
							South facing surface angle with horiz.				
	AM	PM	Alt	Azm	Normal	Horiz.	22	32	42	52	90
Jul 21	6	6	10.7	107.7	113	37	22	14	13	12	8
	7	5	23.1	100.6	203	107	87	75	60	44	14
	8	4	35.7	93.6	241	174	158	143	125	104	16
	9	3	48.4	85.5	261	231	220	205	185	159	31
	10	2	60.9	74.3	271	274	269	254	232	204	54
	11	1	72.4	53.3	277	302	300	285	262	232	69
	12		78.6	0.0	279	311	310	296	273	242	74
	Surface daily totals				3012	2558	2422	2250	2030	1754	458
Aug 21	6	6	6.5	100.5	59	14	9	7	6	6	4
	7	5	19.1	92.8	190	85	77	69	60	50	12
	8	4	31.8	84.7	240	156	152	144	132	116	33
	9	3	44.3	75.0	263	216	220	212	197	178	65
	10	2	56.1	61.3	276	262	272	264	249	226	91
	11	1	66.0	38.4	282	292	305	298	281	257	107
	12		70.3	0.0	284	302	317	309	292	268	113
	Surface daily totals				2902	2352	2388	2296	2144	1934	736
Sep 21	7	5	12.7	81.9	163	51	56	56	55	52	30
	8	4	25.1	73.0	240	124	140	141	138	131	75
	9	3	36.8	62.1	272	188	213	215	211	201	114
	10	2	47.3	47.5	287	237	270	273	268	255	145
	11	1	55.0	26.8	294	268	306	309	303	289	164
	12		58.0	0.0	296	278	318	321	315	300	171
	Surface daily totals				2808	2014	2288	2308	2264	2154	1226
Oct 21	7	5	6.8	73.1	99	19	29	32	34	36	32
	8	4	18.7	64.0	229	90	120	128	133	134	104
	9	3	29.5	53.0	273	155	198	208	213	212	153
	10	2	38.7	39.1	293	204	257	269	273	270	188
	11	1	45.1	21.1	302	236	294	307	311	306	209
	12		47.5	0.0	304	247	306	320	324	318	217
	Surface daily totals				2696	1654	2100	2208	2252	2232	1588
Nov 21	7	5	1.5	65.4	2	0	0	0	1	1	1
	8	4	12.7	56.6	196	55	91	104	113	119	111
	9	3	22.6	46.1	263	118	173	190	202	208	176
	10	2	30.8	33.2	289	166	233	252	265	270	217
	11	1	36.2	17.6	301	197	270	291	303	307	241
	12		38.2	0.0	304	207	282	304	316	320	249
	Surface daily totals				2406	1280	1816	1980	2084	2130	1742
Dec 21	8	4	10.3	53.8	176	41	77	90	101	108	107
	9	3	19.8	43.6	257	102	161	180	195	204	183
	10	2	27.6	31.2	288	150	221	244	259	267	226
	11	1	32.7	16.4	301	180	258	282	298	305	251
	12		34.6	0.0	304	190	271	295	311	318	259
	Surface daily totals				2348	1136	1704	1888	2016	2086	1794

1 BTUH/SQ. FT. = 3.152 W/m²

NOTE: 1) Based on data in Table 1, p. 387 in ref. [3]; 0% ground reflectance; 1.0 clearness factor.
2) See Fig. 4, p. 394 in [3] for typical regional clearness factors.
3) Ground reflection not included on normal or horizontal surfaces.

TABLE A.3 Solar Position and Insolation Values for 40 Degrees North Latitude

Date	Solar time AM	Solar time PM	Solar position Alt	Solar position Azm	Normal	Horiz.	30	40	50	60	90
							South facing surface angle with horiz.				
Jan 21	8	4	8.1	55.3	142	28	65	74	81	85	84
	9	3	16.8	44.0	239	83	155	171	182	187	171
	10	2	23.8	30.9	274	127	218	237	249	254	223
	11	1	28.4	16.0	289	154	257	277	290	293	253
	12		30.0	0.0	294	164	270	291	303	306	263
	Surface daily totals				2182	948	1660	1810	1906	1944	1726
Feb 21	7	5	4.8	72.7	69	10	19	21	23	24	22
	8	4	15.4	62.2	224	73	114	122	126	127	107
	9	3	25.0	50.2	274	132	195	205	209	208	167
	10	2	32.8	35.9	295	178	256	267	271	267	210
	11	1	38.1	18.9	305	206	293	306	310	304	236
	12		40.0	0.0	308	216	306	319	323	317	245
	Surface daily totals				2640	1414	2060	2162	2202	2176	1730
Mar 21	7	5	11.4	80.2	171	46	55	55	54	51	35
	8	4	22.5	69.6	250	114	140	141	138	131	89
	9	3	32.8	57.3	282	173	215	217	213	202	138
	10	2	41.6	41.9	297	218	273	276	271	258	176
	11	1	47.7	22.6	305	247	310	313	307	293	200
	12		50.0	0.0	307	257	322	326	320	305	208
	Surface daily totals				2916	1852	2308	2330	2284	2174	1484
Apr 21	6	6	7.4	98.9	89	20	11	8	7	7	4
	7	5	18.9	89.5	206	87	77	70	61	50	12
	8	4	30.3	79.3	252	152	153	145	133	117	53
	9	3	41.3	67.2	274	207	221	213	199	179	93
	10	2	51.2	51.4	286	250	275	267	252	229	126
	11	1	58.7	29.2	292	277	308	301	285	260	147
	12		61.6	0.0	293	287	320	313	296	271	154
	Surface daily totals				3092	2274	2412	2320	2168	1956	1022
May 21	5	7	1.9	114.7	1	0	0	0	0	0	0
	6	6	12.7	105.6	144	49	25	15	14	13	9
	7	5	24.0	96.6	216	214	89	76	60	44	13
	8	4	35.4	87.2	250	175	158	144	125	104	25
	9	3	46.8	76.0	267	227	221	206	186	160	60
	10	2	57.5	60.9	277	267	270	255	233	205	89
	11	1	66.2	37.1	283	293	301	287	264	234	108
	12		70.0	0.0	284	301	312	297	274	243	114
	Surface daily totals				3160	2552	2442	2264	2040	1760	724
Jun 21	5	7	4.2	117.3	22	4	3	3	2	2	1
	6	6	14.8	108.4	155	60	30	18	17	16	10
	7	5	26.0	99.7	216	123	92	77	59	41	14
	8	4	37.4	90.7	246	182	159	142	121	97	16
	9	3	48.8	80.2	263	233	219	202	179	151	47
	10	2	59.8	65.8	272	272	266	248	224	194	74
	11	1	69.2	41.9	277	296	296	278	253	221	92
	12		73.5	0.0	279	304	306	289	263	230	98
	Surface daily totals				3180	2648	2434	2224	1974	1670	610

Date	Solar time		Solar position		BTUH/sq. ft. total insolation on surfaces						
	AM	PM	Alt	Azm			South facing surface angle with horiz.				
					Normal	Horiz.	30	40	50	60	90
Jul 21	5	7	2.3	115.2	2	0	0	0	0	0	0
	6	6	13.1	106.1	138	50	26	17	15	14	9
	7	5	24.3	97.2	208	114	89	75	60	44	14
	8	4	35.8	87.8	241	174	157	142	124	102	24
	9	3	47.2	76.7	259	225	218	203	182	157	58
	10	2	57.9	61.7	269	265	266	251	229	200	86
	11	1	66.7	37.9	275	290	296	281	258	228	104
	12		70.6	0.0	276	298	307	292	269	238	111
	Surface daily totals				3062	2534	2409	2230	2006	1728	702
Aug 21	6	6	7.9	99.5	81	21	12	9	8	7	5
	7	5	19.3	90.9	191	87	76	69	60	49	12
	8	4	30.7	79.9	237	150	150	141	129	113	50
	9	3	41.8	67.9	260	205	216	207	193	173	89
	10	2	51.7	52.1	272	246	267	259	244	221	120
	11	1	59.3	29.7	278	273	300	292	276	252	140
	12		62.3	0.0	280	282	311	303	287	262	147
	Surface daily totals				2916	2244	2354	2258	2104	1894	978
Sep 21	7	5	11.4	80.2	149	43	51	51	49	47	32
	8	4	22.5	69.6	230	109	133	134	131	124	84
	9	3	32.8	57.3	263	167	206	208	203	193	132
	10	2	41.6	41.9	280	211	262	265	260	247	168
	11	1	47.7	22.6	287	239	298	301	295	281	192
	12		50.0	0.0	290	249	310	313	307	292	200
	Surface daily totals				2708	1788	2210	2228	2182	2074	1416
Oct 21	7	5	4.5	72.3	48	7	14	15	17	17	16
	8	4	15.0	61.9	204	68	106	113	117	118	100
	9	3	24.5	49.8	257	126	185	195	200	198	160
	10	2	32.4	35.6	280	170	245	257	261	257	203
	11	1	37.6	18.7	291	199	283	295	299	294	229
	12		39.5	0.0	294	208	295	308	312	306	238
	Surface daily totals				2454	1348	1962	2060	2098	2074	1654
Nov 21	8	4	8.2	55.4	136	28	63	72	78	82	81
	9	3	17.0	44.1	232	82	152	167	178	183	167
	10	2	24.0	31.0	268	126	215	233	245	249	219
	11	1	28.6	16.1	283	153	254	273	285	288	248
	12		30.2	0.0	288	163	267	287	298	301	258
	Surface daily totals				2128	942	1636	1778	1870	1908	1686
Dec 21	8	4	5.5	53.0	89	14	39	45	50	54	56
	9	3	14.0	41.9	217	65	135	152	164	171	163
	10	2	20.,	29.4	261	107	200	221	235	242	221
	11	1	25.0	15.2	280	134	239	262	276	283	252
	12		26.6	0.0	285	143	253	275	290	296	263
	Surface daily totals				1978	782	1480	1634	1740	1796	1646

1 BTUH/SQ. FT. = 3.152 W/m²

NOTE: 1) Based on data in Table 1, p. 387 in ref. [3]; 0% ground reflectance; 1.0 clearness factor.
2) See Fig. 4, p. 394 in [3] for typical regional clearness factors.
3) Ground reflection not included on normal or horizontal surfaces.

TABLE A.4 Solar Position and Insolation Values for 48 Degrees North Latitude

Date	Solar time AM	PM	Solar position Alt	Azm	Normal	Horiz.	South facing surface angle with horiz. 38	48	58	68	90
Jan 21	8	4	3.5	54.6	37	4	17	19	21	22	22
	9	3	11.0	42.6	185	46	120	132	140	145	139
	10	2	16.9	29.4	239	83	190	206	216	220	206
	11	1	20.7	15.1	261	107	231	249	260	263	243
	12		22.0	0.0	267	115	245	264	275	278	255
	Surface daily totals				1710	596	1360	1478	1550	1578	1478
Feb 21	7	5	2.4	72.2	12	1	3	4	4	4	4
	8	4	11.6	60.5	188	49	95	102	105	106	96
	9	3	19.7	47.7	251	100	178	187	191	190	167
	10	2	26.2	33.3	278	139	240	251	255	251	217
	11	1	30.5	17.2	290	165	278	290	294	288	247
	12		32.0	0.0	293	173	291	304	307	301	258
	Surface daily totals				2330	1080	1880	1972	2024	1978	1720
Mar 21	7	5	10.0	78.7	153	37	49	49	47	45	35
	8	4	19.5	66.8	236	96	131	132	129	122	96
	9	3	28.2	53.4	270	147	205	207	203	193	152
	10	2	35.4	37.8	287	187	263	266	261	248	195
	11	1	40.3	19.8	295	212	300	303	297	283	223
	12		42.0	0.0	298	220	312	315	309	294	232
	Surface daily totals				2780	1578	2208	2228	2182	2074	1632
Apr 21	6	6	8.6	97.8	108	27	13	9	8	7	5
	7	5	18.6	86.7	205	85	76	69	59	48	21
	8	4	28.5	74.9	247	142	149	141	129	113	69
	9	3	37.8	61.2	268	191	216	208	194	174	115
	10	2	45.8	44.6	280	228	268	260	245	223	152
	11	1	51.5	24.0	286	252	301	294	278	254	177
	12		53.6	0.0	288	260	313	305	289	264	185
	Surface daily totals				3076	2106	2358	2266	2114	1902	1262
May 21	5	7	5.2	114.3	41	9	4	4	4	3	2
	6	6	14.7	103.7	162	61	27	16	15	13	10
	7	5	24.6	93.0	219	118	89	75	60	43	13
	8	4	34.7	81.6	248	171	156	142	123	101	45
	9	3	44.3	68.3	264	217	217	202	182	156	86
	10	2	53.0	51.3	274	252	265	251	229	200	120
	11	1	59.5	28.6	279	274	296	281	258	228	141
	12		62.0	0.0	280	281	306	292	269	238	149
	Surface daily totals				3254	2482	2418	2234	2010	1728	982
Jun 21	5	7	7.9	116.5	77	21	9	9	8	7	5
	6	6	17.2	106.2	172	74	33	19	18	16	12
	7	5	27.0	95.8	220	129	93	77	59	39	15
	8	4	37.1	84.6	246	181	157	140	119	95	35
	9	3	46.9	71.6	261	225	216	198	175	147	74
	10	2	55.8	54.8	269	259	262	244	220	189	105
	11	1	62.7	31.2	274	280	291	273	248	216	126
	12		65.5	0.0	275	287	301	283	258	225	133
	Surface daily totals				3312	2626	2420	2204	1950	1644	874

Date	Solar time AM	Solar time PM	Solar position Alt	Solar position Azm	BTUH/sq. ft. total insolation on surfaces Normal	Horiz.	South facing surface angle with horiz. 38	48	58	68	90
Jul 21	5	7	5.7	114.7	43	10	5	5	4	4	3
	6	6	15.2	104.1	156	62	28	18	16	15	11
	7	5	25.1	93.5	211	118	89	75	59	42	14
	8	4	35.1	82.1	240	171	154	140	121	99	43
	9	3	44.8	68.8	256	215	214	199	178	153	83
	10	2	53.5	51.9	266	250	261	246	224	195	116
	11	1	60.1	29.0	271	272	291	276	253	223	137
	12		62.6	0.0	272	279	301	286	263	232	144
	Surface daily totals				3158	2474	2386	2200	1974	1694	956
Aug 21	6	6	9.1	98.3	99	28	14	10	9	8	6
	7	5	19.1	87.2	190	85	75	67	58	47	20
	8	4	29.0	75.4	232	141	145	137	125	109	65
	9	3	38.4	61.8	254	189	210	201	187	168	110
	10	2	46.4	45.1	266	225	260	252	237	214	146
	11	1	52.2	24.3	272	248	293	285	268	244	169
	12		54.3	0.0	274	256	304	296	279	255	177
	Surface daily totals				2898	2086	2300	2200	2046	1836	1208
Sep 21	7	5	10.0	78.7	131	35	44	44	43	40	31
	8	4	19.5	66.8	215	92	124	124	121	115	90
	9	3	28.2	53.4	251	142	196	197	193	183	143
	10	2	35.4	37.8	269	181	251	254	248	236	185
	11	1	40.3	19.8	278	205	287	289	284	269	212
	12		42.0	0.0	280	213	299	302	296	281	221
	Surface daily totals				2568	1522	2102	2118	2070	1966	1546
Oct 21	7	5	2.0	71.9	4	0	1	1	1	1	1
	8	4	11.2	60.2	165	44	86	91	95	95	87
	9	3	19.3	47.4	233	94	167	176	180	178	157
	10	2	25.7	33.1	262	133	228	239	242	239	207
	11	1	30.0	17.1	274	157	266	277	281	276	237
	12		31.5	0.0	278	166	279	291	294	288	247
	Surface daily totals				2154	1022	1774	1860	1890	1866	1626
Nov 21	8	4	3.6	54.7	36	5	17	19	21	22	22
	9	3	11.2	42.7	179	46	117	129	137	141	135
	10	2	17.1	29.5	233	83	186	202	212	215	201
	11	1	20.9	15.1	255	107	227	245	255	258	238
	12		22.2	0.0	261	115	241	259	270	272	250
	Surface daily totals				1668	596	1336	1448	1518	1544	1442
Dec 21	9	3	8.0	40.9	140	27	87	98	105	110	109
	10	2	13.6	28.2	214	63	164	180	192	197	190
	11	1	17.3	14.4	242	86	207	226	239	244	231
	12		18.6	0.0	250	94	222	241	254	260	244
	Surface daily totals				1444	446	1136	1250	1326	1364	1304

1 BTUH/SQ. FT. = 3.152 W/m^2

NOTE: 1) Based on data in Table 1, p. 387 in ref. [3]; 0% ground reflectance; 1.0 clearness factor.
2) See Fig. 4, p. 394 in [3] for typical regional clearness factors.
3) Ground reflection not included on normal or horizontal surfaces.

TABLE A.5 Solar Position and Insolation Values for 56 Degrees North Latitude

Date	Solar time AM	PM	Solar position Alt	Azm	Normal	Horiz.	South facing surface angle with horiz. 46	56	66	76	90
Jan 21	9	3	5.0	41.8	78	11	50	55	59	60	60
	10	2	9.9	28.5	170	39	135	146	154	156	153
	11	1	12.9	14.5	207	58	183	197	206	208	201
	12		14.0	0.0	217	65	198	214	222	225	217
	Surface daily totals				1126	282	934	1010	1058	1074	1044
Feb 21	8	4	7.6	59.4	129	25	65	69	72	72	69
	9	3	14.2	45.9	214	65	151	159	162	161	151
	10	2	19.4	31.5	250	98	215	225	228	224	208
	11	1	22.8	16.1	266	119	254	265	268	263	243
	12		24.0	0.0	270	126	268	279	282	276	255
	Surface daily totals				1986	740	1640	1716	1742	1716	1598
Mar 21	7	5	8.3	77.5	128	28	40	40	39	37	32
	8	4	16.2	64.4	215	75	119	120	117	111	97
	9	3	23.3	50.3	253	118	192	193	189	180	154
	10	2	29.0	34.9	272	151	249	251	246	234	205
	11	1	32.7	17.9	282	172	285	288	282	268	236
	12		34.0	0.0	284	179	297	300	294	280	246
	Surface daily totals				2586	1268	2066	2084	2040	1938	1700
Apr 21	5	7	1.4	108.8	0	0	0	0	0	0	0
	6	6	9.6	96.5	122	32	14	9	8	7	6
	7	5	18.0	84.1	201	81	74	66	57	46	29
	8	4	26.1	70.9	239	129	143	135	123	108	82
	9	3	33.6	56.3	260	169	208	200	186	167	133
	10	2	39.9	39.7	272	201	259	251	236	214	174
	11	1	44.1	20.7	278	220	292	284	268	245	200
	12		45.6	0.0	280	227	303	295	279	255	209
	Surface daily totals				3024	1892	2282	2186	2038	1830	1458
May 21	4	8	1.2	125.5	0	0	0	0	0	0	0
	5	7	8.5	113.4	93	25	10	9	8	7	6
	6	6	16.5	101.5	175	71	28	17	15	13	11
	7	5	24.8	89.3	219	119	88	74	58	41	16
	8	4	33.1	76.3	244	163	153	138	119	98	63
	9	3	40.9	61.6	259	201	212	197	176	151	109
	10	2	47.6	44.2	268	231	259	244	222	194	146
	11	1	52.3	23.4	273	249	288	274	251	222	170
	12		54.0	0.0	275	255	299	284	261	231	178
	Surface daily totals				3340	2374	2374	2188	1962	1682	1218
Jun 21	4	8	4.2	127.2	21	4	2	2	2	2	1
	5	7	11.4	115.3	122	40	14	13	11	10	8
	6	6	19.3	103.6	185	86	34	19	17	15	12
	7	5	27.6	91.7	222	132	92	76	57	38	15
	8	4	35.9	78.8	243	175	154	137	116	92	55
	9	3	43.8	64.1	257	212	211	193	170	143	98
	10	2	50.7	46.4	265	240	255	238	214	184	133
	11	1	55.6	24.9	269	258	284	267	242	210	156
	12		57.5	0.0	271	264	294	276	251	219	164
	Surface daily totals				3438	2526	2388	2166	1910	1606	1120

Date	Solar time		Solar position		BTUH/sq. ft. total insolation on surfaces						
	AM	PM	Alt	Azm			South facing surface angle with horiz.				
					Normal	Horiz.	46	56	66	76	90
Jul 21	4	8	1.7	125.8	0	0	0	0	0	0	0
	5	7	9.0	113.7	91	27	11	10	9	8	6
	6	6	17.0	101.9	169	72	30	18	16	14	12
	7	5	25.3	89.7	212	119	88	74	58	41	15
	8	4	33.6	76.7	237	163	151	136	117	96	61
	9	3	41.4	62.0	252	201	208	193	173	147	106
	10	2	48.2	44.6	261	230	254	239	217	189	142
	11	1	52.9	23.7	265	248	283	268	245	216	165
	12		54.6	0.0	267	254	293	278	255	225	173
	Surface daily totals				3240	2372	2342	2152	1926	1646	1186
Aug 21	5	7	2.0	109.2	1	0	0	0	0	0	0
	6	6	10.2	97.0	112	34	16	11	10	9	7
	7	5	18.5	84.5	187	82	73	65	56	45	28
	8	4	26.7	71.3	225	128	140	131	119	104	78
	9	3	34.3	56.7	246	168	202	193	179	160	126
	10	2	40.5	40.0	258	199	251	242	227	206	166
	11	1	44.8	20.9	264	218	282	274	258	235	191
	12		46.3	0.0	266	225	293	285	269	245	200
	Surface daily totals				2850	1884	2218	2118	1966	1760	1392
Sep 21	7	5	8.3	77.5	107	25	36	36	34	32	28
	8	4	16.2	64.4	194	72	111	111	108	102	89
	9	3	23.3	50.3	233	114	181	182	178	168	147
	10	2	29.0	34.9	253	146	236	237	232	221	193
	11	1	32.7	17.9	263	166	271	273	267	254	223
	12		34.0	0.0	266	173	283	285	279	265	233
	Surface daily totals				2368	1220	1950	1962	1918	1820	1594
Oct 21	8	4	7.1	59.1	104	20	53	57	59	59	57
	9	3	13.8	45.7	193	60	138	145	148	147	138
	10	2	19.0	31.3	231	92	201	210	213	210	195
	11	1	22.3	16.0	248	112	240	250	253	248	230
	12		23.5	0.0	253	119	253	263	266	261	241
	Surface daily totals				1804	688	1516	1586	1612	1588	1480
Nov 21	9	3	5.2	41.9	76	12	49	54	57	59	58
	10	2	10.0	28.5	165	39	132	143	149	152	148
	11	1	13.1	14.5	201	58	179	193	201	203	196
	12		14.2	0.0	211	65	194	209	217	219	211
	Surface daily totals				1094	284	914	986	1032	1046	1016
Dec 21	9	3	1.9	40.5	5	0	3	4	4	4	4
	10	2	6.6	27.5	113	19	86	95	101	104	103
	11	1	9.5	13.9	166	37	141	154	163	167	164
	12		10.6	0.0	180	43	159	173	182	186	182
	Surface daily totals				748	156	620	678	716	734	722

1 BTUH/SQ. FT. = 3.152 W/m²

NOTE: 1) Based on data in Table 1, p. 387 in ref. [3]; 0% ground reflectance; 1.0 clearness factor.
2) See Fig. 4, p. 394 in [3] for typical regional clearness factors.
3) Ground reflection not included on normal or horizontal surfaces.

TABLE A.6 Solar Position and Insolation Values for 64 Degrees North Latitude

Date	Solar time AM	PM	Solar position Alt	Azm	BTUH/sq. ft. total insolation on surfaces Normal	Horiz.	South facing surface angle with horiz. 54	64	74	84	90
Jan 21	10	2	2.8	28.1	22	2	17	19	20	20	20
	11	1	5.2	14.1	81	12	72	77	80	81	81
	12		6.0	0.0	100	16	91	98	102	103	103
	Surface daily totals				306	45	268	290	302	306	304
Feb 21	8	4	3.4	58.7	35	4	17	19	19	19	19
	9	3	8.6	44.8	147	31	103	108	111	110	107
	10	2	12.6	30.3	199	55	170	178	181	178	173
	11	1	15.1	15.3	222	71	212	220	223	219	213
	12		16.0	0.0	228	77	225	235	237	232	226
	Surface daily totals				1432	400	1230	1286	1302	1282	1252
Mar 21	7	5	6.5	76.5	95	18	30	29	29	27	25
	8	4	20.7	62.6	185	54	101	102	99	94	89
	9	3	18.1	48.1	227	87	171	172	169	160	153
	10	2	22.3	32.7	249	112	227	229	224	213	203
	11	1	25.1	16.6	260	129	262	265	259	246	235
	12		26.0	0.0	263	134	274	277	271	258	246
	Surface daily totals				2296	932	1856	1870	1830	1736	1656
Apr 21	5	7	4.0	108.5	27	5	2	2	2	1	1
	6	6	10.4	95.1	133	37	15	9	8	7	6
	7	5	17.0	81.6	194	76	70	63	54	43	37
	8	4	23.3	67.5	228	112	136	128	116	102	91
	9	3	29.0	52.3	248	144	197	189	176	158	145
	10	2	33.5	36.0	260	169	246	239	224	203	188
	11	1	36.5	18.4	266	184	278	270	255	233	216
	12		97.6	0.0	268	190	289	281	266	243	225
	Surface daily totals				2982	1644	2176	2082	1936	1736	1594
May 21	4	8	5.8	125.1	51	11	5	4	4	3	3
	5	7	11.6	112.1	132	42	13	11	10	9	8
	6	6	17.9	99.1	185	79	29	16	14	12	11
	7	5	24.5	85.7	218	117	86	72	56	39	28
	8	4	30.9	71.5	239	152	148	133	115	94	80
	9	3	36.8	56.1	252	182	204	190	170	145	128
	10	2	41.6	38.9	261	205	249	235	213	186	167
	11	1	44.9	20.1	265	219	278	264	242	213	193
	12		46.0	0.0	267	224	288	274	251	222	201
	Surface daily totals				3470	2236	2312	2124	1898	1624	1436
Jun 21	3	9	4.2	139.4	21	4	2	2	2	2	1
	4	8	9.0	126.4	93	27	10	9	8	7	6
	5	7	14.7	113.6	154	60	16	15	13	11	10
	6	6	21.0	100.8	194	96	34	19	17	14	13
	7	5	27.5	87.5	221	132	91	74	55	36	23
	8	4	34.0	73.3	239	166	150	133	112	88	73
	9	3	39.9	57.8	251	195	204	187	164	137	119
	10	2	44.9	40.4	258	217	247	230	206	177	157
	11	1	48.3	20.9	262	231	275	258	233	202	181
	12		49.5	0.0	263	235	284	267	242	211	189
	Surface daily totals				3650	2488	2342	2118	1862	1558	1356

Date	Solar time		Solar position		BTUH/sq. ft. total insolation on surfaces						
	AM	PM	Alt	Azm			South facing surface angle with horiz.				
					Normal	Horiz.	54	64	74	84	90
Jul 21	4	8	6.4	125.3	53	13	6	5	5	4	4
	5	7	12.1	112.4	128	44	14	13	11	10	9
	6	6	18.4	99.4	179	81	30	17	16	13	12
	7	5	25.0	86.0	211	118	86	72	56	38	28
	8	4	31.4	71.8	231	152	146	131	113	91	77
	9	3	37.3	56.3	245	182	201	186	166	141	124
	10	2	42.2	39.2	253	204	245	230	208	181	162
	11	1	45.4	20.2	257	218	273	258	236	207	187
	12		46.6	0.0	259	223	282	267	245	216	195
	Surface daily totals				3372	2248	2280	2090	1864	1588	1400
Aug 21	5	7	4.6	108.8	29	6	3	3	2	2	2
	6	6	11.0	95.5	123	39	16	11	10	8	7
	7	5	17.6	81.9	181	77	69	61	52	42	35
	8	4	23.9	67.8	214	113	132	123	112	97	87
	9	3	29.6	52.6	234	144	190	182	169	150	138
	10	2	34.2	36.2	246	168	237	229	215	194	179
	11	1	37.2	18.5	252	183	268	260	244	222	205
	12		38.3	0.0	254	188	278	270	255	232	215
	Surface daily totals				2808	1646	2108	1008	1860	1662	1522
Sep 21	7	5	6.5	76.5	77	16	25	25	24	23	21
	8	4	12.7	72.6	163	51	92	92	90	85	81
	9	3	18.1	48.1	206	83	159	159	156	147	141
	10	2	22.3	32.7	229	108	212	213	209	198	189
	11	1	25.1	16.6	240	124	246	248	243	230	220
	12		26.0	0.0	244	129	258	260	254	241	230
	Surface daily totals				2074	892	1726	1736	1696	1608	1532
Oct 21	8	4	3.0	58.5	17	2	9	9	10	10	10
	9	3	8.1	44.6	122	26	86	91	93	92	90
	10	2	12.1	30.2	176	50	152	159	161	159	155
	11	1	14.6	15.2	201	65	193	201	203	200	195
	12		15.5	0.0	208	71	207	215	217	213	208
	Surface daily totals				1238	358	1088	1136	1152	1134	1106
Nov 21	10	2	3.0	28.1	23	3	18	20	21	21	21
	11	1	5.4	14.2	79	12	70	76	79	80	79
	12		6.2	0.0	97	17	89	96	100	101	100
	Surface daily totals				302	46	266	286	298	302	300
Dec 21	11	1	1.8	13.7	4	0	3	4	4	4	4
	12		2.6	0.0	16	2	14	15	16	17	17
	Surface daily totals				24	2	20	22	24	24	24

1 BTUH/SQ. FT. = 3.152 W/m²
NOTE: 1) Based on data in Table 1, p. 387 in ref. [3]; 0% ground reflectance; 1.0 clearness factor.
2) See Fig. 4, p. 394 in [3] for typical regional clearness factors.
3) Ground reflection not included on normal or horizontal surfaces.

TABLE A.7 Latitude 24°N, Incident Angles for Horizontal and South-facing Tilted Surfaces

		Horiz.	L − 10	Lat.	Lat. + 10	Lat. + 20	Vert.
7	5	86.8	80.5	76.3	72.4	68.9	62.6
8	4	75.1	67.5	62.7	58.6	55.4	56.6
9	3	64.5	55.3	49.6	44.9	41.8	51.1
10	2	55.7	44.5	37.4	31.6	28.0	46.6
11	1	49.6	36.5	27.6	19.6	14.3	43.6
12		47.4	33.4	23.5	13.5	3.4	42.5

Dates: (Decl.) Dec. 21 (−23.45)

		Horiz.	L − 10	Lat.	Lat. + 10	Lat. + 20	Vert.
7	5	85.2	79.6	75.9	72.6	69.8	65.7
8	4	73.1	66.2	62.0	58.5	56.0	59.8
9	3	62.1	53.5	48.4	44.5	42.2	54.4
10	2	52.8	42.1	35.5	30.6	28.2	50.0
11	1	46.4	33.4	24.8	17.6	14.1	47.0
12		44.0	30.0	20.0	10.0	0.0	46.0

Jan. 21 (−19.9)
Nov. 21 (−19.9)

		Horiz.	L − 10	Lat.	Lat. + 10	Lat. + 20	Vert.
7	5	80.7	77.2	75.2	73.7	72.6	74.8
8	4	67.7	62.9	60.5	59.0	58.5	69.0
9	3	55.6	49.0	45.9	44.3	44.5	63.8
10	2	44.9	35.9	31.5	29.5	30.6	59.6
11	1	37.0	25.0	18.0	14.8	17.6	56.9
12		34.0	20.0	10.0	0.0	10.0	56.0

Feb. 21 (−10.6)
Oct. 21 (−10.7)

		Horiz.	L − 10	Lat.	Lat. + 10	Lat. + 20	Vert.
7	5	76.3	75.2	75.0	75.2	75.9	84.0
8	4	62.8	60.5	60.0	60.5	62.0	78.3
9	3	49.8	45.9	45.0	45.9	48.4	73.3
10	2	37.7	31.5	30.0	31.5	35.5	69.4
11	1	28.1	18.0	15.0	18.0	24.8	66.9
12		24.0	10.0	0.0	10.0	20.0	66.0

Mar. 21 (0.0)
Sep. 21 (0.0)

		Horiz.	L − 10	Lat.	Lat. + 10	Lat. + 20	Vert.
6	6	85.3	88.0	90.0	92.0	93.9	100.6
7	5	71.7	73.5	75.3	77.6	80.2	94.6
8	4	58.0	58.9	60.7	63.4	67.0	89.1
9	3	44.4	44.2	46.2	49.7	54.4	84.4
10	2	31.0	29.5	32.0	36.8	43.2	80.7
11	1	18.9	14.8	18.9	26.2	34.9	78.4
12		12.4	1.6	11.6	21.6	31.6	77.6

Apr. 21 (+11.9)
Aug. 21 (+12.1)

		Horiz.	L − 10	Lat.	Lat. + 10	Lat. + 20	Vert.
6	6	82.0	86.0	90.0	93.4	96.7	108.2
7	5	68.8	72.6	75.9	79.6	83.6	102.3
8	4	55.4	58.5	62.0	66.2	71.1	97.0
9	3	41.7	44.5	48.4	53.5	59.5	92.4
10	2	28.0	30.6	35.5	42.1	49.6	88.9
11	1	14.5	17.6	24.8	33.4	42.6	86.7
12		4.0	10.0	20.0	30.0	40.0	86.0

May 21 (+20.3)
Jul 21 (+20.5)

		Horiz.	L − 10	Lat.	Lat. + 10	Lat. + 20	Vert.
6	6	80.7	86.0	90.0	94.0	97.8	111.3
7	5	67.7	72.4	76.3	80.5	85.0	105.5
8	4	54.5	58.6	62.7	67.5	72.8	100.2
9	3	41.0	44.9	49.6	55.8	61.7	95.7
10	2	27.4	31.6	37.4	44.5	52.4	92.3
11	1	13.7	19.7	27.6	36.5	45.9	90.2
12		0.6	13.4	23.4	33.4	43.4	89.4

Jun. 21 (+23.45)

TABLE A.8 Latitude 32° N, Incident Angles for Horizontal and South-facing Tilted Surfaces

		Horiz.	L − 10	Lat.	Lat. + 10	Lat. + 20	Vert.
8	4	79.7	67.5	62.7	58.6	55.4	54.5
9	3	70.2	55.3	49.6	44.9	41.8	47.1
10	2	62.4	44.5	37.4	31.6	28.0	40.7
11	1	57.3	36.5	27.6	19.6	14.3	36.2
12		55.4	33.4	23.4	13.5	3.5	34.5

Dates (Decl.)
Dec. 21 (−23.45)

		Horiz.	L − 10	Lat.	Lat. + 10	Lat. + 20	Vert.
7	5	88.6	79.6	75.9	72.6	69.8	65.2
8	4	77.5	66.2	62.0	58.5	56.0	57.4
9	3	67.5	53.5	48.4	44.5	42.2	50.0
10	2	59.4	42.1	35.5	30.6	28.2	43.8
11	1	53.9	33.4	24.8	17.6	14.1	39.6
12		52.0	30.0	20.0	10.0	0.0	38.0

Jan. 21 (−19.9)
Nov. 21 (−19.9)

		Horiz.	L − 10	Lat.	Lat. + 10	Lat. + 20	Vert.
7	5	82.9	77.2	75.2	73.7	72.6	73.6
8	4	71.0	62.9	60.5	59.0	58.5	65.9
9	3	60.1	49.0	45.9	44.3	44.5	58.9
10	2	50.9	35.9	31.5	29.5	30.6	53.2
11	1	44.4	25.0	18.0	14.8	17.6	49.4
12		42.0	20.0	10.0	0.0	10.0	48.0

Feb. 21 (−10.6)
Oct. 21 (−10.7)

		Horiz.	L − 10	Lat.	Lat. + 10	Lat. + 20	Vert.
7	5	77.3	75.2	75.0	75.2	75.9	82.1
8	4	64.9	60.5	60.0	60.5	62.0	74.6
9	3	53.2	45.9	45.0	45.9	48.4	68.0
10	2	42.7	31.5	30.0	31.5	35.5	62.7
11	1	35.0	18.0	15.0	18.0	24.8	59.2
12		32.0	10.0	0.0	10.0	20.0	58.0

Mar. 21 (0.0)
Sep. 21 (0.0)

		Horiz.	L − 10	Lat.	Lat. + 10	Lat. + 20	Vert.
6	6	83.9	88.0	90.0	92.0	93.9	99.8
7	5	71.2	73.5	75.3	77.6	80.2	92.1
8	4	58.5	58.9	60.7	63.4	67.0	84.9
9	3	46.1	44.2	46.2	49.7	54.4	78.7
10	2	34.3	29.5	32.0	36.8	43.2	73.8
11	1	24.6	14.8	18.9	26.2	34.9	70.7
12		20.4	1.6	11.6	21.6	31.6	69.6

Apr. 21 (+11.9)
Aug. 21 (+12.1)

		Horiz.	L − 10	Lat.	Lat. + 10	Lat. + 20	Vert.
6	6	79.6	86.6	90.0	93.4	96.7	106.9
7	5	67.2	72.6	75.9	79.6	83.6	99.3
8	4	54.6	58.5	62.0	66.2	71.1	92.4
9	3	41.9	44.5	48.4	53.5	59.5	86.4
10	2	29.4	30.6	35.5	42.1	49.6	81.9
11	1	18.0	17.6	24.8	33.4	42.6	79.0
12		12.0	10.0	20.0	30.0	40.0	78.0

May 21 (+20.3)
Jul. 21 (+20.5)

		Horiz.	L − 10	Lat.	Lat. + 10	Lat. + 20	Vert.
6	6	77.8	86.0	90.0	94.0	97.8	109.7
7	5	65.7	72.4	76.3	80.5	85.0	102.2
8	4	53.1	58.6	62.7	67.5	72.8	95.4
9	3	40.4	44.9	49.6	55.3	61.7	89.6
10	2	27.8	31.6	37.4	44.5	52.4	85.2
11	1	15.8	19.6	27.6	36.5	45.8	82.4
12		8.6	13.4	23.4	33.4	43.4	81.4

Jun. 21 (+23.45)

TABLE A.9 Latitude 40°N, Incident Angles for Horizontal and South-facing Tilted Surfaces

Dates: (Decl.)
Dec. 21 (−23.45)

		Horiz.	L − 10	Lat.	Lat. + 10	Lat. + 20	Vert.
8	4	84.5	67.5	62.7	58.6	55.4	53.2
9	3	76.0	55.3	49.6	44.9	41.8	43.8
10	2	69.3	44.5	37.4	31.6	28.0	35.4
11	1	65.0	36.5	27.6	19.6	14.3	29.0
12		63.4	33.4	23.4	13.5	3.5	26.6

Jan. 21 (−19.9)
Nov. 21 (−19.9)

		Horiz.	L − 10	Lat.	Lat. + 10	Lat. + 20	Vert.
8	4	81.9	66.2	62.0	58.5	56.0	55.7
9	3	73.2	53.5	48.4	44.5	42.2	46.4
10	2	66.2	42.1	35.5	30.6	28.2	38.3
11	1	61.6	33.4	24.8	17.6	14.1	32.3
12		60.0	30.0	20.0	10.0	0.0	30.0

Feb. 21 (−10.6)
Oct. 21 (−10.7)

		Horiz.	L − 10	Lat.	Lat. + 10	Lat. + 20	Vert.
7	5	85.2	77.2	75.2	73.7	72.6	72.7
8	4	74.6	62.9	60.5	59.0	58.5	63.3
9	3	65.0	49.0	45.9	44.3	44.5	59.5
10	2	57.2	35.9	31.5	29.5	30.6	47.1
11	1	51.9	25.0	18.0	14.8	17.6	41.9
12		50.0	20.0	10.0	0.0	10.0	40.0

Mar. 21 (0.0)
Sep. 21 (0.0)

		Horiz.	L − 10	Lat.	Lat. + 10	Lat. + 20	Vert.
7	5	78.6	75.2	75.0	75.2	75.9	80.4
8	4	67.5	60.5	60.0	60.5	62.0	71.3
9	3	57.2	45.9	45.0	45.9	48.4	63.0
10	2	48.4	31.5	30.0	31.5	35.5	56.2
11	1	42.3	18.0	15.0	18.0	24.8	51.6
12		40.0	10.0	0.0	10.0	20.0	50.0

Apr. 21 (+11.9)
Aug. 21 (+12.1)

		Horiz.	L − 10	Lat.	Lat. + 10	Lat. + 20	Vert.
6	6	82.6	88.0	90.0	92.0	93.9	98.9
7	5	71.1	73.5	75.3	77.6	80.2	89.5
8	4	59.7	58.9	60.7	63.4	67.0	80.7
9	3	48.7	44.2	46.2	49.7	54.4	73.1
10	2	38.8	29.5	32.0	36.8	43.2	67.0
11	1	31.3	14.8	18.9	26.2	34.9	63.0
12		28.4	1.6	11.6	21.6	31.6	61.6

May 21 (+20.3)
Jul. 21 (+20.5)

		Horiz.	L − 10	Lat.	Lat. + 10	Lat. + 20	Vert.
5	7	88.1	100.4	104.1	107.4	110.2	114.7
6	6	77.3	86.6	90.0	93.4	96.7	105.2
7	5	66.0	72.6	75.9	79.6	83.6	96.1
8	4	54.6	58.5	62.0	66.2	71.1	87.7
9	3	43.2	44.5	48.4	53.5	59.5	80.5
10	2	32.5	30.6	35.5	42.1	49.6	74.9
11	1	23.8	17.6	24.8	33.4	42.6	71.2
12		20.0	10.0	20.0	30.0	40.0	70.0

Jun. 21 (+23.45)

		Horiz.	L − 10	Lat.	Lat. + 10	Lat. + 20	Vert.
5	7	85.8	99.5	103.7	107.6	111.1	117.2
6	6	75.2	86.0	90.0	94.0	97.8	107.7
7	5	64.0	72.4	76.3	80.5	85.0	98.8
8	4	52.6	58.6	62.7	67.5	72.8	90.6
9	3	41.2	44.9	49.6	55.3	61.7	83.6
10	2	30.2	31.6	37.4	44.5	52.4	78.1
11	1	20.8	19.6	27.6	36.5	45.8	74.6
12		16.6	13.4	23.4	33.4	43.4	73.4

TABLE A.10 Latitude 48°N, Incident Angles for Horizontal and South-facing Tilted Surfaces

		Horiz.	L − 10	Lat.	Lat. + 10	Lat. + 20	Vert.
9	3	82.0	55.3	49.6	44.9	41.8	41.6
10	2	76.4	44.5	37.4	31.6	28.0	31.1
11	1	72.7	36.5	27.6	19.6	14.3	22.4
12		71.4	33.4	23.4	13.5	3.5	18.5

Dec. 21 (−23.45)

		Horiz.	L − 10	Lat.	Lat. + 10	Lat. + 20	Vert.
8	4	86.5	66.2	62.0	58.5	56.0	54.7
9	3	79.0	53.5	48.4	44.5	42.2	43.7
10	2	73.1	42.1	35.5	30.6	28.2	33.5
11	1	69.3	33.4	24.8	17.6	14.1	25.4
12		68.0	30.0	20.0	10.0	0.0	22.0

Jan. 21 (−19.9)
Nov. 21 (−19.9)

		Horiz.	L − 10	Lat.	Lat. + 10	Lat. + 20	Vert.
7	5	87.6	77.2	75.2	73.7	72.6	72.2
8	4	78.4	62.9	60.5	59.0	58.5	61.2
9	3	70.3	49.0	45.9	44.3	44.5	50.7
10	2	63.8	35.9	31.5	29.5	30.6	41.4
11	1	59.5	25.0	18.0	14.8	17.6	34.6
12		58.0	20.0	10.0	0.0	10.0	32.0

Feb. 21 (−10.6)
Oct. 21 (−10.7)

		Horiz.	L − 10	Lat.	Lat. + 10	Lat. + 20	Vert.
7	5	80.0	75.2	75.0	75.2	75.9	78.9
8	4	70.5	60.5	60.0	60.5	62.0	68.2
9	3	61.8	45.9	45.0	45.9	48.4	58.3
10	2	54.6	31.5	30.0	31.5	35.5	49.9
11	1	49.7	18.0	15.0	18.0	24.8	44.1
12		48.0	10.0	0.0	10.0	20.0	42.0

Mar. 21 (0.0)
Sep. 21 (0.0)

		Horiz.	L − 10	Lat.	Lat. + 10	Lat. + 20	Vert.
6	6	81.4	88.0	90.0	92.0	93.9	97.7
7	5	71.4	73.5	75.3	77.6	80.2	86.9
8	4	61.5	58.9	60.7	63.4	67.0	76.7
9	3	52.2	44.2	46.2	49.7	54.4	67.7
10	2	44.2	29.5	32.0	36.8	43.2	60.3
11	1	38.5	14.8	18.9	26.2	34.9	55.3
12		36.4	1.6	11.6	21.6	31.6	53.6

Apr. 21 (+11.9)
Aug. 21 (+12.1)

		Horiz.	L − 10	Lat.	Lat. + 10	Lat. + 20	Vert.
5	7	84.8	100.4	104.1	107.4	110.2	114.2
6	6	75.3	86.6	90.0	93.4	96.7	103.2
7	5	65.4	72.6	75.9	79.6	83.6	92.8
8	4	55.4	58.5	62.0	66.2	71.1	83.1
9	3	45.7	44.5	48.4	53.5	59.5	74.6
10	2	37.0	30.6	35.5	42.1	49.6	67.9
11	1	30.5	17.6	24.8	33.4	42.6	63.5
12		28.0	10.0	20.0	30.0	40.0	62.0

May 21 (+20.3)
Jul. 21 (+20.5)

		Horiz.	L − 10	Lat.	Lat. + 10	Lat. + 20	Vert.
5	7	82.1	99.5	103.7	107.6	111.1	116.3
6	6	72.8	86.0	90.0	94.0	97.3	105.4
7	5	63.0	72.4	76.3	80.5	85.0	95.2
8	4	52.9	58.6	62.7	67.5	72.8	85.7
9	3	43.1	44.9	49.6	55.3	61.7	77.5
10	2	34.2	31.6	37.4	44.5	52.4	71.1
11	1	27.3	19.6	27.6	36.5	45.8	66.9
12		24.6	13.4	23.4	33.4	43.4	65.4

Jun. 21 (+23.45)

TABLE A.11 Latitude 56° N, Incident Angles for Horizontal and South-facing Tilted Surfaces

		Horiz.	L − 10	Lat.	Lat. + 10	Lat. + 20	Vert.
9	3	88.1	55.3	49.6	44.9	41.8	40.5
10	2	83.4	44.5	37.4	31.6	28.0	28.2
11	1	80.5	36.5	27.6	19.6	14.3	16.8
12		79.4	33.4	23.4	13.5	3.4	10.5

Dec. 21 (−23.45)

		Horiz.	L − 10	Lat.	Lat. + 10	Lat. + 20	Vert
9	3	85.0	53.5	48.4	44.5	42.2	42.1
10	2	80.1	42.1	35.5	30.6	28.2	30.0
11	1	77.1	33.4	24.8	17.6	14.1	19.3
12		76.0	30.0	20.0	10.0	0.0	14.0

Jan. 21 (−19.9)
Nov. 21 (−19.9)

		Horiz.	L − 10	Lat.	Lat. + 10	Lat. + 20	Vert.
8	4	82.5	62.9	60.5	59.0	58.5	59.6
9	3	75.8	49.0	45.9	44.3	44.5	47.6
10	2	70.6	35.9	31.5	29.5	30.6	36.5
11	1	67.2	25.0	18.0	14.8	17.6	27.7
12		66.0	20.0	10.0	0.0	10.0	24.0

Feb. 21 (−10.6)
Oct. 21 (−10.7)

		Horiz.	L − 10	Lat.	Lat. + 10	Lat. + 20	Vert.
7	5	81.7	75.2	75.0	75.2	75.9	77.6
8	4	73.9	60.5	60.0	60.5	62.0	65.0
9	3	66.7	45.9	45.0	45.9	48.4	54.1
10	2	61.0	31.5	30.0	31.5	35.5	44.1
11	1	57.3	18.0	15.0	18.0	24.8	36.8
12		56.0	10.0	0.0	10.0	20.0	34.0

Mar. 21 (0.0)
Sep. 21 (0.0)

		Horiz.	L − 10	Lat.	Lat. + 10	Lat. + 20	Vert.
5	7	88.6	102.4	104.7	106.5	107.9	108.8
6	6	80.4	88.0	90.0	92.0	93.9	96.5
7	5	72.0	73.5	75.3	77.6	80.2	84.4
8	4	63.9	58.9	60.7	63.4	67.0	72.9
9	3	56.4	44.2	46.2	49.7	54.4	62.5
10	2	50.1	29.5	32.0	36.8	43.2	53.8
11	1	45.9	14.8	18.9	26.2	34.9	47.8
12		44.4	1.6	11.6	21.6	31.6	45.6

Apr. 21 (+11.9)
Aug. 21 (+12.1)

		Horiz.	L − 10	Lat.	Lat. + 10	Lat. + 20	Vert.
4	8	88.8	113.8	118.0	121.5	124.0	125.5
5	7	81.5	100.4	104.1	107.4	110.2	113.1
6	6	73.5	86.6	90.0	93.4	96.7	101.0
7	5	65.2	72.6	75.9	79.6	83.6	89.4
8	4	56.9	58.5	62.0	66.2	71.1	78.6
9	3	49.1	44.5	48.4	53.5	59.5	68.9
10	2	42.4	30.6	35.5	42.1	49.6	61.1
11	1	37.7	17.6	24.8	33.4	42.6	55.9
12		36.0	10.0	20.0	30.0	40.0	54.0

May 21 (+20.3)
Jul. 21 (+20.5)

		Horiz.	L − 10	Lat.	Lat. + 10	Lat. + 20	Vert.
4	8	85.8	112.5	117.3	121.4	124.6	127.1
5	7	78.6	99.5	103.7	107.6	111.1	114.8
6	6	70.7	86.0	90.0	94.0	97.8	102.9
7	5	62.4	72.4	76.3	80.5	85.0	91.5
8	4	54.1	58.6	62.7	67.5	72.8	80.9
9	3	46.2	44.9	49.6	55.3	61.7	71.6
10	2	39.3	31.6	37.4	44.5	52.4	64.1
11	1	34.4	19.6	27.6	36.5	45.8	59.2
12		32.6	13.4	23.4	33.4	43.4	57.4

Jun. 21 (+23.45)

TABLE A.12 Latitude 64°N, Incident Angles for Horizontal and South-facing Tilted Surfaces

		Horiz.	L − 10	Lat.	Lat. + 10	Lat. + 20	Vert.
11	1	88.2	36.5	27.6	19.6	14.3	13.9
	12	87.4	33.4	23.4	13.5	3.4	2.5

Dec. 21 (−23.45)

		Horiz.	L − 10	Lat.	Lat. + 10	Lat. + 20	Vert.
10	2	87.2	42.1	35.5	30.6	28.2	28.2
11	1	84.8	33.4	24.8	17.6	14.1	15.0
	12	84.0	30.0	20.0	10.0	0.0	6.0

Jan. 21 (−19.9)
Nov. 21 (−19.9)

		Horiz.	L − 10	Lat.	Lat. + 10	Lat. + 20	Vert.
8	4	86.6	62.9	60.5	59.0	58.5	58.8
9	3	81.4	49.0	45.9	44.3	44.5	45.4
10	2	77.4	35.9	31.5	29.5	30.6	32.6
11	1	74.9	25.0	18.0	14.8	17.6	21.4
	12	74.0	20.0	10.0	0.0	10.0	16.0

Feb. 21 (−10.6)
Oct. 21 (−10.7)

		Horiz.	L − 10	Lat.	Lat. + 10	Lat. + 20	Vert.
7	5	83.5	75.2	75.0	75.2	75.9	76.5
8	4	77.3	60.5	60.0	60.5	62.0	63.3
9	3	71.9	45.9	45.0	45.9	48.4	50.5
10	2	67.7	31.5	30.0	31.5	35.5	38.9
11	1	64.9	18.0	15.0	18.0	24.8	29.8
	12	64.0	10.0	0.0	10.0	20.0	26.0

Mar. 21 (0.0)
Sep. 21 (0.0)

		Horiz.	L − 10	Lat.	Lat. + 10	Lat. + 20	Vert.
5	7	86.0	102.4	104.7	106.5	107.9	108.4
6	6	79.6	88.0	90.0	92.0	93.9	95.1
7	5	73.0	73.5	75.3	77.6	80.2	82.0
8	4	66.7	58.9	60.7	63.4	67.0	69.4
9	3	61.0	44.2	46.2	49.7	54.4	57.7
10	2	56.5	29.5	32.0	36.8	43.2	47.6
11	1	53.5	14.8	18.9	26.2	34.9	40.3
	12	52.4	1.6	11.6	21.6	31.6	37.6

Apr. 21 (+11.9)
Aug. 21 (+12.1)

		Horiz.	L − 10	Lat.	Lat. + 10	Lat. + 20	Vert.
4	8	84.2	113.8	118.0	121.5	124.0	124.9
5	7	78.4	100.4	104.1	107.4	110.2	111.6
6	6	72.1	86.6	90.0	93.4	96.7	98.6
7	5	65.5	72.6	75.9	79.6	83.6	86.1
8	4	59.1	58.5	62.0	66.2	71.1	74.2
9	3	53.2	44.5	48.4	53.5	59.5	63.4
10	2	48.4	30.6	35.5	42.1	49.6	54.4
11	1	45.1	17.6	24.8	33.4	42.6	48.3
	12	44.0	10.0	20.0	30.0	40.0	46.0

May 21 (+20.3)
Jul. 21 (+20.5)

		Horiz.	L − 10	Lat.	Lat. + 10	Lat. + 20	Vert.
3	9	85.8	124.7	130.4	135.1	138.2	139.2
4	8	81.0	112.5	117.3	121.4	124.6	125.9
5	7	75.3	99.5	103.7	107.6	111.1	112.8
6	6	69.0	86.0	90.0	94.0	97.8	100.0
7	5	62.5	72.4	76.3	80.5	85.0	87.8
8	4	56.0	58.6	62.7	67.5	72.8	76.2
9	3	50.1	44.9	49.6	55.6	61.7	65.9
10	2	45.1	31.6	37.4	44.5	52.4	57.3
11	1	41.7	19.6	27.6	36.5	45.8	51.5
	12	40.6	13.5	23.4	33.4	43.4	49.4

Jun. 21 (+23.45)

either be equipped with a build-in temperature compensation circuit and have a temperature sensitivity of less than ±1 percent over the range of ambient temperature encountered during the test(s) or have been tested in a temperature-controlled chamber over the same temperature range so that its temperature coefficient has been determined in accordance with reference [5].

6.1.1.2 *Variation in spectral response.* Errors caused by a departure from the required spectral response of the sensor shall not exceed ±2 percent over the range of interest.*

6.1.1.3 *Nonlinearity of response.* Unless the pyranometer was supplied with a calibration curve relating the output to the insolation, its response shall be within ±1 percent of being linear over the range of insolation existing during the tests.

6.1.1.4 *Time response of pyranometer.* The time constant of the pyranometer shall be less than 5 s.

6.1.1.5 *Variation of response with attitude.* The calibration factor of a pyranometer can change when the instrument is used in other than the orientation for which it was calibrated. The instruments' calibration factor (including corrections) shall change less than ±0.5 percent compared with the calibrated orientation when placed in the orientation used during the test(s).

6.1.1.6 *Variation of response with angle of incidence.* Ideally the response of the receiver is proportional to the cosine of the zenith angle

---

*Pyranometer thermopiles which are "all black" and which are coated with Parson's black or 3M 101C10 velvet black paint and which have selected optical grade hemispheres usually satisfy this requirement [5]. Note: Identification of commercial materials does not imply recommendation or endorsement by the National Bureau of Standards.

of the solar beam and is constant at all azimuth angles. The pyranometer's deviation from a true cosine response shall be less than ±1 percent for the incident angles encountered during the test(s).

6.1.2 The pyranometer shall be calibrated within six months of the collector test(s) against other pyranometers whose calibration uncertainty relative to recognized measurement standards is known.*

6.2 Temperature Measurements

6.2.1 Temperature measurements shall be made in accordance with ASHRAE Standard 41-66, Part 1 [6].

6.2.2 *Temperature measurements.* The temperature of the transfer fluid entering and/or leaving the solar collector shall be measured with one of the following:

a. Thermocouples

b. Resistance Thermometers

6.2.3 *Temperature difference measurements across the solar collector.* The temperature difference of the transfer fluid across the solar collector shall be measured with:

a. Thermopile (air or water as the transfer fluid)

b. Calibrated resistance thermometers connected in two arms of a bridge circuit (only when a liquid is the transfer fluid)

6.2.4 The accuracy and precision of the instruments and their associated readout devices shall be within the limits as follows:

*One nationally recognized calibration center is the Eppley Laboratory in Newport, Rhode Island. The calibration data are commonly expressed in cal/(cm² · min) or in langleys/min. In some meteorological services, calibration data are supplied in milliwatt/cm². The following equivalent units shall be used:

1 cal/cm² · min) = 1 ly/min = .001434 W/m²
1 mW/cm² = 0.1 W/m²

	Instrument Accuracy*	Instrument Precision**
Temperature	±0.5°C (±0.9°F)	±0.2°C (±0.4°F)
Temperature Difference	±0.1°C (±0.2°F)	±0.1°C (±0.2°F)

6.2.5 In no case should the smallest scale division of the instrument or instrument system exceed 2 1/2 times the specified precision. For example, if the specified precision is ±0.1°C (±0.2°F), the smallest scale division shall not exceed 0.25°C (0.5°F).

6.2.6 The instruments shall be configured and used in accordance with Section 7 of this standard.

6.2.7 When using thermopiles, they shall be constructed in accordance with ANSI Standard C96.1-1964 (R 1969) [7].

6.3   Liquid Flow Measurements

6.3.1 The flow rate of the transfer fluid when a liquid shall be measured with one of the following:

   a. weigh tank (where proper precaution has been taken to minimize errors due to evaporation)

   b. positive displacement flow meter

   c. turbine flow meter (only when the transfer fluid is pure water or it has been specifically calibrated with the transfer fluid)

   d. magnetic flowmeter

6.3.2 The accuracy of the liquid flow rate measurement using the calibration, if furnished shall be equal to or better than ±1.0% of the measured value.

*The ability of the instrument to indicate the true value of the measured quantity.
**Closeness of agreement among repeated measurements of the same physical quantity.

6.4   Integrators and Recorders

6.4.1 Strip chart recorders used shall have an accuracy equal to or better than ±0.5% of the temperature difference and/or voltage measured and have a time constant of 1 s or less.

6.4.2 Electronic integrators used shall have an accuracy equal to or better than ±1.0% of the measured value.

6.5   Air Flow Measurements

When air is used as the transfer fluid, air flow rate shall be determined as described in Section 7.

6.6   Pressure Measurements

6.6.1 *Nozzle throat pressure.* The pressure measurement at the nozzle throat shall be made with instruments which shall permit measurements of pressure to within ±2.0% absolute and whose smallest scale division shall not exceed 2 1/2 times the specified accuracy [11].

6.6.2 *Air flow measurements.* The static pressure across the nozzle and the velocity pressure at the nozzle throat shall be measured with manometers which have been calibrated against a standard manometer to ±1.0% of the reading. The smallest manometer scale division shall not exceed 2.0% of the reading [11].

6.6.3 *Pressure drop across collector.* The static pressure drop across the solar collector shall be measured with a manometer having an accuracy of 2.49 N/m² (0.01 in. of water).

6.7   Time and Weight Measurements

Time measurements and weight measurements shall be made to an accuracy of ±0.20% [11].

6.8   Wind Velocity

The wind velocity shall be measured with an instrument and associated readout device that can determine the integrated average wind velocity for each 15 minute test period to an accuracy of ±0.8 m/s (1.8 mph).

SECTION 7. APPARATUS AND METHOD OF TESTING

7.1　Liquid as the Transfer Fluid

The test configuration for the solar collector employing liquid as the transfer fluid is shown in Figure A.1.*

7.1.1 *Solar collector.* The solar collector should be mounted in its rigid frame at the predetermined tilt angle (for stationary collectors) or movable frame (for movable collectors) and anchored rigidly enough to a foundation so that the collector can hold its selected angular position against a strong gust of wind.

7.1.2 *Ambient temperature.* The ambient temperature sensor shall be housed in a well-ventilated instrumentation shelter with its bottom 1.25 m (4.1 ft) above the ground and with its door facing north, so that the sun's direct beam cannot fall upon the sensor when the door is opened. The instrument shelter shall be painted white outside and shall not be closer to any obstruction than twice the height of the obstruction itself (i.e., trees, fences, buildings, etc.) [15].

7.1.3 *Pyranometer.* The pyranometer shall be mounted on the surface parallel to the collector surface in such a manner that it does not cast a shadow onto the collector plate. Precautions should be always taken to avoid subjecting the instrument to mechanical shocks or vibration during the installation. The pyranometer should be oriented so that the emerging leads or the connector are located north of the receiving surface (in the Northern Hemisphere) or are in some other manner shaded. This minimizes heating of the electrical connections by the sun.

Care should also be taken to minimize reflected and reradiated energy from the solar collector onto the

---

*The recommended apparatus consists of a closed loop configuration. An open loop configuration is an acceptable alternative provided that the test conditions specified herein can be satisfied.

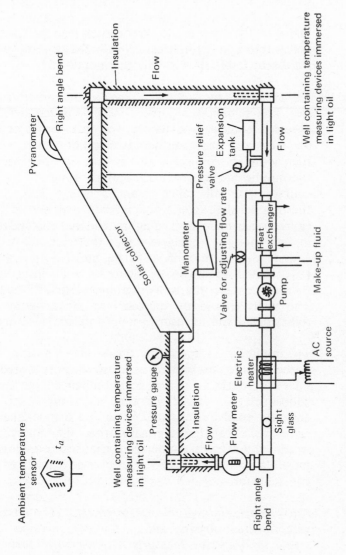

FIG. A.1 Testing configuration for the solar collector when the transfer fluid is liquid.

pyranometer. Some pyranometers come supplied with shields. This should be adjusted so that the highest point on the shield lies parallel to and just below the plane of the thermopile. Some pyranometers not supplied with a shield may be susceptible to error due to reflections by radiation that originates below the plane of the thermopile. Precautions can be taken by constructing a cylindrical shield, the top of which should be coplaner with the thermopile [5].

7.1.4 *Temperature measurement across the solar collector.* The temperature difference of the transfer fluid between entering and leaving the solar collector shall be measured using either two calibrated resistance thermometers connected in two arms of a bridge or a thermopile made from calibrated, type T thermocouple wire all taken from a single spool. The thermopile shall contain any even number of junctions constructed according to the recommendations in reference [7]. Each resistance thermometer or each end of the thermopile is to be inserted into a well [8] located as shown in Figure A.1. To insure good thermal contact, the wells shall be filled with light oil. The wells should be located just downstream of a right angle bend to insure proper mixing [6].

To minimize temperature measurement error, each probe should be located as close as possible to the inlet or outlet of the solar collector device. In addition, the piping between the wells and the collector shall be insulated in such a manner that the calculated heat loss or gain from the ambient air would not cause a temperature change for any test of more than 0.05°C (0.09°F) between each well and the collector.

7.1.5 *Additional temperature measurements.* The temperature of the transfer fluid at the two positions cited above shall also be measured by inserting appropriate sensors into the wells. Reference [6] should be followed in making these measurements.

7.1.6 *Pressure drop across the solar collector.* The pressure drop across the solar collector shall be measured using static pressure tap holes and a manometer. The edges of the holes on the inside surface of the pipe should be free of burrs and should be as small as practicable and not exceeding 1.6 mm (1/16 inch) diameter [12]. The thickness of the pipe wall should be 2 1/2 times the hole diameter [12].

7.1.7 *Reconditioning apparatus.* As shown in Figure A.1, a heat exchanger is used to cool the transfer fluid to simulate the building load and an adjustable electric resistance heater is used to control the inlet temperature to the prescribed test value. This combination of equipment or equivalent shall control the temperature of the fluid entering the solar collector to within ±0.5°C (±0.9°F) at all times during the tests.

7.1.8 *Additional equipment.* A pressure gauge, a pump, and a means of adjusting the flow rate of the transfer fluid shall be provided at the relative locations shown in Figure A.1. Depending upon the test apparatus design, an additional throttle valve may be required in the line just preceding the solar collector for proper control. An expansion tank and a pressure relief valve should be installed to allow the transfer fluid to freely expand and contract in the apparatus.* In addition, filters should be installed within the apparatus as well as a sight glass to insure that the transfer fluid passing through the collector is free of contaminants including air bubbles.

7.2   Air as the Transfer Fluid

The test configuration for the solar collector employing air as the transfer fluid is shown in Figure A.2.**

7.2.1 *Solar collector.* The solar collector should be mounted in its rigid frame at the predetermined tilt angle (for stationary collectors) or movable frame (for movable collectors) and anchored rigidly enough to a

*Figure A.1 should not be interpreted to mean that the relief valve and expansion tank necessarily be located below the solar collector.
**The recommended apparatus consists of a closed loop configuration. An open loop configuration is an acceptable alternative provided that the test conditions specified herein can be satisfied.

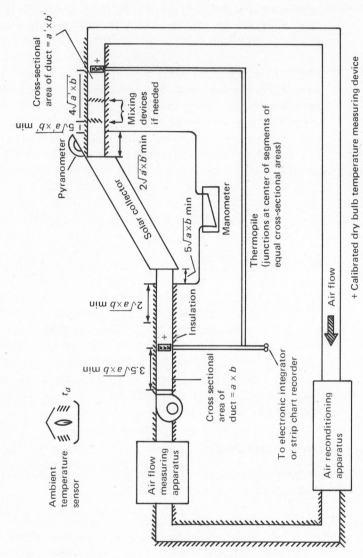

FIG. A.2 Testing configuration for the solar collector when the transfer fluid is air.

Cross-sectional area of duct = $a' \times b'$

Pyranometer

$5\sqrt{a \times b}$ min

Mixing devices if needed

$4\sqrt{a' \times b'}$

$2\sqrt{a \times b}$ min

Solar collector

$5\sqrt{a \times b}$ min

Manometer

Thermopile (junctions at center of segments of equal cross-sectional areas)

Insulation

$2\sqrt{a \times b}$ min

$3.5\sqrt{a \times b}$ min

Cross sectional area of duct = $a \times b$

To electronic integrator or strip chart recorder

Air flow

Air reconditioning apparatus

+ Calibrated dry bulb temperature measuring device

Ambient temperature sensor

t_a

Air flow measuring apparatus

foundation so that the collector can hold its selected angular position against a strong gust of wind.

7.2.2 *Ambient temperature.* The ambient temperature sensor shall be housed in a well-ventilated instrumentation shelter with its bottom 1.25 m (4.1 ft) above the ground and with its door facing north, so that the sun's direct beam cannot fall upon the sensor when the door is opened. The instrument shelter shall be painted white outside and shall not be closer to any obstruction than twice the height of the obstruction itself (i.e., trees, fences, buildings, etc.) [15].

7.2.3 *Pyranometer.* The pyranometer shall be mounted on the surface parallel to the collector surface in such a manner that it does not cast a shadow onto the collector plate. Precuations should be always taken to avoid subjecting the instrument to mechanical shocks or vibration during the installation. The pyranometer should be oriented so that the emerging leads or the connector are located north of the receiving surface (in the Northern Hemisphere) or are in some other manner shaded. This minimizes heating of the electrical connections by the sun.

Care should also be taken to minimize reflected and reradiated energy from the solar collector onto the pyranometer. Some pyranometers come supplied with shields. This should be adjusted to be parallel to and to lie just below the plane of the thermopile. Some pyranometers not supplied with a shield may be susceptible to error due to reflections by radiation that originates below the plane of the thermopile. Precautions can be taken by constructing a cylindrical shield, the top of which should be coplaner with the thermopile [5].

7.2.4 *Test ducts.* The air inlet duct, between the air flow measuring apparatus and the solar collector, shall have the same cross-sectional dimensions as the inlet manifold to the solar collector. The air outlet duct, between the solar collector and the reconditioning

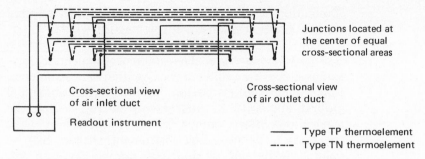

FIG. A.3 Schematic of the thermopile arrangement used to measure the temperature difference across the solar collector.

apparatus, shall have the same cross-sectional dimensions as the outlet manifold from the solar collector.*

7.2.5 *Temperature measurement across the solar collector.* A thermopile shall be used to measure the difference between the inlet air temperature and outlet air temperature of the solar collector. It shall be constructed from calibrated type T thermocouple wire all taken from a single spool. No extension wires are to be used in either its fabrication or installation. The wire diameter must be no larger than 0.51 mm (24 AWG) and the thermopile shall be fabricated as shown in Figure A.3. There shall be a minimum of six junctions in the air inlet test duct and six junctions in the air outlet test duct. These junctions shall be located at the center of equal cross-sectional areas.

During all tests, the variation in temperature at a given cross section of the air inlet and air outlet test ducts shall be less than ±0.5°C (±0.9°F) at the location of the thermopile junctions. The variation shall be checked prior to testing utilizing instrumentation and procedures outlined in reference [6]. If the variation exceeds the limits above, mixing devices shall be installed to achieve this degree of temperature uniformity. Reference [16] discusses the positioning and performance of several types of air mixers.

*The performance of air heaters is expected to be affected by the ductwork entering and leaving the solar collector considerably more so than in the case of solar collectors using a liquid as the transfer fluid.

The ends of the thermopile should be located as near as possible to the inlet and outlet of the solar collector. The air inlet and air outlet ducts shall be insulated in such a manner that the calculated heat loss or gain to or from the ambient air would not cause a temperature change for any test of more than 0.05°C (0.09°F) between the temperature measuring locations and the collector.

7.2.6 *Temperature measurements.* Sensors and read-out devices meeting the accuracy requirements of Section 6 and giving a continuous reading shall be used to measure the temperature at the locations in the air inlet and air outlet ducts shown in Figure A.2. Reference [6] should be followed in making these measurements.

7.2.7 *Duct pressure measurements.* The static pressure drop across the solar collector shall be measured using a manometer as shown in Figures A.2 and A.4 [12]. Each side of the manometer shall be connected to four externally manifolded pressure taps on the air inlet and air outlet ducts. The pressure taps should consist of 6.4 mm (1/4 inch) nipples soldered to the duct and centered over 1 mm (0.040 inch) diameter holes. The edges of these holes on the inside surfaces

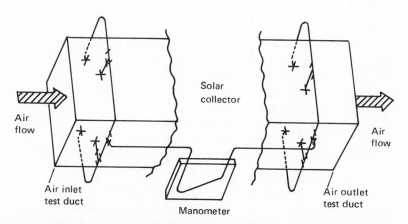

FIG. A.4 Schematic representation of the measurement of pressure drop across the solar collector when air is the transfer fluid.

of the ducts should be free of burrs and other surface irregularities [11].

7.2.8 *Air flow measuring apparatus.* Where the air flow rate is sufficiently large, it shall be measured with the nozzle apparatus discussed in section 7 of reference [11]. As shown in Figure A.5, this apparatus consists basically of a receiving chamber, a discharge chamber and an air flow measuring nozzle. The distance from the center of the nozzle to the side walls shall not be less than 1 1/2 times the nozzle throat diameter, and the diffusion baffles shall be installed in the receiving chamber at least 1 1/2 nozzle throat diameters upstream of the nozzle and 2 1/2 nozzle throat diameters downstream of the nozzle. The apparatus should be designed so that the nozzle can be easily changed and the nozzle used on each test shall be selected so that the throat velocity is between 15 m/s (2960 fpm) and 35 m/s (6900 fpm). When nozzles are constructed in accordance with Figure A.6 and

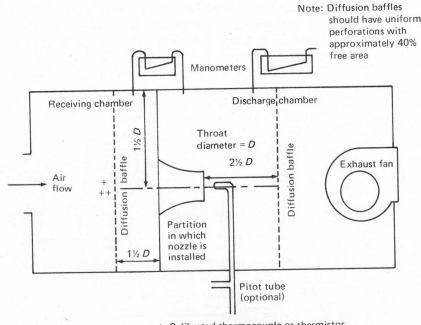

FIG. A.5 Nozzle apparatus for measuring air flow rate.

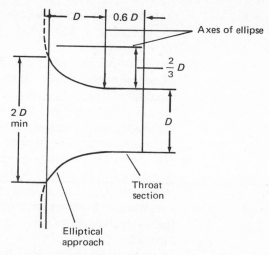

FIG. A.6 Air flow measuring nozzle.

installed in accordance with Section 7.2.9 of this Standard, the discharge coefficient may be assumed to be as follows:

Reynolds Number, N_{Re}	Coefficient of Discharge, C
20,000	0.96
50,000	0.97
100,000	0.98
150,000	0.98
200,000	0.99
250,000	0.99
300,000	0.99
400,000	0.99
500,000	0.99

If the throat diameter of the nozzle is 0.13 m (5 in.) or larger, the discharge coefficient may be assumed to be 0.99. For nozzles smaller than 0.05 m (2 in.) and where a more precise discharge coefficient than given above is desired, the nozzle should be calibrated. The area of the nozzle shall be determined by measuring its diameter to an accuracy of ±0.20% in four places approximately 45 degrees apart around the nozzle in each of two planes through the nozzle throat, one at the outlet and the other in the straight section near the radius [11].

Where the nozzle apparatus is used, an exhaust fan capable of providing the desired flow rates through the solar collector shall be installed in the end wall of the discharge chamber rather than separate from the air flow measuring apparatus as shown in Figure A.2. The dry and wet bulb temperature of the air entering the nozzle shall be measured in accordance with reference [6]. The velocity of the air passing through the nozzle shall be determined by either measuring the velocity head by means of a commercially available pitot tube or by measuring the static pressure drop across the nozzle with a manometer. If the latter method is used, one end of the manometer shall be connected to a static pressure tap located flush with the inner wall of the discharge chamber, or preferably, several taps in each chamber should be manifolded to a single manometer. A means shall also be provided for measuring the absolute pressure of the air in the nozzle throat.

Where the air flow rate is sufficiently small so that a nozzle constructed and installed in accordance with the requirements above would have a throat diameter of smaller than 0.025 m (1 in.), the above configuration should not be used and the air flow measuring apparatus as shown in Figure A.2 should consist of a calibrated flow element* where at least 10 pipe diameters of upstream and downstream pipe section have been included in the calibration.**

7.2.9 *Air leakage.* Air leakage through the air flow measuring apparatus, air inlet test duct, the solar collector and the air outlet test duct shall not exceed ±1.0% of the measured air flow.

---

*Usually an orifice, venturi, or flow nozzle.

**For small flow elements, the discharge coefficients associated with elements varies considerably from those associated with the larger elements. In addition, for small pipe or duct sizes, the ratio of pipe circumference to pipe area becomes large and the characteristics of the upstream and downstream pipe sections affect the behavior of the element itself.

7.2.10 *Air reconditioning apparatus.* The reconditioning apparatus shall not control the dry bulb temperature of the transfer medium entering the solar collector to within ±1.0°C (±1.8°F) of the desired test values at all times during the tests. Its heating and cooling capacity shall be selected so that dry bulb temperature of the air entering the reconditioning apparatus may be raised or lowered the required amount to meet the applicable test conditions in Section 8.

SECTION 8. TEST PROCEDURE AND CALCULATIONS

8.1   General

The performance of the solar collector is determined by obtaining values of instantaneous efficiency for a large combination of values of incident insolation, ambient temperature, and inlet fluid temperature. This requires experimentally measuring the rate of incident solar radiation onto the solar collector as well as the rate of energy addition to the transfer fluid as it passes through the collector, all under quasi-steady conditions.

8.2   Instantaneous Efficiency

It has been shown and discussed by a number of investigators [17, 18, 19 and 20] that the performance of flat plate solar collector operating under steady conditions can be successfully described by the following relationship:

$$\frac{q_u}{A} = I(\tau\alpha)_e - U_L(t_p - t_a) \tag{1}$$

A very similar equation can be used to describe the performance of concentrating collectors [21, 22 and 23]. Equation (1) becomes modified as follows [21]:

$$\frac{q_u}{A_a} = I(\tau\alpha)_e \rho\gamma - U_L \frac{A_r}{A_a}(t_r - t_a) \tag{2}$$

To assist in obtaining detailed information about the performance of collectors and to prevent the necessity of determining some average surface temperature, it has been convenient to introduce a parameter F' where

$F' \equiv$ actual useful energy collected/
useful energy collected if the
entire collector surface were
at the average fluid temperature

Introducing this factor into equation (1) results in

$$\frac{q_u}{A} = F'\left[I(\tau\alpha)_e - U_L\left(\frac{t_{f,i} + t_{f,e}}{2} - t_a\right)\right] \tag{3}$$

If the solar collector efficiency can be defined as

$\eta \equiv$ actual useful energy collected/
solar energy incident upon or
intercepted by the collector

or in equation form

$$\eta = \frac{q_u/A}{I}, \tag{4}$$

then the efficiency of the flat-plate collector is given by:

$$\eta = F'(\tau\alpha)_e - F'U_L\frac{[(t_{f,i} + t_{f,e}) - t_a]/2}{I} \tag{5}$$

Equation (5) indicates that if the efficiency is plotted against an appropriate $\Delta t/I$, a straight line will result where the slope is some function of U_L and the y intercept is some function of $(\tau\alpha)_e$. In reality U_L is not a constant but rather a function of the temperature of the collector and of the ambient weather conditions. In addition, the product $(\tau\alpha)_e$ varies with incident angle to the collector.

The procedures outlined in this document have been developed in an attempt to control the test conditions so that a well defined efficiency "curve" can be obtained with a minimum of scatter. Figure A.7 shows typical test results taken from reference [24] for two flat-plate collectors using air as the transfer fluid. The collector tests were conducted outside and the scatter about the two lines in each figure indicates ". . . apart from experimental errors, the order of variation on account of the variations in heat loss coefficient U_L, and the parameter F' due to variations in ambient wind

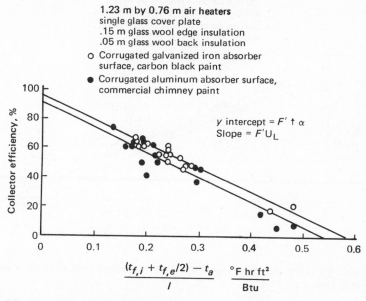

FIG. A.7 Efficiency curves for two flat-plate collectors using air as the transfer fluid (reference [24]).

speed and sky temperatures." Figure A.10 was taken from reference [25] and is for a flat-plate collector using water as the transfer fluid. There is less scatter due to the fact that the tests were conducted indoors using a "solar simulator."

The curves shown in Figures A.7 and A.10 are duplicates of those reported in references [24] and [25], respectively. The abscissa in the first case is in metric units and in the second, English units. The curves to be presented in the test report described herein should be done so the abscissa is either in the SI units of $(°C \cdot m^2)/W$ (as in Figures A.8 and A.11) or as shown in Figures A.9 and A.12. Here the experimentally determined temperature difference has been divided by the difference in temperature between the boiling point and freezing point on the respective scale (100°C, 180°F) and the insolation has been divided by the solar constant, I_{SC} (1353 W/m^2), in appropriate units [26]. The result is an abscissa whose units are dimensionless.

It is expected that a "straight-line" representation will suffice for most conventional flat-plate collectors but that an

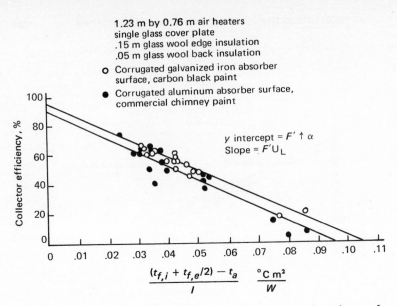

FIG. A.8 Efficiency curves for two flat-plate collectors using air as the transfer fluid (reference [24]).

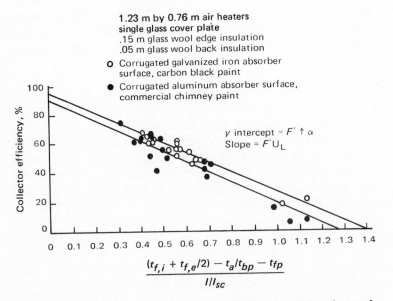

FIG. A.9 Efficiency curves for two flat-plate collectors using air as the transfer fluid (reference [24]).

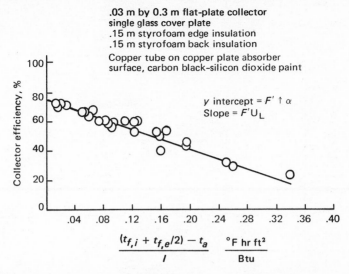

FIG. A.10 Efficiency curve for a flat-plate collector using water as the transfer fluid (reference [25]).

attempt to represent the performance of a concentrating collector on such a plot will require the use of a "higher-order fit" due to the larger variation in U_L and the product $(\tau\alpha)_e$.

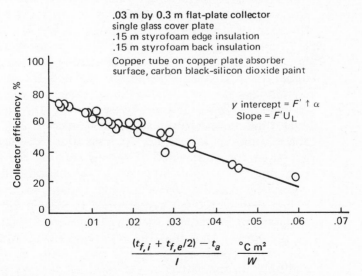

FIG. A.11 Efficiency curve for a flat-plate collector using water as the transfer fluid (reference [25]).

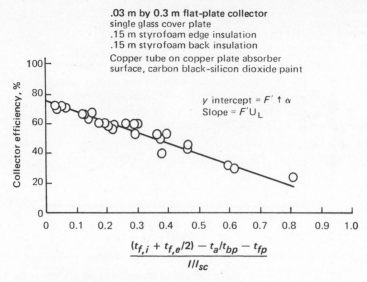

.03 m by 0.3 m flat-plate collector
single glass cover plate
.15 m styrofoam edge insulation
.15 m styrofoam back insulation

Copper tube on copper plate absorber
surface, carbon black-silicon dioxide paint

y intercept = $F' \uparrow \alpha$
Slope = $F'U_L$

$$\frac{(t_{f,i} + t_{f,e}/2) - t_a/t_{bp} - t_{fp}}{I/I_{sc}}$$

FIG. A.12 Efficiency curve for a flat-plate collector using water as the transfer fluid (reference [25]).

8.3   Testing Procedure

The testing of the solar collector shall be conducted in such a way that an "efficiency curve" is determined for the collector under test conditions described in Section 5. and 8.3. At least four different values of inlet fluid temperature shall be used to obtain the values of $\Delta t/I$. Ideally the inlet fluid temperature should correspond to 10, 30, 50, and 70°C (18, 54, 90, and 126°F) *above the ambient temperature*; however the values that can realistically be used will depend upon the particular collector design and the environmental conditions at the location and time of year when the collector is being tested. Consequently, the four different inlet fluid temperatures selected should be as close to the above values as is feasible. At least four "data points" shall be taken for each value of $t_{f,i}$; two during the time period preceding *solar noon* and two in the period following *solar noon*, the specific periods being chosen so that the data points represent times symmetrical to solar noon. This latter requirement is made so that any "transient effects" that may be present will not bias the test results when they are used for design purposes. All test data shall be reported in addition to the fitted curve (see Section 9) so that any difference in efficiency due solely to the operating temperature level of the collector can be discerned in the test report.

The curve shall be established by "data points" that represent 15 minute integrated efficiency values. In other words, the integrated value of incident solar energy will be divided into the integrated value of energy obtained from the collector to obtain the efficiency value for that "instant". Care should be taken to insure that the incident solar energy is steady for each 15 minute segment during which an efficiency value is calculated. Either electronic integrators or continuous pen strip chart recorders may be used to determine the integrated values of incident solar radiation and temperature rise across the collector. However, a strip chart recorder with a recommended chart speed of 30 cm/hr must always be used to monitor the output of the pyranometer to insure that the incident radiation has remained steady during the 15 minute segment. Figures A.13 and A.14 show a strip chart recording of incident solar radiation on a horizontal surface at the National Bureau of Standards site in Gaithersburg, Maryland. Whereas the conditions of Figure A.13 would be perfectly acceptable for obtaining efficiency values, those of Figure A.14 would not be.*

*One or two "blips" of 10 s or less occurring during the 15 minute period such as at 12:18 in Figure A.11 is acceptable.

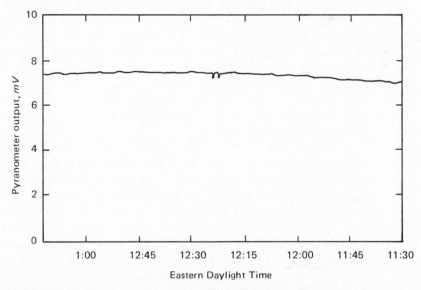

FIG. A.13 Incident solar radiation on a horizontal surface at the National Bureau of Standards site in Gaithersburg, Maryland, March 13, 1974.

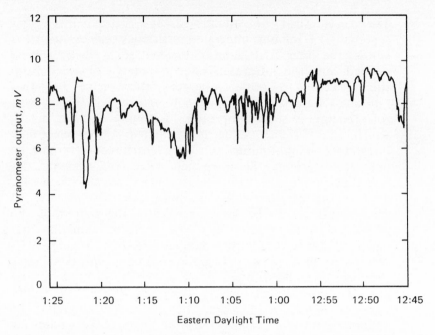

FIG. A.14 Incident solar radiation on a horizontal surface at the National Bureau of Standards site in Gaithersburg, Maryland, March 11, 1974.

The surface of the collector cover plate (if present) as well as exposed envelope of the pyranometer should be wiped clean and dry prior to the tests. If local pollution or sand has formed a deposit on the transparent surfaces, the wiping should be carried out very gently, preferably after blowing off most of the loose material or after wetting it a little, in order to prevent scratching of the surface. This is particularly important for the pyranometer since such abrasive action can appreciably alter the original transmission properties of the enclosing envelope.

The pyranometer shall be checked prior to testing to see if there is any accumulation of water vapor enclosed within the glass cover. The use of "wet" pyranometers (where moisture is visible) shall not be allowed.

In order to obtain sufficiently good "quasi-steady" conditions for the solar collection process, the collector should stand in the sun under no flow conditions until the contained fluid heats up to a temperature equivalent to or slightly greater than the inlet fluid temperature for the test.

The transfer fluid should then be circulated through the collector at the appropriate temperature level for at least 30 minutes* prior to the period in which data will be taken to calculate the efficiency values. During this period, a check should be made to insure that the flow rate of the transfer fluid does not vary by more than ±1% and that the incident solar radiation is steady as described above.

The flow rate of transfer fluid through the collector shall be standardized at one value for all data points. The recommended value of flow rate per unit area (transparent frontal or aperture) for tests are 0.02 kg/(s · m²) (14.7 lbm/(h · ft²)) when a liquid is the transfer fluid and 0.01 m³/(s · m²) (1.96 cfm per ft²) of standard are when the transfer fluid is air. It is recognized that in some cases the collector will have been designed for a flow rate much different than specified above. In such cases, the design flow rate should be used.

In order to determine and report the fraction of the incident solar radiation that is diffuse for each efficiency value, the sensing element of the pyranometer shall be shaded from the direct beam of sun just prior and just following each 15 minute testing period and the value of the incident radiation determined.** This shall be accomplished by using a small disk attached to a slender rod held on a direct line between the pyranometer and the sun. The disk should be just large enough to shade the sensing element alone. In reference [5], this is accomplished by a disk 100 mm in diameter and held at a distance of 1 m from the sensing element.***

8.4   Calculation of Instantaneous Efficiency

For each 15 minute segment for which an efficiency value is to be determined, the value is calculated using the equation:

---

*30 minutes is felt to be sufficient for typical tube and sheet type solar collectors using water as the transfer fluid. For those collectors having higher thermal capacity, a longer time period may be necessary.

**A normal incidence pyrheliometer can be used in lieu of shading the sensing element of the pyranometer.

***This was when using a Moll-Gorszynski Pyranometer made by Kipp and Zonen.

$$\eta = \frac{\left[\dot{m}c_p \int_{\tau_1}^{\tau_2} (t_{f,e} - t_{f,i})\, d\tau\right] \Big/ A_a}{\int_{\tau_1}^{\tau_2} I\, d\tau} \tag{6}$$

The quantities $\dot{m}$ and c_p have been taken out of the integration in the numerator since they remain essentially constant during the test. Note that the collector area used for the calculation is not the absorbing surface area but rather the transparent frontal area or aperture area.

At least sixteen data points shall be obtained for the establishment of the "efficiency curve" and an equation for the curve shall be obtained using the standard technique of a least-squares fit to a second-order polynomial.*

8.5   An Experimental Check

As an independent check on the experimental results, the inlet temperature, $t_{f,i}$, and the outlet temperature, $t_{f,e}$, of the collector shall be recorded on continuous pen strip chart recorders. The quantity $\tau_1 \int \tau_2 \, (t_{f,e} - t_{f,i})\, d\tau$ shall be approximated using these recordings and compared with the identical quantity obtained by using the primary method which measures the temperature difference directly.

8.6   Calculation of Air Flow Rate

The air flow rate through the nozzle is calculated by the following equations:

$$Q_{mi} = 1.41 C A_n (P_v v_n')^{0.5} \tag{7}$$

$$v_n' = 10.1 \times 10^4 \frac{V_n}{P_n} (1 + W_n) \tag{8}$$

The air flow rate of standard air is then:

$$Q_s = \frac{Q_{mi}}{1.2 v_n'} \tag{9}$$

*One should consult any standard text discussing analysis of experimental data for a presentation of this technique (i.e., [27] and [28]).

8.7 Calculation of Nozzle Reynolds Number

The Reynolds number is calculated as follows:

$$N_{Re} = fV_a D \qquad (10)$$

The temperature factor is as follows:

Temperature, $°C$	Factor, f
−6.7	78275
+4.4	72075
+15.6	67425
+26.7	62775
+37.8	58125
+48.9	55025
+60.0	51925
+71.1	48825

8.8 Calculation of Theoretical Power Requirements

In order to calculate the theoretical power required to move the transfer fluid through the solar collector, the following equation shall be used:

$$P_{th} = \frac{\dot{m}\Delta P}{\rho} \qquad (11)$$

SECTION 9. DATA TO BE RECORDED AND TEST REPORT

9.1 Test Data

Table A.13 lists the measurements which are to be made at the beginning of the testing day and during the individual tests to obtain an efficiency "data point".

9.2 Test Report

Table A.14 specifies the data and information that shall be reported in testing the solar collector.

TABLE A.13 Test Data to Be Recorded

Item	Test involving air as the transfer fluid	Test involving a liquid as the transfer fluid
Date	X	X
Observer(s)	X	X
Equipment name plate data	X	X
Collector tilt angle	X	X
Collector azimuth angle (as a function of time if movable)	X	X
Collector aperture area or frontal transparent area	X	X
Local standard time, at the beginning of collector warm-up and at the beginning and end of each 15 minute test period	X	X
Barometric pressure	X	
Ambient air temperature (at the beginning and end of each 15 minute test period)	X	X
$\Delta t = t_{f,e} - t_{f,i}$ across solar collector (either as a continuous function of time or as a 15 minute integrated quantity)	X	X
Inlet temperature, $t_{f,i}$ (as a continuous function of time)	X	X
Outlet temperature, $t_{f,e}$ (as a continuous function of time)	X	X
Liquid flow rate		X
Gauge pressure at solar collector inlet		X
Gauge pressure at nozzle throat	X	
Nozzle throat diameter	X	
Velocity pressure at nozzle throat or static pressure difference across nozzle	X	
Dry bulb temperature at nozzle throat	X	
Wet bulb temperature at nozzle throat	X	
Pressure drop across solar collector	X	X
Height of the collector outlet above the collector inlet	X	X
Wind velocity near the collector surface or aperture (15 minute average)	X	X
I, the incident solar radiation onto the collector (as a continuous function of time and as a 15 minute integrated quantity if desired)	X	X
I_d, the diffuse component of the solar radiation onto the collector (at the beginning of the 15 minute period and after the completion of the 15 minute period)	X	X

238

TABLE A.14 Data to Be Reported

General Information

Manufacturer or Project Name

Collector Model No.

Construction details of the collector:

Gross dimensions and area
Area of absorbing surface
Cover plate,* dimensions, materials, optical properties (if known)
Reflector,* dimensions and shape, materials, optical properties (if known)
Absorber plate, dimensional layout and configuration of flow path, absorptivity to short
   wave radiation (if known), emissivity for long wave radiation (if known), description of
   coating (including maximum allowable temperature if known)
Air space(s),* thickness and description of contained gas or construction
Insulation,* material, thickness, thermal properties

Transfer fluid used and its properties

Weight of collector per m² of gross cross-sectional area

Volumetric capacity of the collector per m² of gross cross-sectional area if designed to operate with a liquid as the transfer fluid

Normal operating temperature range

Minimum transfer fluid flow rate

Maximum transfer fluid flow rate

Maximum operating pressure

Description of apparatus, including flow configuration and instrumentation used in testing
   (include photographs)

Description of the mounting of the collector for testing

Location of tests (longitude, latitude, and elevation above sea level)

Efficiency Tests

A plot of the efficiency versus $(t_{f,i} + t_{f,e} - t_a/2)/I$

An equation for the efficiency curve

For each "data point"

$\dot{m}$
c_p
$\int_{\tau_1}^{\tau_2} (t_{f,e} - t_{f,i})\, d\tau$
$\int_{\tau_1}^{\tau_2} I\, d\tau$
pressure drop across the solar collector
collector tilt angle
collector azimuth angle (as a function of time if movable)
incident angle
inlet fluid temperature, $t_{f,i}$
percentage of incident radiation that is diffuse
wind speed near the collector surface or aperture

*If applicable

USCOMM-NBS-DC-243

SECTION 10. NOMENCLATURE

A Cross-sectional area, m^2

A_a Transparent frontal area for a flat-plate collector or aperture for a concentrating collector, m^2

A_n Area of nozzle, m^2

A_r Absorbing or receiving area of the concentrating solar collector, m^2

C Nozzle throat diameter, m

c_p Specific heat of the transfer fluid, $J/(kg \cdot {}^\circ C)$

D Nozzle throat diameter, m

f Temperature factor for the calculation of nozzle N_{Re}

F' Solar collector efficiency factor

h Outside surface heat transfer coefficient (includes radiation and convection) for the solar collector, $W/(m^2 \cdot {}^\circ C)$

I Total solar energy incident upon the plane of the solar collector per unit time per unit area, W/m^2

I_d Diffuse solar energy incident upon the plane of the solar collector per unit time per unit area, W/m^2

I_{sc} Solar constant, 1353 W/m^2

$\dot{m}$ Mass flow rate of the transfer fluid, kg/s

N_{Re} Reynolds number

P_n Absolute pressure at the nozzle throat, N/m^2

P_{th} Theoretical power required to move the transfer fluid through the solar collector, W

P_v Velocity pressure at the nozzle throat or the static pressure difference across the nozzle, N/m^2

ΔP Pressure drop across the solar collector, N/m^2

$Q_{mi.}$ Measured air flow rate, m^3/s

Q_s Standard air flow rate, m^3/s

q_u Rate of useful energy extraction from the solar collector, W

t_a Ambient air temperature, ${}^\circ C$

t_{bp} Temperature of the boiling point on a temperature scale, ${}^\circ C$ or ${}^\circ F$

$t_{f,e}$ Temperature of the fluid leaving the collector, ${}^\circ C$

$t_{f,i}$ Temperature of the fluid entering the collector, ${}^\circ C$

t_{fp} Temperature of freezing point on a temperature scale, ${}^\circ C$ or ${}^\circ F$

t_p Average temperature of the absorber surface of the solar collector, ${}^\circ C$

t_r Average temperature of the absorber surface of the concentrating solar collector, ${}^\circ C$

Δt Temperature difference, ${}^\circ C$

U_L Heat transfer loss coefficient for the solar collector, $W/(m^2 \cdot {}^\circ C)$

V_a Velocity of the air at the nozzle throat, m/s

v_n Specific volume of the air at dry and wet bulb temperature conditions existing at the nozzle but at standard barometric pressure, m^3/kg dry air

v_n' Specific volume of the air at the nozzle, m^3/kg dry air

W_n Humidity ratio at the nozzle, kg H_2O/kg dry air

α Absorptance of the solar collector absorbing surface to solar radiation

γ The fraction of specularly reflected radiation from the reflector which is intercepted by the solar collector absorbing surface

η Solar collector efficiency, %

ρ Specular reflectance of the solar collector reflector, or density, kg/m^3

τ Time, s, or transmittance of the solar collector cover plate

$(\tau\alpha)_e$ Effective transmission-absorptance factor for the solar collector

τ_1 Time at the beginning of a 15 minute test period, s

τ_2 Time at the end of a 15 minute test period, s

SECTION 11. REFERENCES

1. Morrison, C. A., and E. A. Farber, "Development and Use of Insolation Data for South Facing Surfaces in Northern Latitudes", paper presented at the ASHRAE Symposium, Solar Energy Applications, Montreal, Canada, June 22, 1974.

2. Personal communication with J. I. Yellott, Arizona State University, Tempe, Arizona, data to be published in 1975 revised edition of the ASHRAE publication, *Low Temperature Engineering Application of Solar Energy.*

3. *ASHRAE Handbook of Fundamentals*, American Society of Heating, Refrigerating and Air-Conditioning Engineers, Inc., 345 East 47th Street, New York, N. Y. 10017, 1972.

4. Threlkeld, J. L., *Thermal Environmental Engineering*, Prentice-Hall, Englewood Cliffs, New Jersey, Second Edition, 1970.

5. Latimer, J. R., "Radiation Measurement", *International Field Year for the Great Lakes, Technical Manual Series, No. 2*, The Secretariat, Canadian National Committee for the International Hydrological Decade, No. 8 Building, Carling Avenue, Ottawa, Canada, 1971.

6. "Standard Measurements Guide: Section on Temperature Measurements", *ASHRAE Standard 41-66, Part 1*, American Society of Heating, Refrigerating and Air-Conditioning Engineers, Inc., 345 East 47th Street, New York, N. Y. 10017, January, 1966.

7. "American Standard for Temperature Measurement Thermocouples C96.1-1964" (R 1969), American National Standards Institute, 1969.

8. "Instruments and Apparatus, Part 3, Temperature Measurement", Supplement to the ASME Power Test Codes, American Society of Mechanical Engineers, 345 East 47th Street, New York, N. Y. 10017, March, 1961.

9. Davis, J. C., "Radiation Errors in Air Ducts Under Isothermal Conditions Using Thermocouples, Thermistors, and a Resistance Thermometer", *NBS Building Science Series 26*, November, 1969. (Available from the Superintendent of Documents, U. S. Government Printing Office, Washington, D. C. 20402—order by SD Catalog No. C 13.29/2:26, $0.25.)

10. Davis, J. C., Faison, T. K., and P. R. Achenbach, "Errors in Temperature Measurement of Moving Air Under Isothermal Conditions Using Thermocouples, Thermistors, and Thermometers, *ASHRAE Transactions*, Vol. 72, Part 1, 1966.

11. "Methods of Testing for Rating Unitary Air Conditioning and Heat Pump Equipment", *ASHRAE Standard 37-69*, American Society of Heating, Refrigerating and Air-Conditioning Engineers, Inc., 345 East 47th Street, New York, N. Y. 10017, April, 1969.

12. "Instruments and Apparatus, Part 2, Pressure Measurement", Supplement to the ASME Power Test Codes, American Society of Mechanical Engineers, 345 East 47th Street, New York, N. Y. 10017, July, 1964.

13. "Instruments and Apparatus, Part 5, Measurement of Quantity of Materials, Chapter 4, Flow Measurement", Supplement to the ASME Power Test Codes, American Society of Mechanical Engineers, 345 East 47th Street, New York, N. Y. 10017, February, 1959.

14. "Fluid Meters, Their Theory and Application", 5th edition, American Society of Mechanical Engineers, 345 East 47th Street, New York, N. Y. 10017, 1959.

15. "Instructions for Climatological Observations, Circular B", United States Weather Bureau Service, 11th edition, January, 1962.

16. Faison, T. K., Davis, J. C., and P. R. Achenbach, "Performance of Louvered Devices as Air Mixers, *NBS Building Science Series 27*, March, 1970. (Available from the Superintendent of Documents, U. S. Government Printing Office, Washington, D. C. 20402—order by SD Catalog No. C 13.29/2:27, $0.30.)

17. Hottel, H. C., and B. B. Woertz, "The Performance of Flat-Plate Solar-Heat Collectors", *ASME Transactions*, Vol. 64, p. 91, 1942.

18. Hottel, H. C., and A. Whillier, "Evaluation of Flat-Plate Collector Performance", *Transactions of the Conference on the Use of Solar Energy*, Vol. 2, Part I, p. 74, University of Arizona Press, 1958.

19. Bliss, R. W., "The Derivation of Several 'Plate-Efficiency Factors' Useful in the Design of Flat-Plate Solar Heat Collectors", *Solar Energy*, Vol. 3, No. 4, p. 55, 1959.

20. Whillier, A., "Design Factors Influencing Collector Performance", *Low Temperature Engineering Application of Solar Energy*, American Society of Heating, Refrigerating and Air-Conditioning Engineers, Inc., 345 East 47th Street, New York, N. Y. 10017, 1967.

21. Duffie, J. A., and W. A. Beckman, *Solar Energy Thermal Processes*, John Wiley and Sons, 1974.

22. Nevins, R. G., and P. E. McNall, "A High-Flux Low-Temperature Solar Collector", *ASHRAE Transactions*, Vol. 64, pp. 69–82, 1958.

23. Löf, G. O. G., Fester, D. A., and J. A. Duffie, "Energy Balance on a Parabolic Cylinder Solar Reflector", *ASME Transactions*, Vol. 84A, p. 24, 1962.

24. Gupta, C. L., and H. P. Garg, "Performance Studies on Solar Air Heaters", *Solar Energy*, Vol. 11, No. 1, 1967.

25. Simon, F. F., and P. Harlament, "Flat-Plate Collector Performance Evaluation: The Case for a Solar Simulator Approach", *NASA TM X-71427*, October, 1973.

26. Thekaekara, M. P., "Solar Energy Outside the Earth's Atmosphere", *Solar Energy*, Vol. 14, No. 2, p. 109, 1973.

27. Holman, J. P., *Experimental Methods for Engineers*, McGraw-Hill, Second Edition, 1971.

28. Bevington, P. R., *Data Reduction and Error Analysis for the Physical Sciences*, McGraw-Hill, 1969.

Example Determination of Spectral Radiation Properties

This appendix discusses one of several means of determining spectral-radiation properties of an opaque surface. This calculation example uses tabular values of radiation functions to evaluate the effective reflectance of a surface for 1000°R and 10,000°R sources. To illustrate the method, the reflectance at room temperature of a white paint irradiated by a black-body source at 1000°R is compared with the reflectance at the same temperature of the same paint to radiation from an anodized aluminum source at 1000°R. Reflectance of the same paint for solar radiation may also be computed, assuming the sun is a black body at 10,000°R and $\rho_\lambda = 0.75$ below λ of $0.4\,\mu$. The monochromatic reflectance of the paint and the aluminum source are plotted as functions of wavelength in Fig. B.1.

RADIATION FUNCTIONS AND ALUMINUM EMITTANCE

Table B.1 lists λT in increments of $400\mu°R$ in the first column. This λT value, $400\mu°R$, corresponds to wavelength increments of 0.4 for a 1000°R source. Since only 0.03 percent of the total emissive power of a 1000°R source lies below a wavelength of $1.8\,\mu$ (see Table 2.4), the first column starts with a λT value of 1,800. Values of $E_{b,\lambda}/\sigma T^5$ from Table 2.4 are tabulated in the second column, values of λ for $T = 1000°R$ in the third. Then values of spectral emittance of $\epsilon_\lambda (= 1 - \rho_\lambda)$ for anodized aluminum are read from Fig. B.1 and tabulated in the fourth column. Values of ϵ_λ above $15\,\mu$ are assumed equal to 0.8. This extrapolation is necessary because no long-wave emittance data are available. A change of 0.10 in the extrapolated value would not cause an error of more than 0.013 in the total emittance, however, because only 13.2 percent of the total emissive power lies at wavelengths greater than $15\,\mu(\lambda T = 15,000)$ for a 1000°R source.

The total emittance $\bar{\epsilon}_T$ of the aluminum is now calculated. The sum of column 5 to the point where λT is 15,000 is 147.80×10^{-5}. To the sum of column 5 is added the estimated

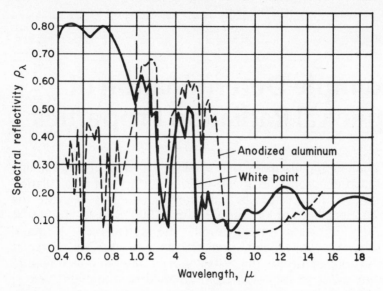

FIG. B.1 Spectral distribution of the monochromatic reflectivity of a white paint and of anodized aluminum. (Extracted from "Thermal Radiation Tables and Application" by R. V. Dunkle, published in *Trans. ASME*, vol. 76, 1955, with permission of the publishers, The American Society of Mechanical Engineers.)

value of ϵ_λ between $\lambda T = 15{,}000$ and $\lambda T = \infty$, denoted by $\bar{\epsilon}_{15{,}000-\infty}$, times the portion of the total emissive power at wavelengths greater than $\lambda = 15\ \mu$. The entire manipulation can be written in the form

$$\bar{\epsilon}_T = \Delta(\lambda T)\left(\sum_0^{\lambda T} \frac{\epsilon_\lambda E_{b,\lambda T}}{\sigma T^5}\right) + \bar{\epsilon}_{15{,}000-\infty}\left(1 - \int_0^{\lambda T} \frac{E_{b,\lambda T}\,d(\lambda T)}{\sigma T^5}\right)$$

$$= 400 \times 147.8 \times 10^{-5} + 0.8 \times 0.132 = 0.697$$

WHITE PAINT REFLECTANCE—BLACK BODY AND ALUMINUM SOURCES

The reflectance of white paint is to be determined for a black-body source at 1000°R. Reflectance values of flat, white paint, taken from Fig. B.1 at appropriate wavelengths, are tabulated in column 6 of Table B.1. Data are available to 21 μ, and the fraction of the total emissive power at wavelengths greater than 21 μ for a black body at 1000°R is 0.059, according to Table 2.4. In each row of column 7 the product of the corresponding values from columns 2 and 6 is listed. These spectral reflected power results shown in column 7 for a black body are also plotted in

TABLE B.1 Computing Effective Emissivity and Reflectivity from Spectral Data

λT	$\dfrac{E_{b,\lambda} \times 10^5}{\sigma T^5}$	λ	Anodized aluminum		White paint		
			ϵ	$\dfrac{\epsilon E_{b,\lambda} \times 10}{\sigma T^5}$	ρ_λ	$\dfrac{\rho_\lambda E_{b,\lambda} \times 10^5}{\sigma T^5}$	$\dfrac{\epsilon \rho_\lambda E_{b,\lambda} \times 10}{\sigma T^5}$
1,800	0.21	1.8	0.34	0.07	0.57	0.12	0.04
2,200	1.04	2.2	0.32	0.33	0.50	0.52	0.17
2,600	2.75	2.6	0.60	1.65	0.40	1.10	0.66
3,000	5.08	3.0	0.90	4.57	0.19	0.97	0.87
3,400	7.50	3.4	0.65	4.88	0.10	0.75	0.48
3,800	9.59	3.8	0.52	4.99	0.34	3.26	1.69
4,200	11.14	4.2	0.44	4.90	0.48	5.35	2.35
4,600	12.10	4.6	0.42	5.08	0.42	5.08	2.13
5,000	12.53	5.0	0.39	4.89	0.51	6.39	2.49
5,400	12.55	5.4	0.42	5.27	0.44	5.52	2.32
5,800	12.26	5.8	0.48	5.88	0.10	1.22	0.59
6,200	11.77	6.2	0.54	6.36	0.15	1.77	0.96
6,600	11.13	6.6	0.48	5.34	0.18	2.00	0.96
7,000	10.43	7.0	0.52	5.42	0.09	0.94	0.49
7,400	9.71	7.4	0.80	7.77	0.10	0.97	0.78
7,800	8.98	7.8	0.93	8.35	0.09	0.81	0.75
8,200	8.28	8.2	0.945	7.82	0.06	0.50	0.47
8,600	7.61	8.6	0.945	7.19	0.07	0.53	0.50
9,000	6.98	9.0	0.940	6.56	0.10	0.70	0.66
9,400	6.40	9.4	0.945	6.05	0.14	0.90	0.85
9,800	5.86	9.8	0.940	5.51	0.13	0.76	0.71
10,200	5.37	10.2	0.94	5.06	0.13	0.70	0.66
10,600	4.92	10.6	0.94	4.62	0.14	0.69	0.65
11,000	4.50	11.0	0.935	4.21	0.16	0.72	0.67
11,400	4.13	11.4	0.93	3.84	0.19	0.78	0.73
11,800	3.79	11.8	0.92	3.49	0.22	0.83	0.76
12,200	3.48	12.2	0.92	3.20	0.225	0.78	0.72
12,600	3.20	12.6	0.885	2.83	0.22	0.70	0.62
13,000	2.94	13.0	0.885	2.60	0.20	0.59	0.52
13,400	2.71	13.4	0.890	2.41	0.17	0.46	0.41
13,800	2.50	13.5	0.86	2.15	0.14	0.35	0.30
14,200	2.31	14.2	0.85	1.96	0.14	0.32	0.27
14,600	2.13	14.6	0.825	1.76	0.13	0.28	0.23
15,000	1.97	15.0	$0.80 \times \frac{1}{2}$	0.79	0.115	0.23	0.10
15,400	1.83	15.4	...	...	0.12	0.22	0.18
15,800	1.69	15.8	...	...	0.135	0.23	0.18
16,200	1.57	16.2	...	...	0.15	0.24	0.19
16,600	1.46	16.6	...	...	0.17	0.25	0.20
17,000	1.36	17.0	...	...	0.18	0.24	0.19
17,400	1.26	17.4	...	...	0.18	0.23	0.18
17,800	1.18	17.8	...	...	0.18	0.21	0.17
18,200	1.10	18.2	...	...	0.18	0.20	0.16
18,600	1.03	18.6	...	...	0.18	0.19	0.15
19,000	0.96	19.0	...	...	0.20	0.19	0.15
19,400	0.90	19.4	...	...	0.19	0.17	0.14
19,800	0.84	19.8	...	...	0.18	0.15	0.12
20,200	0.79	20.2	...	...	0.17	0.13	0.10
20,600	0.75	20.6	...	...	0.15	0.11	0.09
21,000	0.70	21.0	...	...	$0.15 \times \frac{1}{2}$	0.05	0.04
Sum				147.80		50.45	29.88

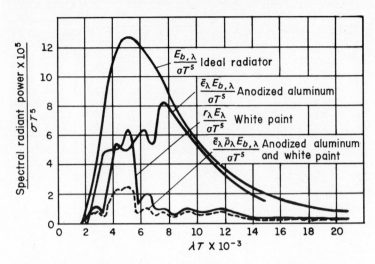

FIG. B.2 Spectral distribution of emissive power for anodized aluminum at 1000°R and of reflectivity of a white paint irradiated by a black source and by an anodized aluminum source at 1000°R. (Extracted from "Thermal Radiation Tables and Application" by R. V. Dunkle, published in *Trans. ASME*, vol. 76, 1955, with permission of the publishers, The American Society of Mechanical Engineers.)

Fig. B.2. If an estimated reflectance of 0.15 at wavelengths greater than 21 μ is used, the total effective reflectance $\bar{\rho}_T$ of the paint for a black-body source at 1000°R is

$$\bar{\rho}_T = 400 \times 50.45 \times 10^{-5} + 0.15 \times 0.059 = 0.211$$

Finally, the reflectance of the same paint is calculated when irradiated by an anodized aluminum source at 1000°R. In column 8 the products of columns 7 and 4 are tabulated. The sum of column 8, with the same extrapolation as used before, divided by the total emittance of the aluminum source (0.697), gives the effective reflectance of the paint for radiation from an anodized aluminum source at 1000°R to be 0.181. Under these conditions the reflectance is 15 percent less than the radiation from a black-body source at the same temperature.

The results of this exercise for 1000°R sources are:

Material	Average property value, $\bar{\rho}_T$ or $\bar{\epsilon}_T$
Anodized aluminum	$\bar{\epsilon}_T = 0.697$
White paint	Black body source: $\bar{\rho}_T = .211$
	Aluminum source: $\bar{\rho}_T = .181$

Reference Tables and Charts

TABLE C.1 Weather Data for Selected U.S. Cities*

Location	Yearly averages and totals					Precipi-tation, days	Snow per season, in.	32°F or less, days	Particulate pollution, $\mu g/m^3$
	Sun total, %	Wind, avg, mi/hr	Rain total, in.	Relative humidity, %					
				7 A.M.	1 P.M.				
AL Mobile	61	9.5	68.13	85	57	123	0.4	21	
AK Juneau	31	8.6	54.62	87	81	220	102.7	149	
AZ Phoenix	86	5.8	7.20	55	33	35		16	123
AR Little Rock	62	8.2	48.66	84	56	102	5.5	68	
CA Los Angeles	73	7.3	12.63	77	61	36			104
Sacramento	78	8.5	16.29	83	65	57		20	
San Francisco	67	10.5	18.69	84	69	62		4	59
CO Denver	70	9.1	14.81	68	40	87	58.3	164	135
CT Hartford	57	9.1	42.92	78	53	125	56.3	138	69
DE Wilmington	58	9.0	44.56	79	55	116	22.6	104	127
DC Washington	58	9.4	40.78	73	51	111	18.2	82	77
FL Jacksonville	60	8.9	53.36	87	55	116	0.1	11	
Miami	66	9.0	59.76	84	61	128			69
GA Atlanta	61	9.2	47.14	83	57	115	1.8	62	85
HI Honolulu	69	11.5	21.89	75	71	101			42
ID Boise	67	9.0	11.43	70	53	90	21.7	127	105
IL Chicago	57	10.3	33.18	75	58	120	38.4	119	145
Peoria	58	10.4	34.84	83	62	109	22.4	133	
IN Indianapolis	59	9.8	39.25	83	61	121	20.6	122	118
IA Des Moines	60	11.3	30.37	82	63	105	31.9	140	103
KS Wichita	65	12.8	28.41	80	55	83	14.3	112	67
KY Louisville	58	8.3	41.32	82	59	122	18.0	98	127
LA New Orleans	61	8.4	55.90	88	63	112	0.2	12	75
ME Portland	59	8.7	42.85	80	60	126	73.7	162	
MD Baltimore	58	9.8	43.05	77	53	111	25.1	103	118
MA Boston	60	13.2	42.77	72	55	128	42.2	100	
MI Detroit	54	10.1	30.95	78	58	131	31.6	125	129
Sault Ste. Marie	47	9.8	31.22	86	67	164	104.0	179	
MN Duluth	55	11.8	28.97	79	62	134	76.8	186	76
Minneapolis-St. Paul	58	10.7	24.78	81	61	112	44.4	160	74
MS Jackson	59	7.9	50.82	90	58	106	1.8	54	80
MO Kansas City	64	10.0	34.07	77	57	99	20.1	100	116
St. Louis	59	9.5	35.31	81	57	105	17.4	107	213

*Federal government data, published 1971. For monthly data see *Statistical Abstract of the United States: 1971*, 92nd ed., U.S. Bureau of the Census, 1971.

TABLE C.1   Weather Data for Selected U.S. Cities *(continued)*

| | | | | | Relative humidity, % | | | Snow | | |
	Location	Sun total, %	Wind, avg, mi/hr	Rain total, in.	7 A.M.	1 P.M.	Precipitation, days	per season, in.	32°F or less, days	Particulate pollution, $\mu g/m^3$
MT	Great Falls	64	13.2	14.07	66	50	99	55.2	150	
NE	Omaha	62	11.0	27.56	79	59	98	32.0	137	115
NV	Reno	80	6.4	7.15	69	44	48	25.3	188	
NH	Concord	54	6.7	38.80	80	53	125	62.7	169	35
NJ	Atlantic City	56	10.9	42.36	80	54	112	17.8	121	
NM	Albuquerque	77	8.8	8.13	57	37	58	9.8	118	
NY	Albany	54	8.8	35.08	79	56	133	63.9	146	
	Buffalo	53	12.5	35.65	79	63	165	87.2	138	
	New York	59	9.5	42.37	72	56	121	29.9	82	131
NC	Charlotte	66	7.5	43.38	84	53	110	5.4	76	108
	Raleigh	61	7.9	43.58	84	52	113	7.4	88	
ND	Bismarck	62	10.8	15.15	77	55	96	38.1	188	82
OH	Cincinnati	58	7.1	39.51	80	57	132	19.0	99	117
	Cleveland	52	10.9	35.35	79	62	154	50.3	128	128
	Columbus	55	8.5	36.67	81	59	134	28.0	125	106
OK	Oklahoma City	67	13.2	30.82	82	56	82	9.5	85	70
OR	Portland	46	7.7	37.18	87	73	153	8.4	44	84
PA	Philadelphia	58	9.6	42.48	76	54	116	21.9	108	129
	Pittsburgh	53	9.4	36.14	79	57	149	46.0	128	165
PR	San Juan	64	8.5	64.21	82	66	205			
RI	Providence	57	11.0	42.13	73	54	124	40.9	126	80
SC	Columbia	64	7.0	46.82	86	50	109	1.2	55	
SD	Sioux Falls	63	11.1	25.16	81	59	93	41.6	173	69
TN	Memphis	65	9.2	49.73	82	57	103	6.1	60	76
	Nashville	58	7.8	45.15	85	57	118	11.8	75	98
TX	Dallas	65	10.9	34.55	79	55	80	2.1	37	84
	El Paso	83	9.9	7.89	52	35	44	4.4	63	
	Houston	62	10.8	45.95	87	60	103	0.4	13	88
UT	Salt Lake City	70	8.6	13.90	67	47	87	55.1	141	90
VT	Burlington	52	8.9	33.21	77	59	149	73.8	167	46
VA	Norfolk	63	10.6	44.94	79	58	115	7.5	56	94
	Richmond	61	7.7	44.21	82	53	113	15.0	86	
WA	Seattle	45	9.5	38.94	84	75	161	15.8	32	61
	Spokane	57	8.4	17.19	76	62	115	54.6	139	
WV	Charleston	48	6.6	44.43	82	55	147	28.4	101	213
WI	Milwaukee	56	11.8	29.51	81	64	120	43.4	149	124
WY	Cheyenne	64	12.7	15.06	63	40	99	51.6	171	32

TABLE C.2 Weather Conditions for Environmental Design*

Location	Lati-tude, deg†	Eleva-tion, ft	Degree-days heating‡	Winter§ 99%	Winter§ 97.5%	Summer¶ 1% DB	Summer¶ 1% WB	Summer¶ 5% DB	Summer¶ 5% WB
USA									
Alabama, Birmingham	33	610	2,551	19	22	97	94	93	77
Alaska, Anchorage	61	90	10,864	−25	−20	73	70	67	59
Alaska, Juneau	58	17	9,075	−7	−4	75	71	68	62
Arizona, Phoenix	34	1,117	1,765	31	34	108	106	104	75
Arkansas, Little Rock	35	257	3,219	19	23	99	96	94	78
California, Los Angeles	34	312	2,061	42	44	94	90	87	69
California, Sacramento	39	17	2,502	30	32	100	97	94	69
California, San Francisco	38	52	3,015	42	44	80	77	73	61
Colorado, Denver	40	5,283	6,283	−2	3	92	90	89	63
Connecticut, Hartford	42	15	6,235	1	5	90	88	85	74
Delaware, Wilmington	40	78	4,930	12	15	93	90	87	76
District of Columbia	39	14	4,224	16	19	94	92	90	76
Florida, Jacksonville	31	24	1,239	29	32	96	94	92	79
Florida, Miami	26	7	214	44	47	92	90	89	79
Florida, Tampa	28	19	683	36	39	92	91	90	79
Georgia, Atlanta	34	1,005	2,961	18	23	95	92	90	76
Hawaii, Honolulu	21	7	0	60	62	87	85	84	73
Idaho, Boise	44	2,842	5,809	4	10	96	93	91	65
Illinois, Chicago	42	594	6,639	−3	1	94	91	88	75
Indiana, Indianapolis	40	793	5,699	0	4	93	91	88	76
Iowa, Des Moines	42	948	6,588	−7	−3	95	92	89	76
Kansas, Topeka	39	877	5,182	3	6	99	96	94	77
Kentucky, Louisville	38	474	4,660	8	12	96	93	91	77
Louisiana, New Orleans	30	3	1,385	32	35	93	91	90	79
Maine, Portland	44	61	7,511	−5	0	88	85	81	71
Maryland, Baltimore	39	146	4,654	12	15	94	91	89	77
Massachusetts, Boston	42	15	5,634	6	10	91	88	85	73
Michigan, Detroit, (Met.)	42	633	6,232	4	8	92	88	85	74
Michigan, Sault Ste. Marie	47	721	9,048	−12	−8	83	81	78	69
Minnesota, Duluth	47	1,426	10,000	−19	−15	85	82	79	69
Minnesota, Minneapolis/ St. Paul	45	822	8,382	−14	−10	92	89	86	74
Mississippi, Jackson	32	330	2,239	21	24	98	96	94	78
Missouri, Kansas City	39	742	4,711	4	8	100	97	94	76
Missouri, St. Louis	39	465	4,900	7	11	96	94	92	77
Montana, Billings	46	3,567	7,049	−10	−6	94	91	88	65
Montana, Butte	46	5,526	−	−24	−16	86	83	80	57
Nebraska, Omaha	41	978	6,612	−5	−1	97	94	91	76
Nevada, Las Vegas	36	2,162	2,709	23	26	108	106	104	70
New Hampshire, Man-chester	43	253	−	−5	1	92	89	86	73
New Jersey, Atlantic City	40	11	4,812	14	18	91	88	85	76

*Condensed with permission from "ASHRAE Handbook of Fundamentals" and "ASHRAE Guide and Data Book," American Society of Heating, Refrigerating and Air-Conditioning Engineers, 1972 and 1968, respectively. Extended data for over 1,000 locations are given in these sources.

†Latitudes are given to the nearest degree.

‡Degree-days are the yearly totals, 65°F base, i.e., for all days when the mean temperature was below 65°F. For monthly and yearly degree-days see the "ASHRAE Guide and Data Book."

§Winter temperatures are the temperatures equaled or exceeded 99% (and 97.5%) of the time during the coldest months.

¶Summer temperatures represent the highest 1% (or 2.5 or 5%) hourly dry-bulb (DB) or wet-bulb (WB) temperatures during the warmest months. In a normal season there would be about 30 hr above the 1% design temperature and 150 hr above the 5% dry-bulb design temperature. In most locations the 1% design wet-bulb temperature is about 2° above the 5% dry-bulb value.

TABLE C.2 Weather Conditions for Environmental Design *(continued)*

Location	Lati-tude, deg	Eleva-tion, ft	Degree-days heating	Temperatures, deg F					
				Winter		Summer			
				99%	97.5%	1% DB	1% WB	5% DB	5% WB
USA *(cont.)*									
New Mexico, Albuquerque	35	5,310	4,348	14	17	96	94	92	64
New York, Albany	43	277	6,875	−5	0	91	88	85	73
New York, Buffalo	43	705	7,062	3	6	88	86	83	72
New York, NYC-LaGuardia	41	19	4,811	12	16	93	90	87	75
North Carolina, Durham	36	406	—	15	19	94	92	89	76
North Dakota, Bismarck	47	1,647	8,851	−24	−19	95	91	88	70
Ohio, Cincinnati	39	761	4,410	8	12	94	92	90	76
Ohio, Cleveland	41	777	6,351	2	7	91	89	86	74
Oklahoma, Tulsa	36	650	3,860	12	16	102	99	96	77
Oregon, Portland	46	57	4,635	26	29	91	88	84	67
Pennsylvania, Philadelphia	40	7	5,144	11	15	93	90	87	76
Pennsylvania, Pittsburgh	41	749	5,987	7	11	90	88	85	73
Rhode Island, Providence	42	55	5,954	6	10	89	86	83	74
South Carolina, Columbia	34	217	2,484	20	23	98	96	94	78
South Dakota, Sioux Falls	44	1,420	7,839	−14	−10	95	92	89	74
Tennessee, Memphis	35	263	3,232	17	21	98	96	94	78
Tennessee, Nashville	36	577	3,578	12	16	97	95	92	77
Texas, Dallas	33	481	2,363	19	24	101	99	97	78
Texas, Houston	30	158	1,396	29	33	96	94	92	79
Texas, San Antonio	30	792	1,546	25	30	99	97	96	76
Utah, Salt Lake City	41	4,220	6,052	5	9	97	94	92	65
Vermont, Burlington	45	331	8,269	−12	−7	88	85	83	71
Virginia, Richmond	38	162	3,865	14	18	96	93	91	77
Washington, Seattle	48	14	4,424	28	32	81	79	76	64
Washington, Spokane	48	2,357	6,655	−2	4	93	90	87	63
West Virginia, Charleston	38	939	4,476	9	14	92	90	88	74
Wisconsin, LaCrosse	44	652	7,589	−12	−8	90	88	85	75
Wisconsin, Milwaukee	43	672	7,635	−6	−2	90	87	84	73
Wyoming, Cheyenne	41	6,126	7,381	−6	−2	89	86	83	61
CANADA									
Alberta, Edmonton	54	2,219	10,268	−29	−26	86	83	80	65
Br. Columbia, Vancouver	49	16	5,515	15	19	80	78	76	65
Manitoba, Winnipeg	50	786	10,679	−28	−25	90	87	84	72
Ontario, Ottawa	45	339	8,735	−17	−13	90	87	84	73
Ontario, Toronto	44	578	6,827	−3	1	90	87	85	73
Quebec, Montreal	46	98	8,203	−16	−10	88	86	84	73
Quebec, Quebec	47	245	9,372	−19	−13	86	82	79	71
OTHER COUNTRIES									
Argentina, Buenos Aires	35	89	—	32	34	91	89	86	75
Australia, Sydney	34	138	—	40	42	89	84	80	72
Brazil, Sao Paulo	24	2,608	—	42	46	86	84	82	74
France, Paris	49	164	—	22	25	89	86	83	67
Germany, Berlin	52	187	—	7	12	84	81	78	66
Hong Kong, Hong Kong	22	109	—	48	50	92	91	90	80
India, Calcutta	23	21	—	52	54	98	97	96	82
Italy, Rome	42	377	—	30	33	94	92	89	72
Japan, Tokyo	36	19	—	26	28	91	89	87	79
Mexico, Mexico City	19	7,575	—	37	39	83	81	79	59
Netherlands, Amsterdam	52	5	—	20	23	79	76	73	63
Soviet Union, Moscow	56	505	—	−11	−6	84	81	78	65
Spain, Madrid	40	2,188	—	25	28	93	91	89	67
Sweden, Stockholm	59	146	—	5	8	78	74	72	60
United Kingdom, London	51	149	—	24	26	82	79	76	65

TABLE C.3 Maximum Solar Altitude, Radiation Intensity, and Heat Gain, 24 to 56°N Latitude*

	N latitude, deg	Jan. 21	Feb. 21	March 21	April 21	May 21	June 21	July 21	Aug. 21	Sept. 21	Oct. 21	Nov. 21	Dec. 21
Solar altitude at noon, deg	24	46.0	55.2	66.0	77.6	86.0	89.5	86.6	78.3	66.0	55.5	46.2	42.6
Max. normal irradiation	24	320	323	317	300	287	280	279	285	301	311	314	317
Max. heat gain, horiz. skylight	24	215	249	275	284	282	279	278	276	267	244	213	199
Max. heat gain, window	24	253	243	234	229	218	212	214	220	222	235	249	252
Max. heat gain half-day	24	1064	1095	981	953	950	936	933	916	939	1050	1046	1019
Solar altitude at noon, deg	32	38.0	47.2	58.0	69.6	78.0	81.5	78.6	70.3	58.0	47.5	38.2	34.6
Max. normal irradiation	32	309	316	312	298	286	280	278	283	296	304	303	304
Max. heat gain. horiz. skylight	32	176	217	252	272	277	276	273	265	244	213	175	158
Max. heat gain, window	32	249	248	227	228	220	214	216	219	218	239	244	246
Max. heat gain, half-day	32	974	1091	1054	965	983	972	964	929	1004	1044	955	947
Solar altitude at noon, deg	40	30.0	39.2	50.0	61.6	70.0	73.5	70.6	62.3	50.0	39.5	30.2	26.6
Max. normal irradiation	40	293	306	306	294	284	278	276	279	290	293	287	284
Max. heat gain, horiz. skylight	40	133	180	223	253	265	267	262	247	215	177	132	113
Max. heat gain, window	40	254	247	236	225	220	215	216	216	226	238	250	253
Max. heat gain, half-day	40	903	1035	1100	1003	1002	1019	986	961	1045	989	884	831
Solar altitude at noon, deg	48	22.0	31.2	42.0	53.6	62.0	65.4	62.6	54.3	42.0	31.5	22.2	18.6
Max. normal irradiation	48	267	291	297	289	280	275	272	273	280	278	260	250
Max. heat gain, horiz. skylight	48	85	138	188	227	247	252	245	222	182	136	85	64
Max. heat gain, window	48	245	251	239	219	218	215	214	211	228	233	240	233
Max. heat gain, half-day	48	735	936	1111	1086	1058	1098	1044	1040	1050	895	719	623
Solar altitude at noon, deg	56	14.0	23.2	34.0	45.6	54.0	57.4	54.6	46.3	34.0	23.5	14.2	10.6
Max. normal irradiation	56	216	267	284	280	275	270	267	265	266	252	210	180
Max. heat gain, horiz. skylight	56	39	91	149	196	222	230	221	193	144	90	40	23
Max. heat gain, window	56	205	244	241	224	215	213	211	215	231	234	200	171
Max. heat gain, half-day	56	489	819	1080	1144	1120	1158	1101	1095	1013	776	476	339

*Adapted with permission from "ASHRAE Handbook of Fundamentals," American Society of Heating, Refrigerating, and Air-Conditioning Engineers, 1972.

TABLE C.4 Free-convection Coefficients*,†,‡

Configuration	Laminar flow, $10^9 > N_{Gr}N_{Pr} > 10^{-4}$	Turbulent flow, $N_{Gr}N_{Pr} > 10^9$
Horizontal cylinders, $\beta < 30°$	$h_c = 0.27 \left(\dfrac{\Delta T}{d}\right)^{\frac{1}{4}} (\cos \beta)^{\frac{1}{4}}$	$h_c = 0.18(\Delta T)^{\frac{1}{3}} (\cos \beta)^{\frac{1}{3}}$
Vertical plates and cylinders, $30° < \beta < 90°$	$h_c = 0.29 \left(\dfrac{\Delta T}{s}\right)^{\frac{1}{4}} (\sin \beta)^{\frac{1}{4}}$	$h_c = 0.19(\Delta T)^{\frac{1}{3}} (\sin \beta)^{\frac{1}{3}}$
Horizontal plates, $\beta < 30°$ Heated plates up Cooled plates down	$h_c = 0.27 \left(\dfrac{\Delta T}{s}\right)^{\frac{1}{4}} (\cos \beta)^{\frac{1}{4}}$	$h_c = 0.22(\Delta T)^{\frac{1}{3}} (\cos \beta)^{\frac{1}{3}}$
Heated plates down Cooled plates up	$h_c = 0.12 \left(\dfrac{\Delta T}{s}\right)^{\frac{1}{4}} (\cos \beta)^{\frac{1}{4}}$	

*In free convection (no wind)

$$N_{Gr} = \frac{\rho^2 g s^3 \, \Delta T}{\mu^2 T_s} \quad \text{and} \quad N_{Pr} = \frac{c_p \mu}{k}$$

where N_{Gr} = Grashof number
ρ = density, lb_m/ft^3
g = acceleration of gravity, ft/sec
s = side, or vertical, dimension, ft
ΔT = surface temperature to fluid temperature difference $(T_s - T_f)$, °R
μ = viscosity, lb/ft-sec
T_s = surface temperature, °R
N_{Pr} = Prandtl number
c_p = specific heat, Btu/(lb)(°F)
k = conductivity, Btu/(hr)(ft)(°F)

†Other symbols used in the table are defined as follows: h_c = convection coefficient, Btu/(hr)(ft²)(°F); d = diameter, ft; β = collector tilt angle, deg.
‡In forced convection (where the windspeed is greater than 0.5 mph), $h_c = 1 + 0.35v$, where v is the windspeed, in knots.

TABLE C.5 Surface Heat-transfer Coefficients $h_{c,in}$ and $h_{c,out}$*†
for Various Types and Orientations of Building Surfaces

Surface description	Wind conditions	Heat-flow direction	Surface emittance $\bar{\epsilon}_{ir}$		
			0.90	0.20	0.05
Horizontal	Still air	Up	1.63	0.91	0.76
Horizontal	Still air	Down	1.08	0.37	0.22
Sloping, 45°	Still air	Up	1.60	0.88	0.73
Sloping, 45°	Still air	Down	1.32	0.60	0.45
Vertical	Still air	Horizontal	1.46	0.74	0.59
Any position	15-mph	Any	6.00	. . .	. . .
Any position	7.5-mph	Any	4.00	. . .	. . .

*Adapted with permission from "ASHRAE Handbook of Fundamentals," 1967, p. 429.
†Values of $h_{c,in}$ and $h_{c,out}$ are in Btu per hour per square foot per degree Fahrenheit.

TABLE C.6 Reciprocal of Capital Recovery Factor

Yr	Rate i_{ann}				
n	0.0025($\frac{1}{4}$%)	0.004167($\frac{5}{12}$%)	0.005($\frac{1}{2}$%)	0.005833($\frac{7}{12}$%)	0.0075($\frac{3}{4}$%)
1	0.99750623	0.99585062	0.99502488	0.99420050	0.99255583
2	1.99252492	1.98756908	1.98509938	1.98263513	1.97772291
3	2.98506227	2.97517253	2.97024814	2.96533732	2.95555624
4	3.97512446	3.95867804	3.95049566	3.94234034	3.92611041
5	4.96271766	4.93310261	4.92586633	4.91367722	4.88943961
6	5.94784804	5.91346318	5.89638441	5.87938083	5.84559763
7	6.93052174	6.88477661	6.86207404	6.83948384	6.79463785
8	7.91074487	7.85205970	7.82295924	7.79401874	7.73661325
9	8.88852357	8.81532916	8.77906392	8.74301780	8.67157642
10	9.86386391	9.77460165	9.73041186	9.68651314	9.59957958
11	10.8367720	10.7298938	10.6770267	10.6245367	10.5206745
12	11.8072538	11.6812220	11.6189321	11.5571201	11.4349127
13	12.7753156	12.6286028	12.5561513	12.4842951	12.3423451
14	13.7409631	13.5720526	13.4887078	13.4060929	13.2430224
15	14.7042026	14.5115877	14.4166246	14.3225447	14.1369950
16	15.6650400	15.4472242	15.3399250	15.2336816	15.0243126
17	16.6234813	16.3789785	16.2586319	16.1395343	15.9050249
18	17.5795325	17.3068665	17.1727680	17.0401335	16.7791811
19	18.5331995	18.2309044	18.0823562	17.9355097	17.6468298
20	19.4844883	19.1511082	18.9874191	18.8256931	18.5080197
21	20.4334048	20.0674936	19.8879793	19.7107140	19.3627987
22	21.3799549	20.9800766	20.7840590	20.5906021	20.2112146
23	22.3241445	21.8888730	21.6756806	21.4653874	21.0533147
24	23.2659796	22.7938984	22.5628662	22.3350993	21.8891461
25	24.2054659	23.6951685	23.4456380	23.1997673	22.7187555
26	25.1426094	24.5926980	24.3240179	24.0594207	23.5421891
27	26.0774158	25.4865052	25.1980278	24.9140885	24.3594929
28	27.0098911	26.3766027	26.0676894	25.7637997	25.1707125
29	27.9400410	27.2630068	26.9330242	26.6085830	25.9758933
30	28.8678713	28.1457329	27.7940540	27.4484669	26.7750802
31	29.7933879	29.0247963	28.6508000	28.2834799	27.5683178
32	30.7165964	29.9002120	29.5032835	29.1136503	28.3556504
33	31.6375026	30.7719954	30.3515259	29.9390061	29.1371220
34	32.5561123	31.6401614	31.1955482	30.7595752	29.9127762
35	33.4724313	32.5047250	32.0353713	31.5753855	30.6826563
36	34.3864651	33.3657013	32.8710162	32.3864645	31.4468053
37	35.2982196	34.2231050	33.7025037	33.1928396	32.2052658
38	36.2077003	35.0769511	34.5298544	33.9945381	32.9580802
39	37.1149130	35.9272542	35.3530890	34.7915872	33.7052905
40	38.0198634	36.7740290	36.1722279	35.5840137	34.4469384
41	38.9225570	37.6172903	36.9872914	36.3718446	35.1830654
42	39.8229995	38.4570526	37.7982999	37.1551065	35.9137126
43	40.7211965	39.2933304	38.6052735	37.9338259	36.6389207
44	41.6171536	40.1261382	39.4082324	38.7080290	37.3587302
45	42.5108764	40.9554903	40.2071964	39.4777422	38.0731814
46	43.4023705	41.7814011	41.0021855	40.2429914	38.7823140
47	44.2916414	42.6038849	41.7932194	41.0038026	39.4861677
48	45.1786946	43.4229559	42.5803178	41.7602014	40.1847819
49	46.0635358	44.2386283	43.3635003	42.5122135	40.8781954
50	46.9461704	45.0509162	44.1427864	43.2598643	41.5664471

TABLE C.6 Reciprocal of Capital Recovery Factor *(continued)*

Yr	Rate i_{ann}				
n	0.025($\frac{1}{4}$%)	0.004167($\frac{5}{12}$%)	0.005($\frac{1}{2}$%)	0.005833($\frac{7}{12}$%)	0.0075($\frac{3}{4}$%)
50	46.9461704	45.0509162	44.1427864	43.2598643	41.5564471
51	47.8266039	45.8598335	44.9181954	44.0031791	42.2495753
52	48.7048418	46.6653944	45.6897466	44.7421830	42.9276181
53	49.5808895	47.4676127	46.4574593	45.4769011	43.6006135
54	50.4547527	48.2665022	47.2213526	46.2073582	44.2685990
55	51.3264366	49.0620769	47.9814454	46.9335789	44.9316119
56	52.1959467	49.8543505	48.7377566	47.6555880	45.5896893
57	53.0632885	50.6433366	49.4903050	48.3734098	46.2428678
58	53.9284673	51.4290489	50.2391095	49.0870686	46.8911839
59	54.7914886	52.2115009	50.9841886	49.7965885	47.5346738
60	55.6523577	52.9907063	51.7255608	50.5019935	48.1733735
61	56.5110800	53.7666785	52.4632445	51.2033075	48.8073186
62	57.3676608	54.5394309	53.1972582	51.9005543	49.4365445
63	58.2221056	55.3089768	53.9276201	52.5937574	50.0610864
64	59.0744195	56.0753296	54.6543484	53.2829402	50.6809791
65	59.9246080	56.8385025	55.3774611	53.9681262	51.2962571
66	60.7726763	57.5985087	56.0969762	54.6493384	51.9069550
67	61.6186297	58.3553614	56.8129117	55.3265999	52.5131067
68	62.4624736	59.1090736	57.5252852	55.9999336	53.1147461
69	63.3042130	59.8596583	58.2341147	56.6693623	53.7119068
70	64.1438534	60.6071286	58.9394176	57.3349087	54.3046221
71	64.9813999	61.3514974	59.6412115	57.9965952	54.8929252
72	65.8168577	62.0927775	60.3395139	58.6544443	55.4768488
73	66.6502322	62.8309817	61.0343422	59.3084781	56.0564256
74	67.4815283	63.5661229	61.7257137	59.9587190	56.6316879
75	68.3107515	64.2982136	62.4136454	60.6051887	57.2026679
76	69.1379067	65.0272667	63.0981547	61.2479092	57.7693975
77	69.9629992	65.7532946	63.7792584	61.8869023	58.3319081
78	70.7860341	66.4763100	64.4569735	62.5221895	58.8902314
79	71.6070166	67.1963253	65.1313169	63.1537924	59.4443984
80	72.4259517	67.9133530	65.8023054	63.7817323	59.9944401
81	73.2428446	68.6274055	66.4699556	64.4060304	60.5403872
82	74.0577003	69.3384951	67.1342842	65.0267080	61.0822702
83	74.8705240	70.0466341	67.7953076	65.6437859	61.6201193
84	75.6813207	70.7518348	68.4530424	66.2572851	62.1539646
85	76.4900955	71.4541094	69.1075049	66.8672262	62.6838358
86	77.2968533	72.1534699	69.7587114	67.4736301	63.2097626
87	78.1015993	72.8499285	70.4066780	68.0765171	63.7317743
88	78.9043385	73.5434973	71.0514209	68.6759076	64.2499000
89	79.7050758	74.2341882	71.6929561	69.2718220	64.7641688
90	80.5038163	74.9220131	72.3312996	69.8642803	65.2746092
91	81.3005649	75.6069840	72.9664672	70.4533027	65.7812498
92	82.0953265	76.2891127	73.5984749	71.0389091	66.2841189
93	82.8881063	76.9684110	74.2273382	71.6211192	66.7832446
94	83.6789090	77.6448906	74.8530728	72.1999528	67.2786547
95	84.4677397	78.3185633	75.4756943	72.7754295	67.7703768
96	85.2546031	78.9894406	76.0952183	73.3475687	68.2584386
97	86.0395044	79.6575342	76.7116600	73.9163897	68.7428671
98	86.8224483	80.3228557	77.3250348	74.4819119	69.2236894
99	87.6034397	80.9854164	77.9353580	75.0441544	69.7009324
100	88.3824835	81.6452280	78.5426448	75.6031361	70.1746227

TABLE C.6 Reciprocal of Capital Recovery Factor *(continued)*

Yr	Rate i_{ann}				
n	0.01(1%)	0.01125($1\frac{1}{8}$%)	0.0125($1\frac{1}{4}$%)	0.015($1\frac{1}{2}$%)	0.0175($1\frac{3}{4}$%)
1	0.99009901	0.98887515	0.98765432	0.98522167	0.98280098
2	1.97039506	1.96674923	1.96311538	1.95588342	1.94869875
3	2.94098521	2.93374460	2.92653371	2.91220042	2.89798403
4	3.90196555	3.88998230	3.87805798	3.85438465	3.83094254
5	4.85343124	4.83558200	4.81783504	4.78264497	4.74785508
6	5.79547647	5.77066205	5.74600992	5.69718717	5.64899762
7	6.72819453	6.69533948	6.66272585	6.59821396	6.53464139
8	7.65167775	7.60973002	7.56812420	7.48592508	7.40505297
9	8.56601758	8.51394810	8.46234498	8.36051732	8.26049432
10	9.47130453	9.40810690	9.34552591	9.22218455	9.10122291
11	10.3676282	10.2923183	10.2178034	10.0711178	9.92749181
12	11.2550775	11.1666930	11.0793120	10.9075052	10.7395497
13	12.1337401	12.0313404	11.9301847	11.7315322	11.5376410
14	13.0037030	12.8863688	12.7705527	12.5433815	12.3220059
15	13.8650525	13.7318851	13.6005459	13.3432330	13.0928805
16	14.7178738	14.5679951	14.4202923	14.1312640	13.8504968
17	15.5622513	15.3948036	15.2299183	14.9076493	14.5950828
18	16.3982686	16.2124139	16.0295489	15.6725609	15.3268627
19	17.2260085	17.0209285	16.8193076	16.4261684	16.0460567
20	18.0455530	17.8204485	17.5993161	17.1686388	16.7528813
21	18.8569831	18.6110739	18.3696949	17.9001367	17.4475492
22	19.6603793	19.3929037	19.1305629	18.6208244	18.1302695
23	20.4558211	20.1660358	19.8820374	19.3308614	18.8012476
24	21.2433873	20.9305669	20.6242345	20.0304054	19.4606856
25	22.0231557	21.6865928	21.3572687	20.7196112	20.1087820
26	22.7952087	22.4342079	22.0812530	21.3986317	20.7457317
27	23.5596076	23.1735060	22.7962993	22.0676175	21.3717264
28	24.3164432	23.9045795	23.5025178	22.7267167	21.9869547
29	25.0657853	24.6275199	24.2000176	23.3760756	22.5916017
30	25.8077082	25.3424177	24.8889062	24.0158380	23.1858403
31	26.5433854	26.0493623	25.5692901	24.6461458	23.7698765
32	27.2695895	26.7484424	26.2412742	25.2671387	24.3438590
33	27.9896925	27.4397452	26.9049622	25.8789544	24.9079695
34	28.7026659	28.1233575	27.5604564	26.4817285	25.4623779
35	29.4085801	28.7993646	28.2078582	27.0755946	26.0072510
36	30.1075050	29.4678513	28.8172674	27.6606843	26.5427528
37	30.7995099	30.1289011	29.4787826	28.2371274	27.0690445
38	31.4846633	30.7825969	30.1025013	28.8050516	27.5862846
39	32.1630330	31.4290204	30.7185198	29.3645829	28.0946286
40	32.8346861	32.0682526	31.3269332	29.9158452	28.5942295
41	33.4996892	32.7003734	31.9278352	30.4589608	29.0852379
42	34.1581081	33.3254620	32.5213187	30.9940500	29.5678014
43	34.8100081	33.9435965	33.1074753	31.5212316	30.0420652
44	35.4554535	34.5548544	33.6863954	32.0406222	30.5081722
45	36.0945084	35.1593121	34.2581683	32.5523372	30.9662626
46	36.7272361	35.7570454	34.8228822	33.0564898	31.4164743
47	37.3536991	36.3481289	35.3806244	33.5531920	31.8589428
48	37.9739595	36.9326367	35.9314809	34.0425536	32.2938013
49	38.5880787	37.5106420	36.4755367	34.5246834	32.7211806
50	39.1961175	38.0822171	37.0128758	34.9996881	33.1412095

TABLE C.6   Reciprocal of Capital Recovery Factor *(continued)*

Yr	Rate i_{ann}				
n	0.01(1%)	0.01125(1⅛%)	0.0125(1¼%)	0.015(1½%)	0.0175(1¾%)
50	39.1961175	38.0822171	37.0128758	34.9996881	33.1412095
51	39.7981362	38.6474335	37.5435810	35.4676730	33.5540142
52	40.3941942	39.2063619	38.0677343	35.9287419	33.9597191
53	40.9843507	39.7590723	38.5854166	36.3829969	34.3584463
54	41.5686641	40.3056339	39.0967078	36.8305388	34.7503158
55	42.1471922	40.8461151	39.6016867	37.2714668	35.1354455
56	42.7199922	41.3805836	40.1004313	37.7058786	35.5139513
57	43.2871210	41.9091061	40.5930186	38.1338706	35.8859473
58	43.8486347	42.4317490	41.0795245	38.5555375	36.2515452
59	44.4045888	42.9485775	41.5600242	38.9709729	36.6108553
60	44.9550384	43.4596563	42.0345918	39.3802689	36.9639855
61	45.5000380	43.9650495	42.5033005	39.7835161	37.3110423
62	46.0396416	44.4648203	42.9662228	40.1808041	37.6521300
63	46.5739026	44.9590312	43.4234299	40.5722208	37.9873514
64	47.1028738	45.4477441	43.8749925	40.9578530	38.3168072
65	47.6266078	45.9310201	44.3209802	41.3377862	38.6405968
66	48.1451562	46.4089197	44.7614619	41.7121046	38.9588175
67	48.6585705	46.8815028	45.1965056	42.0808912	39.2715651
68	49.1669015	47.3488285	45.6261784	42.4442278	39.5789337
69	49.6701995	47.8109553	46.0505466	42.8021949	39.8810160
70	50.1685143	48.2679409	46.4696756	43.1548718	40.1779027
71	50.6618954	48.7198427	46.8836302	43.5023368	40.4694832
72	51.1503915	49.1667171	47.2924743	43.8446668	40.7564454
73	51.6340510	49.6086202	47.6962709	44.1819377	41.0382756
74	52.1129218	50.0456071	48.0950824	44.5142243	41.3152586
75	52.5870512	50.4777326	48.4889703	44.8416003	41.5874777
76	53.0564864	50.9050508	48.8779953	45.1641383	41.8550149
77	53.5212736	51.3276151	49.2622176	45.4819096	42.1179508
78	53.9814590	51.7454785	49.6416964	45.7949848	42.3763644
79	54.4370882	52.1586932	50.0164903	46.1034333	42.6303336
80	54.8882061	52.5673109	50.3866571	46.4073235	42.8799347
81	55.3348575	52.9713829	50.7522539	46.7067227	43.1252430
82	55.7770867	53.3709596	51.1133372	47.0016972	43.3663322
83	56.2149373	53.7660910	51.4699626	47.2923125	43.6032749
84	56.6484528	54.1568267	51.8221853	47.5786330	43.8361424
85	57.0776760	54.5432156	52.1700596	47.8607222	44.0650043
86	57.5026495	54.9253059	52.5136391	48.1386425	44.2899310
87	57.9234154	55.3031455	52.8529769	48.4124557	44.5109887
88	58.3400152	55.6767817	53.1881253	48.6822224	44.7282444
89	58.7524903	56.0462613	53.5191361	48.9481123	44.9417636
90	59.1608815	56.4116304	53.8460604	49.2098545	45.1516104
91	59.5652292	56.7729349	54.1689485	49.4678370	45.3578380
92	59.9655735	57.1302199	54.4878504	49.7220069	45.5605386
93	60.3619539	57.4835302	54.8028152	49.9724206	45.7597431
94	60.7544098	57.8329100	55.1138915	50.2191335	45.9555215
95	61.1429800	58.1784029	55.4211274	50.4622005	46.1479327
96	61.5277030	58.5200523	55.7245703	50.7016754	46.3370345
97	61.9086168	58.8579010	56.0242670	50.9376112	46.5228841
98	62.2857592	59.1919911	56.3202637	51.1700603	46.7055372
99	62.6591676	59.5223645	56.6126061	51.3990742	46.8850488
100	63.0288788	59.8490625	59.9013394	51.6247037	47.0614730

TABLE C.6   Reciprocal of Capital Recovery Factor *(continued)*

Yr	Rate i_{ann}				
n	0.02(2%)	0.0225($2\frac{1}{4}$%)	0.025($2\frac{1}{2}$%)	0.0275($2\frac{3}{4}$%)	0.03(3%)
1	0.98039216	0.97799511	0.97560976	0.97323601	0.97087379
2	1.94156094	1.93446955	1.92742415	1.92042434	1.91346970
3	2.88388327	2.86989687	2.85602356	2.84226213	2.82861135
4	3.80772870	3.78474021	3.76197421	3.73942787	3.71709840
5	4.71345951	4.67945253	4.64582850	4.61258186	4.57970719
6	5.60143089	5.55447680	5.50812536	5.46236678	5.41719144
7	6.47199107	6.41024626	6.34939060	6.28940806	6.23028296
8	7.32548144	7.24718461	7.17013717	7.09431441	7.01969219
9	8.16223671	8.06570622	7.97086553	7.87767826	7.78610892
10	8.98258501	8.86621635	8.75206393	8.64007616	8.53020284
11	9.78684805	9.64911134	9.51420871	9.38206926	9.25262411
12	10.5753412	10.4147788	10.2577646	10.1042037	9.95400399
13	11.3483737	11.1635979	10.9831850	10.8070109	10.6349553
14	12.1062488	11.8959392	11.6909122	11.4910081	11.2960731
15	12.8492635	12.6121655	12.3813777	12.1566989	11.9379351
16	13.5777093	13.3126313	13.0550027	12.8045732	12.5611020
17	14.2918719	13.9976834	13.7121977	13.4351077	13.1661185
18	14.9920313	14.6676611	14.3533636	14.0487666	13.7535131
19	15.6784620	15.3228959	14.9788913	14.6460016	14.3237991
20	16.3514333	15.9637124	15.5891623	15.2272521	14.8774749
21	17.0112092	16.5904277	16.1845486	15.7929461	15.4150241
22	17.6580482	17.2033523	16.7654132	16.3434999	15.9369166
23	18.2922041	17.8027896	17.3321105	16.8793186	16.4436084
24	18.9139256	18.3890362	17.8849858	17.4007967	16.9355421
25	19.5234565	18.9623826	18.4243764	17.9083180	17.4131477
26	20.1210358	19.5231126	18.9506111	18.4022559	17.8768424
27	20.7068978	20.0715038	19.4640109	18.8829741	18.3270315
28	21.2812724	20.6078276	19.9648887	19.3508264	18.7641082
29	21.8443847	21.1323498	20.4535499	19.8061571	19.1884546
30	22.3964556	21.6453298	20.9302926	20.2493013	19.6004413
31	22.9377015	22.1470219	21.3954074	20.6805852	20.0004285
32	23.4683348	22.6376742	21.8491780	21.1003262	20.3887655
33	23.9885636	23.1175298	22.2918809	21.5088333	20.7657918
34	24.4985917	23.5868262	22.7237863	21.9064071	21.1318367
35	24.9986193	24.0457958	23.1451573	22.2933403	21.4872201
36	25.4888425	24.4946658	23.5562511	22.6699175	21.8322525
37	25.9694534	24.9336585	23.9573181	23.0364161	22.1672354
38	26.4406406	25.3629912	24.3486030	23.3931057	22.4924616
39	26.9025888	25.7828765	24.7303444	23.7402488	22.8082151
40	27.3554792	26.1935222	25.1027751	24.0781011	23.1147720
41	27.7994895	26.5951317	25.4661220	24.4069110	23.4124000
42	28.2347936	26.9879039	25.8206068	24.7269207	23.7013592
43	28.6615623	27.3720332	26.1664457	25.0383656	23.9819021
44	29.0799631	27.7477097	26.5038495	25.3414751	24.2542739
45	29.4901599	28.1151195	26.8330239	25.6364721	24.5187125
46	29.8923136	28.4744445	27.1541696	25.9235738	24.7754491
47	30.2865820	28.8258626	27.4674826	26.2029915	25.0247078
48	30.6731196	29.1695478	27.7731537	26.4749309	25.2667066
49	31.0520780	29.5056702	28.0713695	26.7395922	25.5016569
50	31.4236059	29.8343963	28.3623117	26.9971700	25.7297640

TABLE C.6 Reciprocal of Capital Recovery Factor *(continued)*

Yr	Rate i_{ann}				
n	0.02(2%)	0.0225(2¼%)	0.025(2½%)	0.0275(2¾%)	0.03(3%)
50	31.4236059	29.8343963	28.3623117	26.9971700	25.7297640
51	31.7878489	30.1558888	28.6461577	27.2478540	25.9512272
52	32.1449499	30.4703069	28.9230807	27.4918287	26.1662400
53	32.4950489	30.7778062	29.1932495	27.7292737	26.3749903
54	32.8382833	31.0785391	29.4568288	27.9603637	26.5776605
55	33.1747875	31.3726544	29.7139793	28.1852688	26.7744276
56	33.5046936	31.6602977	29.9648578	28.4041545	26.9654637
57	33.8281310	31.9416114	30.2096174	28.6171820	27.1509357
58	34.1452265	32.2167349	30.4484072	28.8245081	27.3310055
59	34.4561044	32.4858043	30.6813729	29.0262852	27.5058306
60	34.7608867	32.7489529	30.9086565	29.2226620	27.6755637
61	35.0596928	33.0063109	31.1303966	29.4137830	27.8403531
62	35.3526400	33.2580057	31.3467284	29.5997888	28.0003428
63	35.6398432	33.5041621	31.5577838	29.7808163	28.1556726
64	35.9214149	33.7449018	31.7636915	29.9569989	28.3064783
65	36.1974655	33.9803440	31.9645771	30.1284661	28.4528915
66	36.4681035	34.2106054	32.1605630	30.2953441	28.5950403
67	36.7334348	34.4357999	32.3517688	30.4577558	28.7330488
68	36.9935635	34.6560391	32.5383110	30.6158207	28.8670377
69	37.2485917	34.8714318	32.7203034	30.7696552	28.9971240
70	37.4986193	35.0820849	32.8978570	30.9193725	29.1234214
71	37.7437444	35.2881026	33.0710800	31.0650827	29.2460401
72	37.9840631	35.4894869	33.2400780	31.2068931	29.3650875
73	38.2196697	35.6866376	33.4049542	31.3449082	29.4806675
74	38.4506566	35.8793521	33.5658089	31.4792294	29.5928811
75	38.6771143	36.0678261	33.7227404	31.6099556	29.7018263
76	38.8991317	36.2521526	33.8758443	31.7371830	29.8075983
77	39.1167958	36.4324231	34.0252140	31.8610054	29.9102896
78	39.3301919	36.6087267	34.1709405	31.9815138	30.0099899
79	39.5394039	36.7811509	34.3131127	32.0987969	30.1067863
80	39.7445136	36.9497808	34.4518172	32.2129410	30.2007634
81	39.9456016	37.1147000	34.5871388	32.3240301	30.2920033
82	40.1427466	37.2759903	34.7191598	32.4321461	30.3805858
83	40.3360261	37.4337313	34.8479607	32.5373685	30.4665881
84	40.5255158	37.5880013	34.9736202	32.6397747	30.5500856
85	40.7112900	37.7383765	35.0962149	32.7394401	30.6311510
86	40.8934216	37.8864318	35.2158194	32.8364380	30.7098554
87	41.0719819	38.0307402	35.3325067	32.9308399	30.7862673
88	41.2470411	38.1718730	35.4463480	33.0227153	30.8604537
89	41.4186677	38.3099003	35.5574127	33.1121317	30.9324794
90	41.5869292	38.4448902	35.6657685	33.1991549	31.0024071
91	41.7518913	38.5769098	35.7714814	33.2838490	31.0702982
92	41.9136190	38.7060242	35.8746160	33.3662764	31.1362118
93	42.0721754	38.8322975	35.9752352	33.4464978	31.2002057
94	42.2276230	38.9557922	36.0734002	33.5245720	31.2623356
95	42.3800225	39.0765694	36.1691709	33.6005567	31.3226559
96	42.5294339	39.1946889	36.2626057	33.6745078	31.3812193
97	42.6759155	39.3102092	36.3537617	33.7464796	31.4380770
98	42.8195250	39.4231875	36.4426943	33.8165251	31.4932787
99	42.9603187	39.5336797	36.5294579	33.8846960	31.5468725
100	43.0983516	39.6417405	36.6141053	33.9510423	31.5989053

TABLE C.6 Reciprocal of Capital Recovery Factor *(continued)*

Yr	Rate i_{ann}				
n	$0.035(3\frac{1}{2}\%)$	$0.04(4\%)$	$0.045(4\frac{1}{2}\%)$	$0.05(5\%)$	$0.055(5\frac{1}{2}\%)$
1	0.96618357	0.96153846	0.95693780	0.95238095	0.94786730
2	1.89969428	1.88609467	1.87266775	1.85941043	1.84631971
3	2.80163698	2.77509103	2.74806435	2.72324803	2.69793338
4	3.67307921	3.62989522	3.58752570	3.54595050	3.50515012
5	4.51505238	4.45182233	4.38997674	4.32947667	4.27028448
6	5.32855302	5.24213686	5.15787248	5.07569207	4.99553031
7	6.11454398	6.00205467	5.89270094	5.78637340	5.68296712
8	6.87395554	6.73274487	6.59588607	6.46321276	6.33456599
9	7.60768651	7.43533161	7.26879050	7.10782168	6.95219525
10	8.31660532	8.11089578	7.91271818	7.72173493	7.53762583
11	9.00155104	8.76047671	8.52891692	8.30641422	8.09253633
12	9.66333433	9.38507376	9.11858078	8.86325164	8.61851785
13	10.3027385	9.98564785	9.68285242	9.39357299	9.11707853
14	10.9205203	10.5631229	10.2228253	9.89864094	9.58964790
15	11.5174109	11.1183874	10.7395457	10.3796580	10.0375809
16	12.0941168	11.6522956	11.2340150	10.8377696	10.4621620
17	12.0513206	12.1656689	11.7071914	11.2740662	10.8646086
18	13.1896817	12.6592970	12.1599918	11.6895869	11.2460745
19	13.7098374	13.1339394	12.5932936	12.0853209	11.6076535
20	14.2124033	13.5903263	13.0079365	12.4622103	11.9503825
21	14.6979742	14.0291599	13.4047239	12.8211527	12.2752441
22	15.1671248	14.4511153	13.7844248	13.1630026	12.5831697
23	15.6204105	14.8568417	14.1477749	13.4885739	12.8750424
24	16.0583676	15.2469631	14.4954784	13.7986418	13.1516990
25	16.4815146	15.6220799	14.8282090	14.0939446	13.4139327
26	16.8903523	15.9827692	15.1466114	14.3751853	13.6624954
27	17.2853645	16.3295857	15.4513028	14.6430336	13.8980999
28	17.6670188	16.6630632	15.7428735	14.8981273	14.1214217
29	18.0357670	16.9837146	16.0218885	15.1410736	14.3331012
30	18.3920454	17.2920333	16.2888885	15.3724510	14.5337452
31	18.7362758	17.5884936	16.5443910	15.5928105	14.7239291
32	19.0688655	17.8735515	16.7888909	15.8026767	14.9041982
33	19.3902082	18.1476457	17.0228621	16.0025492	15.0750694
34	19.7006842	18.4111978	17.2467580	16.1929040	15.2370326
35	20.0006611	18.6646132	17.4610124	16.3741943	15.3905522
36	20.2904938	18.9082820	17.6660406	16.5468517	15.5360684
37	20.5705254	19.1425788	17.8622398	16.7112873	15.6739985
38	20.8410874	19.3678642	18.0499902	16.8678927	15.8047379
39	21.1024999	19.5844848	18.2296557	17.0170407	15.9286615
40	21.3550723	19.7927739	18.4015844	17.1590864	16.0461247
41	21.5991037	19.9930518	18.5661095	17.2943680	16.1574642
42	21.8348828	20.1856267	18.7235498	17.4232076	16.2629992
43	22.0626887	20.3707949	18.8742103	17.5459120	16.3630324
44	22.2827910	20.5488413	19.0183831	17.6627733	16.4578506
45	22.4954503	20.7200397	19.1563474	17.7740698	16.5477257
46	22.7009181	20.8846536	19.2883707	17.8800665	16.6329154
47	22.8994378	21.0429361	19.4147088	17.9810157	16.7136639
48	23.0912443	21.1951309	19.5356065	18.0771578	16.7902027
49	23.2765645	21.3414720	19.6512981	18.1687217	16.8627514
50	23.4556179	21.4821846	19.7621178	18.2559255	16.9315179

TABLE C.6   Reciprocal of Capital Recovery Factor *(continued)*

Yr	Rate i_{ann}				
n	0.06(6%)	0.065(6½%)	0.07(7%)	0.075(7½%)	0.08(8%)
1	0.94339623	0.93896714	0.93457944	0.93023256	0.92592593
2	1.83339267	1.82062642	1.80801817	1.79556517	1.78326475
3	2.67301195	2.64847551	2.62431604	2.60052574	2.57709699
4	3.46510561	3.42579860	3.38721126	3.34932627	3.31212684
5	4.21236379	4.15567944	4.10019744	4.04588490	3.99271004
6	4.91732433	4.84101356	4.76653966	4.69384642	4.62287966
7	5.58238144	5.48451977	5.38928940	5.29660132	5.20637006
8	6.20979381	6.08875096	5.97129851	5.85730355	5.74663894
9	6.80169227	6.65610419	6.51523225	6.37888703	6.24688791
10	7.36008705	7.18883022	7.02358154	6.86408096	6.71008140
11	7.88687458	7.68904246	7.49867434	7.31542415	7.13896426
12	8.38384394	8.15872532	7.94268630	7.73527827	7.53607802
13	8.85268296	8.59974208	8.35765074	8.12584026	7.90377594
14	9.29498393	9.01384233	8.74546799	8.48915373	8.24423698
15	9.71224899	9.40266885	9.10791401	8.82711975	8.55947869
16	10.1058953	9.76776418	9.44664860	9.14150674	8.85136916
17	10.4772597	10.1105767	9.76322299	9.43395976	9.12163811
18	10.8276035	10.4324664	10.0590869	9.70600908	9.37188714
19	11.1581165	10.7347102	10.3355952	9.95907821	9.60359920
20	11.4699212	11.0185072	10.5940142	10.1944914	9.81814741
21	11.7640766	11.2849833	10.8355273	10.4134803	10.0168032
22	12.0415817	11.5351956	11.0612405	10.6171910	10.2007437
23	12.3033790	11.7701367	11.2721874	10.8066893	10.3710589
24	12.5503575	11.9907387	11.4693340	10.9829668	10.5287583
25	12.7833562	12.1978767	11.6535832	11.1469459	10.6747762
26	13.0031662	12.3923725	11.8257787	11.2994845	10.8099780
27	13.2105341	12.5749977	11.9867090	11.4413810	10.9351648
28	13.4061643	12.7464767	12.1371113	11.5733776	11.0510785
29	13.5907210	12.9074898	12.2776741	11.6961652	11.1584060
30	13.7648312	13.0586759	12.4090412	11.8103863	11.2577833
31	13.9290860	13.2006347	12.5318142	11.9166384	11.3497994
32	14.0840434	13.3339293	12.6465553	12.0154776	11.4349994
33	14.2302296	13.4590885	12.7537900	12.1074210	11.5138884
34	14.3681411	13.5766089	12.8540094	12,1929498	11.5869337
35	14.4982464	13.6869567	12.9476723	12.2725114	11.6545682
36	14.6209871	13.7905697	13.0352078	12.3465222	11.7171928
37	14.7367803	13.8878589	13.1170166	12.4153695	11.7751785
38	14.8460192	13.9792102	13.1934735	12.4794135	11.8288690
39	14.9490747	14.0649861	13.2649285	12.5389893	11.8785824
40	15.0462969	14.1455269	13.3317088	12.5944087	11.9246133
41	15.1380159	14.2211520	13.3941204	12.6459615	11.9672346
42	15.2245433	14.2921615	13.4524490	12.6939177	12.0066987
43	15.3061729	14.3588371	13.5069617	12.7385281	12.0432395
44	15.3831820	14.4214433	13.5579081	12.7800261	12.0770736
45	15.4558321	14.4802284	13.6055216	12.8186290	12.1084015
46	15.5243699	14.5354257	13.6500202	12.8545386	12.1374088
47	15.5890282	14.5872542	13.6916076	12.8879429	12.1642674
48	15.6500266	14.6359195	13.7304744	12.9190166	12.1891365
49	15.7075723	14.6816145	13.7667985	12.9479224	12.2121634
50	15.7618606	14.7245207	13.8007463	12.9748116	12.2334846

(From Mathematical Tables From Handbook of Chemistry & Physics, 10th Edition © 1954 by The Chemical Rubber Co. Used by Permission of The Chemical Rubber Co.)

TABLE C.7 Radiation Properties of Metals*

Material and surface description	Solar absorptance $\bar{\alpha}_s$ and infrared emittance $\bar{\epsilon}_{ir}$				Applicable specifications
	Initial		Degradation after 1 yr UV exposure		
	$\bar{\alpha}_s$	$\bar{\epsilon}_{ir}$	$\bar{\alpha}_s$	$\bar{\epsilon}_{ir}$	
Uncoated structural aluminum:[a,b,c,d]	$0.15-0.70$	$0.03-0.06$	0	0	
6061-T6 Cleaned and degreased	0.37 ± 0.10	0.04 ± 0.01	0	0	
1145-H34	0.14 ± 0.04	0.04 ± 0.01	0	0	
2024 As received	0.66 ± 0.10	0.05 ± 0.01	0	0	
1100-0 Mirror finish	0.20 ± 0.05	0.04 ± 0.01	0	0	
3003-0 As received	0.22 ± 0.05	0.04 ± 0.01	0	0	
Treated structural aluminum:[a,d]					
6061 Roughened (grit-blasted)	0.46 ± 0.10	0.30 ± 0.10	0	0	
3003-0 Grit-blasted	0.35 ± 0.10	0.11 ± 0.05	0	0	
6061-T6 Mechanically polished	0.19 ± 0.03	0.04 ± 0.01	0	0	
6061 Chemically polished	0.16 ± 0.03	0.04 ± 0.01	0	0	
Coated structural aluminum					
Chemical conversion coating[e,f,g]	$0.18-0.50$	$0.04-0.11$	0.05 ± 0.05	0	MIL-C-5541
2024 Color chem. film	0.36 ± 0.10	0.08 ± 0.03		0	MIL-C-5541
7075 Polished and chem. film[g]	0.20 ± 0.05	0.05 ± 0.02		0	MIL-C-5541, class 2, grade B
5052 Bleached irridite	0.30 ± 0.05	0.05 ± 0.02		0	
6061 Bleached irridite	0.27 ± 0.05	0.06 ± 0.02		0	
Anodized aluminum:[e,f,h,i,j,k]	$0.12-0.90$	$0.10-0.90$			
0.001-in.-thick anodize:					
1170	0.31 ± 0.05	0.81 ± 0.05		0	MIL-A-8625, Type 2
1199	0.18 ± 0.05	0.82 ± 0.05	0.30^f	0	Type 2
1199	0.28 ± 0.05	0.77 ± 0.05	0.20^f	0	Type 3
2024	0.43 ± 0.05	0.86 ± 0.05	0.40	0	Type 2
2024	0.88 ± 0.05	0.84 ± 0.05	None	0	Type 3
5052	0.40 ± 0.05	0.86 ± 0.05		0	Type 2
5557	0.21 ± 0.05	0.80 ± 0.05	0.20	0	Type 2
5557	0.44 ± 0.05	0.76 ± 0.05	-0.70	0	Type 3
6061	0.52 ± 0.05	0.86 ± 0.05	0.20	0	Type 2
6061	0.86 ± 0.05	0.83 ± 0.05	None	0	Type 3
7075	0.50 ± 0.05	0.88 ± 0.05		0	Type 2
0.0002-in.-thick anodize:					
Alzak, specular	0.77 ± 0.05	0.86 ± 0.05	None		Type 2 and dyed black
0.0003-in.-thick anodize:					
3003 (dyed black)	0.14 ± 0.03	0.73 ± 0.05	0.07 ± 0.05	0	Proprietary anodizing process
Uncoated structural magnesium					
AZ31B	0.27 ± 0.05	0.10 ± 0.05	0	0	
HM21A	0.35 ± 0.05	0.11 ± 0.03	0	0	
Treated structural magnesium[l,m]					
Chemical-conversion coated:					
Dow No. 9 on AZ31 alloy	0.41 ± 0.05	0.40 ± 0.05	0.05 ± 0.05	0	MIL-M-3171, type 4
Dow No. 7 on AZ31B alloy	0.82 ± 0.05	0.70 ± 0.05	0 ± 0.05	0	
Anodized:					
Dow No. 17, heavy, on AS21B alloy	0.89 ± 0.05	0.82 ± 0.05	0 ± 0.05	0	MIL-M-45202, type 2, class D

*Lettered footnotes are grouped at the end of the table.

TABLE C.7 Radiation Properties of Metals *(continued)*

Material and surface description	Solar absorptance $\bar{\alpha}_s$ and infrared emittance $\bar{\epsilon}_{ir}$				Applicable specifications
	Initial		Degradation after 1 yr UV exposure		
	$\bar{\alpha}_s$	$\bar{\epsilon}_{ir}$	$\bar{\alpha}_s$	$\bar{\epsilon}_{ir}$	
Untreated, miscellaneous metals:[a,c,d]					
Beryllium	0.60 ± 0.15	0.07 ± 0.02	0	0	
Copper[n]	0.35 ± 0.05	0.04 ± 0.01	0	0	
Molybdenum foil	0.36 ± 0.05	0.05 ± 0.02	0	0	
Nickel foil	0.40 ± 0.05	0.05 ± 0.02	0	0	
Titanium	0.50 ± 0.05	0.13 ± 0.03	0	0	
Tungsten	0.50 ± 0.05	0.04 ± 0.02	0	0	
TZM molybdenum alloy	0.50 ± 0.05	0.11 ± 0.03	0	0	
Inconel	0.36 ± 0.05	0.14 ± 0.03	0	0	
Stainless steel	0.44 ± 0.12	0.12 ± 0.02	0	0	
Haynes 25 alloy (L605)	0.39 ± 0.05	0.18 ± 0.03	0	0	
Chemically treated metals					
Oxidized-copper Ebanol-C process[e,o,p]	0.26 — 0.93	0.03 — 0.40			MIL-F-495
Orion black on stainless steel	0.81 ± 0.05	0.17 ± 0.03			
Hi-shear 1200-1, type 2, No. 306 spec on titanium	0.92 ± 0.05	0.81 ± 0.05			
Electroplating and electroless plating[d]					
Gold plating, 24-carat	0.24 ± 0.03	0.25 ± 0.01	0	0	PR6-3, type 1
			0	0	MIL-G-45204, type 1
Gold plating, 23+-carat (hard)[q]	0.28 ± 0.03	0.10 ± 0.03	0	0	PR6-3, type 2
					MIL-G-45204, type 2
Electroless gold	0.25 ± 0.03	0.03 ± 0.01	0	0	
Copper plating	0.47 ± 0.05	0.04 ± 0.01	0	0	
Silver plating[n]	0.10 ± 0.04	0.03 ± 0.01	0	0	
Black chrome duramir BK	0.97 ± 0.03	0.73 ± 0.05	0	0	
Vacuum deposited coatings:[c,d]					
Aluminum on resin	0.11 ± 0.02	0.04 ± 0.01	0	0	PR6-11
Aluminum on polymetric film	0.12 ± 0.03	0.04 ± 0.01	0	0	MT3-14
Gold on resin-sputtered	0.23 ± 0.03	0.03 ± 0.01	0	0	
gold[o]	0.25 ± 0.03	0.05 ± 0.02	0	0	

[a]Solar absorptance is a critical function of surface condition for metals. Uncleaned metals tend to have high solar absorptance which is unstable in space.

[b]Solar absorptance of untreated aluminum alloys subject to wide variations.

[c]All uncoated metals must be handled with great care. Vacuum-deposited metals, in particular, are easily removed by rubbing, exposing high absorptance and high emittance substrates.

[d]No degradation in space environment is expected for metals except from micrometeorite bombardment and charged particle sputtering (changes not significant in most applications for a 1-yr lifetime.

[e]Process not normally controlled for thermal properties.

[f]Extrapolated degradation based on short-time tests only (±0.1).

[g]Treatment can be performed in-house.

[h]Properties are strong function of alloy and anodizing parameters.

[i]Solar absorptance is a significant function of thickness and alloy type. Emittance is nearly independent of thickness for thicknesses greater than approximately 0.3 mil.

[j]Low values of emittance are difficult to reproduce except for certain exotic anodizing processes.

[k]No charged-particle degradation data available.

[l]No degradation data available.

[m]Numerous corrosion-prevention coatings (chemical and electrochemical), with a wide range of properties, are available for magnesium.

[n]Tarnishes readily in air; i.e., solar absorptance increases.

[o]No ultraviolet degradation in short-term tests.

[p]Higher values of emittance available.

[q]Solar absorptance estimated.

TABLE C.8 Radiation Properties of Paints

Material and surface description	Solar absorptance $\bar{\alpha}_s$ and infrared emittance $\bar{\epsilon}_{ir}$				Applicable specifications
	Initial		Degradation after 1 yr UV exposure		
	$\bar{\alpha}_s$	$\bar{\epsilon}_{ir}$	$\bar{\alpha}_s$	$\bar{\epsilon}_{ir}$	
White paints[a]					
Potassium zirconium silicate[b,c]	0.13 ± 0.03	0.86 ± 0.03	0.26 ± 0.05	0	MT6-2; PR6-1
IITRI S-13G[d,e]	0.20 ± 0.02	0.88 ± 0.03	0.10 ± 0.05	0	MT6-5C; PR5-25
Dow Corning 92-007[e]	0.21 ± 0.03	0.85 ± 0.03	0.22 ± 0.05	0	MT6-7; PR5-17
3M Co. Velvet White 100[f,g,h]	0.24 ± 0.04	0.85 ± 0.03	0.20	0	
3M Co. Velvet White 400[g,h,i]	0.24 ± 0.04	0.86 ± 0.03	0.20 ± 0.05	0	MT6-7; PR5-13
Cat-a-lac White Gloss 443-3-1	0.24 ± 0.02	0.88 ± 0.03	0.40 ± 0.10	0	MT6-9; PR5-2
IITRI Z-93[j]	0.19 ± 0.03	0.90 ± 0.02	0.02 ± 0.02	0	C115413
Black paints					
3M Co. Velvet Black 401-C10	0.98 ± 0.01	0.90 ± 0.02	−0.02 ± 0.02	0	MT6-8; PR5-13
Cat-a-lac Flat Black 463-3-8	0.97 ± 0.01	0.86 ± 0.03	−0.02 ± 0.02		MT6-9; PR5-2
Sicon Flat Black 8 x 906[k]	0.96 ± 0.02	0.83 ± 0.03	−0.02 ± 0.02		
Parson's Black[l]	0.98 ± 0.01	0.92 ± 0.02	−0.02 ± 0.02		
Colored paints					
Laminar X-500 Chromium Oxide green[f]	0.72 ± 0.03	0.83 ± 0.03	0.04 ± 0.04		
Clear PR5-12, 0.002 in.	Substrate-dependent	0.81 ± 0.05		0	PR5-12
Metallic pigmented paint[m]					
Inter-Chem Alum. Silicone[m,n,o]	0.25 ± 0.03	0.25 ± 0.03	0.06 ± 0.05	−0.02	MT6-4; PR5-19
Silver Chromatone[f,p]	0.24 ± 0.04	0.28 ± 0.06		0.08	
Cat-a-lac Epoxy 443-3-18 Aluminum[m]	0.56 ± 0.05	0.60 ± 0.04	0.06 ± 0.05	−0.10	
UCARSIL R104	0.32 ± 0.03	0.34 ± 0.09			
Trail Alum. Silicon Enamel[o,q]	0.30 ± 0.03	0.38 ± 0.04			
Seidenberg's Alum. Silicon[r]	0.25 ± 0.03	0.30 ± 0.04			

[a]The properties of white paints are functions of thickness. Values given here are for that thickness at which the solar absorptance is minimum (usually 3 to 10 mils).
[b]Difficult to apply.
[c]Extremely difficult to clean.
[d]Very short potentiometer life.
[e]Easy to apply using standard painting procedures.
[f]Extrapolated degradation based on short-time tests only (±0.1).
[g]Physically durable.
[h]Readily available and relatively inexpensive.
[i]The 400 series should have properties similar to the 100 series.
[k]High temperature paint (800°F).
[l]Very fragile surface; outgasses badly.
[m]Aluminum pigmented paints also degrade in emittance; i.e., emittance decreases.
[n]Same as Rinshed Mason Aluminum Silicone. Presently this paint is expensive in small quantities.
[o]Requires 1-hr cure at 400°F.
[p]Difficult to produce consistent property values.
[q]Short-term UV tests indicate degradation comparable to Interchem Aluminum Paint.
[r]Formula reported to be very stable under charged-particle and UV exposure.

TABLE C.9 Radiation Properties of Plastic Films

Material and gauge size	Solar absorptance $\bar{\alpha}_s$ and infrared emittance $\bar{\epsilon}_{ir}$				Applicable specifications
	Initial		Degradation after 1 yr UV exposure		
	$\bar{\alpha}_s$	$\bar{\epsilon}_{ir}$	$\bar{\alpha}_s$	$\bar{\epsilon}_{ir}$	
Transparent polymers, rear-surface-coated					
Aluminized polyethylene terephthalate[d]					
25[b]	0.14 ± 0.03	0.36 ± 0.03	0.22	0	MT3-14, Type 1
100[b]	0.16 ± 0.03	0.54 ± 0.03	0.21	0	
200[b]	0.17 ± 0.03	0.70 ± 0.03	0.22	0	
300[b]	0.18 ± 0.03	0.75 ± 0.03	0.24	0	MT3-14, Type 1
Aluminized polyimide[c,d]					
50[b]	0.33 ± 0.03	0.49 ± 0.03	0.07	0	MT3-14, Type 2
100[b]	0.36 ± 0.03	0.54 ± 0.03	0.05	0	
300[b]	0.44 ± 0.03	0.78 ± 0.03	0.01	0	
500[b]	0.53 ± 0.03	0.80 ± 0.03	0.01	0	MT3-14, Type 2
Aluminized fluorinated ethylene propylene (FEP Teflon, Type A)					
100	0.15 ± 0.03	0.60 ± 0.03	0.05 ± 0.03	0	MT3-14, Type 3
200	0.15 ± 0.03	0.66 ± 0.03	0.05 ± 0.03	0	
500	0.15 ± 0.03	0.78 ± 0.03	0.05 ± 0.03	0	
Silver overcoated with Inconel on FEP Teflon, Type A					
200	0.08 ± 0.02	0.66 ± 0.03	0.05 ± 0.03	0	C122491
Dow Corning 92-007 on rear surface of polyimide					
300	0.48 ± 0.03	0.86 ± 0.03	0.03 ± 0.03	0	

[a]DuPont's tradename for polyethylene terephthalate is "mylar."
[b]This specification presently covers only the metallized side.
[c]DuPont's tradename for polyimide is "Kapton."
[d]Polyimide metallized with silver or gold provides slightly lower values of solar absorptance; e.g., for 300 gauge + gold, $\bar{\alpha}_s = 0.42$.

TABLE C.10 Radiation Properties of Second-surface Mirrors

Material and surface description	Solar absorptance $\bar{\alpha}_s$ and infrared emittance $\bar{\epsilon}_{ir}$				Applicable specifications
	Initial		Degradation after 1 yr UV exposure		
	$\bar{\alpha}_s$	$\bar{\epsilon}_{ir}$	$\bar{\alpha}_s$	$\bar{\epsilon}_{ir}$	
Second-surface metallized glass mirrors					
Silver–fused silica[a,b]	0.07 ± 0.02	0.78 ± 0.02	0.01 ± 0.01	0	
Silver microsheet, 0.006 in.[c,d]	0.07 ± 0.02	0.82 ± 0.02	0.02 ± 0.02	0	C114766

[a]Expensive ($2 to $6/in.²) and laborious to apply but has very stable properties.
[b]Properties are not a significant function of thickness.
[c]Not so expensive as fused silica (~$2 to $4/in.²).
[d]Degree of charged-particle degradation uncertain but greater than for fused silica.

TABLE C.11 Radiation Properties of Miscellaneous Materials

Material and surface description	Solar absorptance $\bar{\alpha}_s$ and infrared emittance $\bar{\epsilon}_{ir}$				Applicable specifications
	Initial		Degradation after 1 yr UV exposure		
	$\bar{\alpha}_s$	$\bar{\epsilon}_{ir}$	$\bar{\alpha}_s$	$\bar{\epsilon}_{ir}$	
Miscellaneous components and materials[a] Solar cells:					
Texas Instruments single crystal 1–3 Ω-cm silicon, OCLI 420-nm blue-filter, microsheet cover	0.63 ± 0.03	0.83 ± 0.02	0.02 ± 0.02	0[b]	
Hoffman 7–14-Ω-cm silicon, 410-nm blue filter, fused-silica cover	0.79 ± 0.03	0.81 ± 0.02	0.02 ± 0.02	0[b]	PT 7-10; TP7-14
Hoffman silicon, blue-red filter, fused-silica cover	0.71 ± 0.03	0.80 ± 0.02	0.02 ± 0.02	0[b]	
Centralab 7–14-Ω-cm silicon, OCLI blue filter, microsheet cover	0.84 ± 0.03	0.82 ± 0.02	0.02 ± 0.02	0[b]	
Fiber glass:	0.20–0.70	0.85–0.90	. . .		

[a]Other cover-slide materials, such as sapphire, produce completely different emittance values.
[b]No degradation data available.

TABLE C.12 Manufacturers of High-performance Flat-plate collectors*

Manufacturer	Address
Beasley Industries, Ltd.	Bolton Avenue Devon Park South Australia 5008
Daylin, Inc.	9606 Santa Monica Blvd. Beverly Hills, Calif. 90210
Energex Corp.	481 Tropicana Las Vegas, Nev. 89109
Miromit, Ltd.	44 Montefiore Street Tel Aviv, Israel
PPG Industries	Pittsburg, Pa. 15238
Sunworks, Inc.	669 Boston Post Road Guilford, Conn. 06437
Solaron Corporation (Air Systems)	4850 Olive Street Denver, Colo. 80022

*Collectors produced by the companies listed in Table C.12 are capable of supplying heated water or water/glycol at 170°F or greater on a continuous basis.

TABLE C.13 Conductance and Resistance Values for External Air Surfaces*

Wind condition / Position of surface	Direction of heat flow	Type of surface					
		Foil		Aluminum-coated paper		Nonreflective building materials	
		Conductance C, Btu/(hr)(ft²)(°F)	Resistance R, 1/[Btu/(hr)(ft²)(°F)]	Conductance C, Btu/(hr)(ft²)(°F)	Resistance R, 1/[Btu/(hr)(ft²)(°F)]	Conductance C, Btu/(hr)(ft²)(°F)	Resistance R, 1/[Btu/(hr)(ft²)(°F)]
Still air:							
Horizontal	Up	0.76	1.32	0.91	1.10	1.63	0.61
45° slope	Up	0.73	1.37	0.88	1.14	1.60	0.62
Vertical	Horizontal	0.59	1.70	0.74	1.35	1.46	0.68
45° slope	Down	0.45	2.22	0.60	1.67	1.32	0.76
Horizontal	Down	0.22	4.55	0.37	2.70	1.08	0.92
7.5-mph wind Any position	Any direction (for summer calculations)	···	···	···	···	4.00	0.25
15-mph wind Any position	Any direction (for winter calculations)	···	···	···	···	6.00	0.17

*Adapted with permission from Johns-Mansville, Denver, Colo.; all values originally taken from "ASHRAE Handbook of Fundamentals," 1972.

TABLE C.14 Conductance and Resistance Values for Internal Air Spaces*

Position of air space	Direction of heat flow†	Thickness, in.	Temp. cond.‡	Foil and nonreflective building materials		Aluminum-coated paper and nonreflective bldg. materials		Both surfaces nonreflective building materials	
				Conductance C, Btu/(hr)(ft²)(°F)	Resistance R, 1/[Btu/(hr)(ft²)(°F)]	Conductance C, Btu/(hr)(ft²)(°F)	Resistance R, 1/[Btu/(hr)(ft²)(°F)]	Conductance C, Btu/(hr)(ft²)(°F)	Resistance R, 1/[Btu/(hr)(ft²)(°F)]
Horizontal	Up	3/4	W	0.45	2.23	0.59	1.71	1.15	0.87
		3/4	S	0.44	2.26	0.61	1.63	1.32	0.76
		4	W	0.37	2.73	0.50	1.99	1.07	0.94
		4	S	0.36	2.75	0.53	1.87	1.24	0.80
45° slope	Up	3/4	W	0.36	2.78	0.50	2.02	1.06	0.94
		3/4	S	0.36	2.81	0.53	1.90	1.24	0.81
		4	W	0.33	3.00	0.47	2.13	1.04	0.96
		4	S	0.33	3.00	0.51	1.98	1.21	0.82
Vertical	Horiz.	3/4	W	0.29	3.48	0.42	2.36	0.99	1.01
		3/4	S	0.31	3.28	0.48	2.10	1.19	0.84
		4	W	0.29	3.45	0.43	2.34	0.99	1.01
		4	S	0.29	3.44	0.46	2.16	1.17	0.91
45° slope	Down	3/4	W	0.28	3.57	0.42	2.40	0.98	1.02
		3/4	S	0.31	3.24	0.48	2.09	1.19	0.84
		4	W	0.23	4.41	0.36	2.75	0.93	1.08
		4	S	0.23	4.36	0.40	2.50	1.11	0.90
Horizontal	Down	3/4	W	0.28	3.55	0.42	2.39	0.98	1.02
		1½	W	0.17	5.74	0.31	3.21	0.88	1.14
		4	W	0.11	8.94	0.25	4.02	0.81	1.23
		3/4	S	0.31	3.25	0.48	2.08	1.19	0.84
		1½	S	0.19	5.24	0.36	2.76	1.07	0.93
		4	S	0.12	8.08	0.30	3.38	1.01	0.99

*Adapted with permission from Johns-Mansville, Denver, Colo.; all values originally derived from "ASHRAE Handbook of Fundamentals," 1972.
†Heat flows from hot to cold. For ceiling installation the direction of heat flow would normally be "up" for winter and "down" for summer. In a floor the direction of heat flow would be "down" in winter and "up" in summer. Heat flow in walls would be in a horizontal direction.
‡W = winter; S = summer.

TABLE C.15 Conductance and Resistance Values for Exterior Siding Materials*

Material	Description	Conductivity k, Btu/(hr)(ft²)(°F/in.)	Thickness, in.	Conductance C, Btu/(hr)(ft²)(°F)	Resistance R, 1/[Btu/(hr)(ft²)(°F)]
Brick	Common	5.0	4	1.25	0.80
Brick	Face	9.0	4	2.27	0.44
Stucco		5.0	1	5.0	0.20
Asbestos cement shingles				4.76	0.21
Wood shingles	16–7$\frac{1}{2}$-in. exposure			1.15	0.87
Wood shingles	Double 16–12-in. exposure			0.84	1.19
Wood shingles	Plus $\frac{5}{16}$ in. Insulated backerboard			0.71	1.40
Asbestos cement siding	$\frac{1}{4}$ in. lapped			4.76	0.21
Asphalt roll siding				6.50	0.15
Asphalt insulating siding			$\frac{1}{2}$	0.69	1.46
Wood	Drop siding, 1 x 8 in.			1.27	0.79
Wood	Bevel, $\frac{1}{2}$ x 8 in. lapped			1.23	0.81
Wood	Bevel, $\frac{3}{4}$ x 10 in. lapped			0.95	1.05
Wood	Plywood, $\frac{3}{8}$ in. lapped			1.59	0.59
Hardboard	Medium density	0.73	$\frac{1}{4}$	2.94	0.34
	Tempered	1.00	$\frac{1}{4}$	4.00	0.25
Plywood lap siding			$\frac{3}{8}$	1.79	0.56
Plywood flat siding			$\frac{3}{8}$	2.33	0.43

*Adapted with permission from Johns-Mansville, Denver, Colo.; all values originally derived from "ASHRAE Handbook of Fundamentals," 1972.

TABLE C.16 Conductance and Resistance Values for Sheathing and Building Paper*

Material	Description	Conductivity k, Btu/(hr)(ft²)(°F/in.)	Thickness, in.	Conductance C, Btu/(hr)(ft²)(°F)	Resistance R, 1/[Btu/(hr)(ft²)(°F)]
Gypsum	...	1.11	$\frac{3}{8}$	3.10	0.32
			$\frac{1}{2}$	2.25	0.45
			$\frac{5}{8}$	1.75	0.57
Plywood	...	0.80	$\frac{1}{4}$	3.20	0.31
			$\frac{3}{8}$	2.13	0.47
			$\frac{1}{2}$	1.60	0.62
			$\frac{5}{8}$	1.28	0.78
			$\frac{3}{4}$	1.07	0.93
Nail-base sheathing	...	0.44	$\frac{1}{2}$	0.88	1.14
Wood sheathing	Fir or pine	0.80	$\frac{3}{4}$	1.06	0.94
Sheathing paper	Vapor-permeable			16.70	0.06
Vapor barrier	2 layers mopped 15-lb felt			8.35	0.12
	Plastic film			Negl.	Negl.

*Adapted with permission from Johns-Mansville, Denver, Colo.; all values originally derived from "ASHRAE Handbook of Fundamentals," 1972..

TABLE C.17 Conductance and Resistance Values for Masonry Materials*

Material	Description	Conductivity k, Btu/(hr)(ft²)(°F/in.)	Thickness, in.	Conductance C, Btu/(hr)(ft²)(°F)	Resistance R, 1/[Btu/(hr)(ft²)(°F)]
Concrete blocks, three-oval core	Sand and gravel aggregate	...	4	1.40	0.71
			8	0.90	1.11
			12	0.78	1.28
	Cinder aggregate	...	4	0.90	1.11
			8	0.58	1.72
			12	0.53	1.89
	Lightweight aggre- gate	...	4	0.67	1.50
			8	0.50	2.00
			12	0.44	2.27
Hollow clay tile	1 cell deep	...	4	0.90	1.11
	2 cells deep	...	8	0.54	1.85
	3 cells deep	...	12	0.40	2.50
Gypsum parti- tion tile	3 × 12 × 30 in. solid	...	3	0.79	1.26
	3 × 12 × 30 in. 4-cell	...	3	0.74	1.35
	4 × 12 × 30 in. 3-cell	...	4	0.60	1.67
Cement mortar		5.0	1	5.0	0.20
Stucco		5.0	1	5.0	0.20
Gypsum	Poured	1.66	1	1.66	0.60
	Precast	2.80	2	1.40	0.71
Concrete	Sand and gravel or stone	12.0	1	12.0	0.08
Lightweight concrete	Perlite or zonolite mixture				
	1:4 mix, 36 lb/ft³	0.72–0.75	1	0.74	1.35
	1:5 mix, 30 lb/ft³	0.61–0.72	1	0.67	1.49
	1:6 mix, 27 lb/ft³	0.54–0.61	1	0.58	1.72
	1:8 mix, 22 lb/ft³	0.47–0.54	1	0.51	1.96
Stone	...	12.5	1	12.5	0.08

*Adapted with permission from Johns-Mansville, Denver, Colo.; all values originally derived from "ASHRAE Handbook of Fundamentals," 1972.

TABLE C.18 Conductance and Resistance Values for Woods*

Material	Description	Conductivity k, Btu/(hr)(ft²)(°F/in.)	Thickness, in.	Conductance C, Btu/(hr)(ft²)(°F)	Resistance R, 1/[Btu/(hr)(ft²)(°F)]
Maple, oak and similar hardwoods	45 lb/ft³	1.10	$\frac{3}{4}$	1.47	0.68
Fir, pine and similar softwoods	32 lb/ft³	C.80	$\frac{3}{4}$	1.06	0.94
			$1\frac{1}{2}$	0.53	1.89
			$2\frac{1}{2}$	0.32	3.12
			$3\frac{1}{2}$	0.23	4.35

*Adapted with permission from Johns-Mansville, Denver, Colo.; all values originally derived from "ASHRAE Handbook of Fundamentals," 1972.

TABLE C.19 Conductance and Resistance Values for Wall-insulation Materials*

Material	Description	Conductivity k, Btu/(hr)(ft^2)($^\circ$F/in.)	Thickness, in.	Conductance C, Btu/(hr)(ft^2)($^\circ$F)	Resistance R, 1/[Btu/(hr)(ft^2)($^\circ$F)]
Fiber glass roof insulation	...	...	$\frac{15}{16}$	0.27	3.70
			$1\frac{1}{16}$	0.24	4.17
			$1\frac{5}{16}$	0.19	5.26
			$1\frac{5}{8}$	0.15	6.67
			$1\frac{7}{8}$	0.13	7.69
			$2\frac{1}{4}$	0.11	9.09
Urethane roof insulation	Thickness includes membrane roofing on both sides	0.13	$\frac{4}{5}$	0.19	5.26
			1	0.15	6.67
			$1\frac{1}{5}$	0.12	8.33
Styrofoam SM & TG	2.1 lb/ft^3	0.19	$\frac{3}{4}$	0.25	3.93
			1	0.19	5.26
			$1\frac{1}{2}$	0.13	7.89
			2	0.95	10.52
Wood shredded	Cemented in pre-formed slabs	0.60	1	0.60	1.67
Insulating board	Building and service board, decorative ceiling panels	0.38	$\frac{3}{8}$	1.01	0.99
			$\frac{1}{2}$	0.76	1.32
			$\frac{9}{16}$	0.68	1.48
			$\frac{3}{4}$	0.51	1.98
Thermal, acoustical fiber glass	...	0.39	$2\frac{3}{4}$	0.14	7.00
		0.36	4	0.09	11.00
		0.34	$6\frac{1}{2}$	0.05	19.00
Corkboard	6.4 lb/ft^3	0.26	1	0.26	3.85
Expanded polystyrene Extruded	1.8 lb/ft^3	0.25	1	0.25	4.00
Molded beads	1.0 lb/ft^3	0.26	1	0.26	3.85
Urethane foam Thurane (Dow Chemical)	1.9 lb/ft^3	0.17	$\frac{3}{4}$	0.23	4.41
			$1\frac{1}{2}$	0.11	8.82
			2	0.09	11.76
Fiberglas perimeter insulation	...	...	1	0.23	4.30
			$1\frac{1}{4}$	0.19	5.40
Fiberglas form board	...	...	1	0.25	4.00

*Adapted with permission from Johns-Mansville, Denver, Colo.; all values originally derived from "ASHRAE Handbook of Fundamentals," 1972.

TABLE C.20 Conductance and Resistance Values for Roofing Materials*

Material	Description	Conductivity k, Btu/(hr)(ft^2)($^\circ$F/in.)	Thickness, in.	Conductance C, Btu/(hr)(ft^2)($^\circ$F)	Resistance R, 1/[Btu/(hr)(ft^2)($^\circ$F)]
Asbestos cement shingles	120 lb/ft^3	...	...	4.76	0.21
Asphalt shingles	70 lb/ft^3	...	...	2.27	0.44
Wood shingles	...	...	...	1.06	0.94
Slate	...	...	$\frac{1}{2}$	20.0	0.05
Asphalt roll roofing	70 lb/ft^3	...	...	6.50	0.15
Built-up roofing	Smooth or gravel surface	...	$\frac{3}{8}$	3.00	0.33
Sheet metal	...	...	...	Negl.	Negl.

*Adapted with permission from Johns-Mansville, Denver, Colo.; all values originally derived from "ASHRAE Handbook of Fundamentals," 1972.

TABLE C.21 Conductance and Resistance Values for Flooring Materials*

Material	Description	Conductivity k, Btu/(hr)(ft²)(°F/in.)	Thickness, in.	Conductance C, Btu/(hr)(ft²)(°F)	Resistance R, 1/[Btu/(hr)(ft²)(°F)]
Asphalt, vinyl, rubber, or linoleum tile	...	...	...	20.0	0.05
Cork tile	...	0.45	$\frac{1}{8}$	3.60	0.28
Terrazzo	...	12.5	1	12.50	0.08
Carpet and fibrous pad	...	...	...	0.48	2.08
Carpet and rubber pad	...	...	...	0.81	1.23
Plywood subfloor	...	0.80	$\frac{5}{8}$	1.28	0.78
Wood subfloor	...	0.80	$\frac{3}{4}$	1.06	0.94
Wood, hardwood finish	...	1.10	$\frac{3}{4}$	1.47	0.68

*Adapted with permission from Johns-Mansville, Denver, Colo.; all values originally derived from "ASHRAE Handbook of Fundamentals," 1972.

TABLE C.22 Conductance and Resistance Values for Interior Finishes*

Material	Description	Conductivity k, Btu/(hr)(ft²)(°F/in.)	Thickness, in.	Conductance C, Btu/(hr)(ft²)(°F)	Resistance R, 1/[Btu/(hr)(ft²)(°F)]
Gypsum board		1.11	$\frac{3}{8}$	3.10	0.32
			$\frac{1}{2}$	2.25	0.45
Cement plaster	Sand aggregate	5.0	$\frac{1}{2}$	10.00	0.10
			$\frac{3}{4}$	6.66	0.15
Gypsum plaster	Sand aggregate	5.6	$\frac{1}{2}$	11.10	0.09
			$\frac{5}{8}$	9.10	0.11
Gypsum plaster	Lightweight aggregate	1.6	$\frac{1}{2}$	3.12	0.32
			$\frac{5}{8}$	2.67	0.39
Gypsum plaster on					
Metal lath	Sand aggregate		$\frac{3}{4}$	7.70	0.13
Metal lath	Lightweight aggregate		$\frac{3}{4}$	2.13	0.47
Gypsum board, $\frac{3}{8}$ in.	Sand aggregate		$\frac{7}{8}$	2.44	0.41
Insulating board		0.38	$\frac{1}{2}$	0.74	1.35
Plywood		0.80	$\frac{3}{8}$	2.13	0.47

*Adapted with permission from Johns-Mansville, Denver, Colo.; all values originally derived from "ASHRAE Handbook of Fundamentals," 1972.

TABLE C.23 Conductance and Resistance Values for Glass*

Material	Description	Conductivity k, Btu/(hr)(ft²)(°F/in.)	Thickness, in.	Conductance C (U), Btu/(hr)(ft²)(°F)	Resistance R (1/U), 1/[Btu/(hr)(ft²)(°F)]
Single-plate				1.13	0.88
Double-plate	Air space, $\frac{3}{16}$ in.	...	...	0.69	1.45
Storm windows	Air space, 1–4 in.	...	...	0.56	1.78
Solid-wood door	Actual thickness, $1\frac{1}{2}$ in.	...	...	0.49	2.04
Storm door, wood and glass	With wood and glass storm door	...	...	0.27	3.70
Storm door, metal and glass	With metal and glass storm door	...	...	0.33	3.00

*Adapted with permission from Johns-Mansville, Denver, Colo.; all values originally derived from "ASHRAE Handbook of Fundamentals," 1972.

The altitude and azimuth of the sun are given by

$$\sin a = \sin \phi \sin \delta + \cos \phi \cos \delta \cos h \qquad (1)$$

and

$$\sin \alpha = -\cos \delta \sin h / \cos a \qquad (2)$$

where

a = altitude of the sun (angular elevation above the horizon),
ϕ = latitude of the observer,
δ = declination of the sun,
h = hour angle of sun (angular distance from the meridian of the observer),
α = azimuth of the sun (measured eastward from north).

From equations (1) and (2) it can be seen that the altitude and azimuth of the sun are functions of the latitude of the observer, the time of day (hour angle) and the date (declination).

Table 170 [*] provides a series of charts, one for each 5 degrees of latitude (except 5°, 15°, 75°, and 85°) giving the altitude and azimuth of the sun as a function of the true solar time and the declination of the sun in a form originally suggested by Hand.[1] Linear interpolation for intermediate latitudes will give results within the accuracy to which the charts can be read.

On these charts, a point corresponding to the projected position of the sun is determined from the heavy lines corresponding to declination and solar time.

To find the solar altitude and azimuth:
1. Select the chart or charts appropriate to the latitude.
2. Find the solar declination δ corresponding to the date.
3. Determine the *true solar time* as follows:
   (a) To the *local standard time* (zone time) add 4 minutes for each degree of longitude the station is east of the standard meridian or subtract 4 minutes for each degree west of the standard meridian to get the *local mean solar time*.
   (b) To the *local mean solar time* add algebraically the equation of time; the sum is the required *true solar time*.
4. Read the required altitude and azimuth at the point determined by the declination and the true solar time. Interpolate linearly between two charts for intermediate latitudes.

It should be emphasized that the solar altitude determined from these charts is the true geometric position of the center of the sun. At low solar elevations terrestrial refraction may considerably alter the apparent position of sun. Under average atmospheric refraction the sun will appear on the horizon when it actually is about 34' below the horizon; the effect of refraction decreases rapidly with increasing solar elevation. Since sunset or sunrise is defined as the time when the upper limb of the sun appears on the horizon, and the semidiameter of the sun is 16', sunset or sunrise occurs under average atmospheric refraction when the sun is 50' below the horizon. In polar regions especially, unusual atmospheric refraction can make considerable variation in the time of sunset or sunrise.

The 90° N. chart is included for interpolation purposes, the azimuths lose their directional significance at the pole.

Altitude and azimuth in southern latitudes.—To compute solar altitude and azimuth for southern latitudes, change the sign of the solar declination and proceed as above. The resulting azimuths will indicate angular distance from *south* (measured eastward) rather than from north.

[*]Fig. C.1 (*a–g*) in this volume.

(a)

FIG. C.1 Description of method for calculating true solar time, together with accompanying meteorological charts, for computing solar altitude and azimuth angles. (*a*) Description of method; (*b*) chart, 25°N latitude; (*c*) chart, 30°N latitude; (*d*) chart, 35°N latitude; (*e*) chart, 40°N latitude; (*f*) chart, 45°N latitude; (*g*) chart, 50°N latitude. (*Description and charts reproduced from the "Smithsonian Meteorological Tables" with permission from the Smithsonian Institute, Washington, D.C.*)

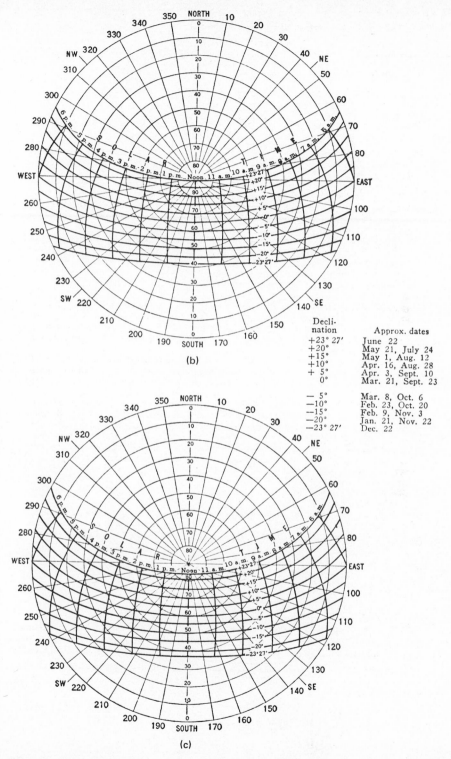

Declination | Approx. dates
+23° 27' | June 22
+20° | May 21, July 24
+15° | May 1, Aug. 12
+10° | Apr. 16, Aug. 28
+ 5° | Apr. 3, Sept. 10
0° | Mar. 21, Sept. 23
− 5° | Mar. 8, Oct. 6
−10° | Feb. 23, Oct. 20
−15° | Feb. 9, Nov. 3
−20° | Jan. 21, Nov. 22
−23° 27' | Dec. 22

(b)

(c)

FIG. C.1 *(continued)*

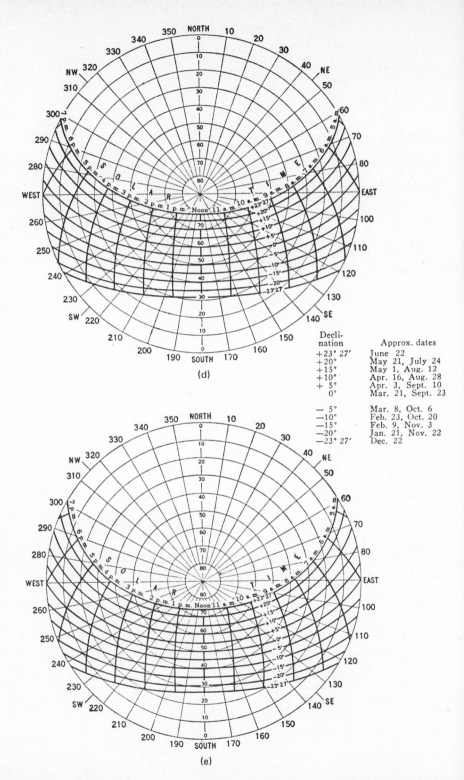

Declination	Approx. dates
+23° 27′	June 22
+20°	May 21, July 24
+15°	May 1, Aug. 12
+10°	Apr. 16, Aug. 28
+5°	Apr. 3, Sept. 10
0°	Mar. 21, Sept. 23
−5°	Mar. 8, Oct. 6
−10°	Feb. 23, Oct. 20
−15°	Feb. 9, Nov. 3
−20°	Jan. 21, Nov. 22
−23° 27′	Dec. 22

(d)

(e)

FIG. C.1 *(continued)*

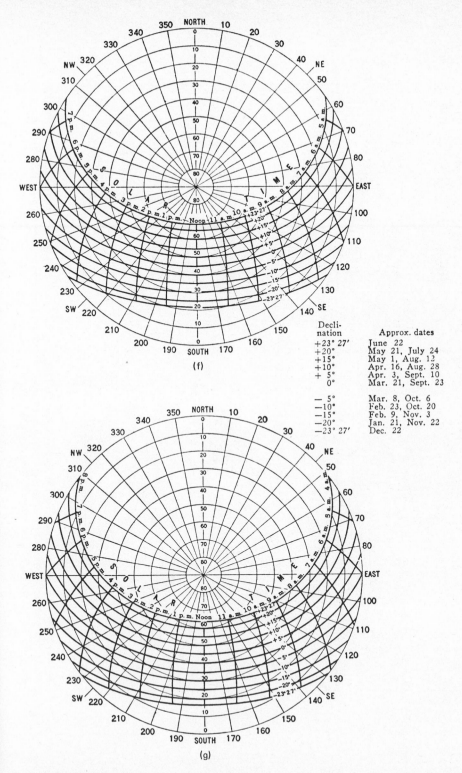

Declination	Approx. dates
+23° 27′	June 22
+20°	May 21, July 24
+15°	May 1, Aug. 13
+10°	Apr. 16, Aug. 28
+ 5°	Apr. 3, Sept. 10
0°	Mar. 21, Sept. 23
— 5°	Mar. 8, Oct. 6
—10°	Feb. 23, Oct. 20
—15°	Feb. 9, Nov. 3
—20°	Jan. 21, Nov. 22
—23° 27′	Dec. 22

(f)

(g)

FIG. C.1 *(continued)*

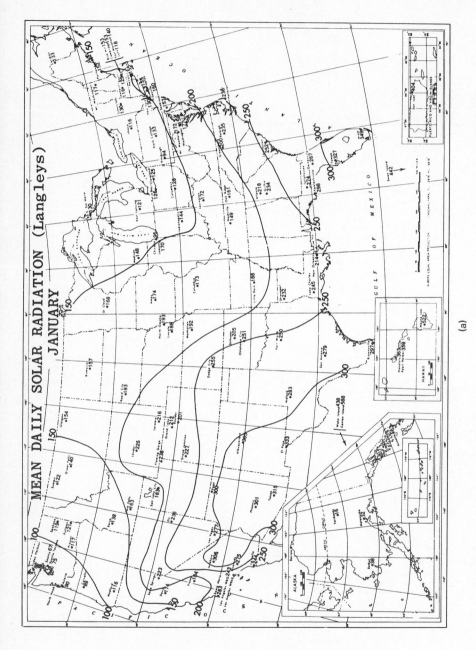

(a)

FIG. C.2 Mean daily insolation data for the United States. (*a*) January; (*b*) February; (*c*) March; (*d*) April; (*e*) May; (*f*) June; (*g*) July; (*h*) August; (*i*) September; (*j*) October; (*k*) November; (*l*) December. (*Reproduced with permission from the Climatic Atlas of the United States, U. S. Government Printing Office, 1968.*)

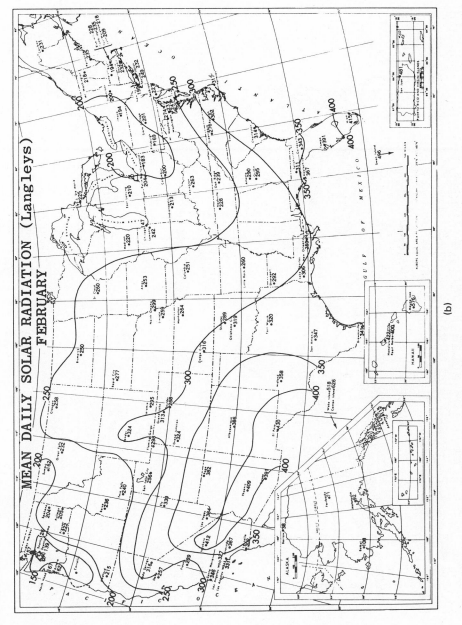

MEAN DAILY SOLAR RADIATION (Langleys)
FEBRUARY

(b)

FIG. C.2 *(continued)*

277

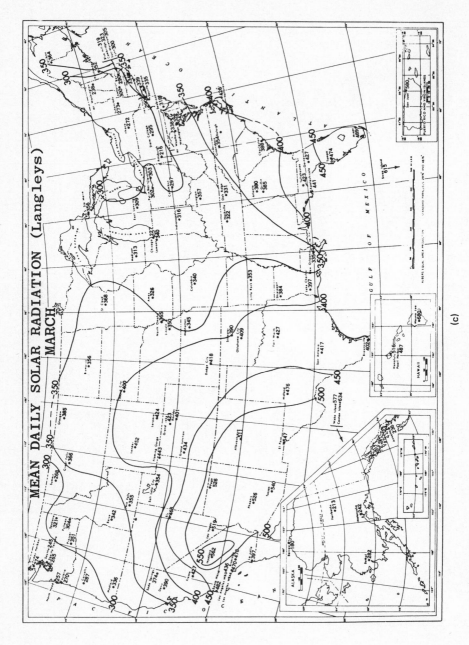

FIG. C.2 (continued)

278

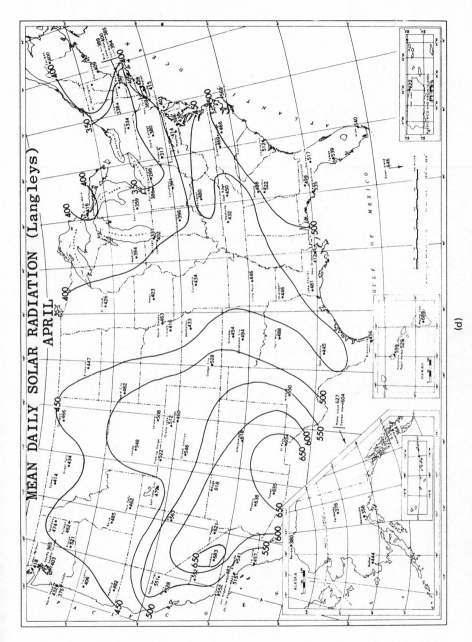

MEAN DAILY SOLAR RADIATION (Langleys)
APRIL

FIG. C.2 *(continued)*

(d)

279

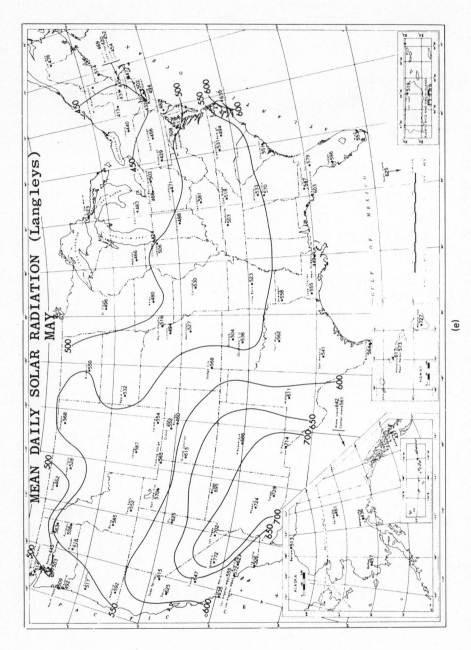

MEAN DAILY SOLAR RADIATION (Langleys)
MAY

FIG. C.2 *(continued)*

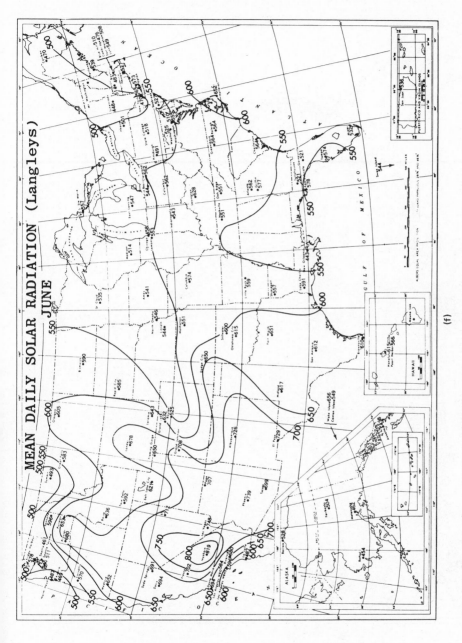

FIG. C.2 *(continued)*

281

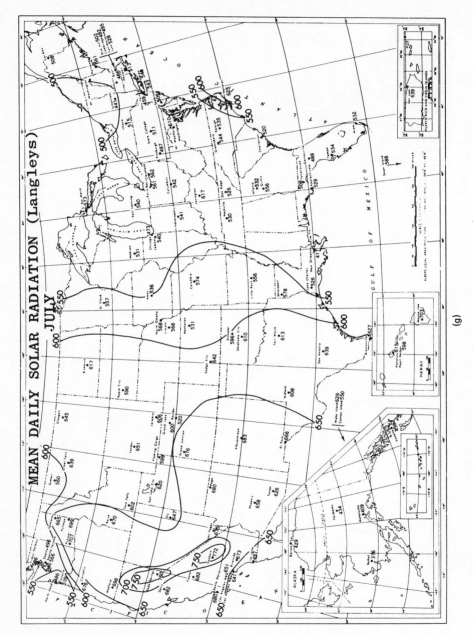

FIG. C.2 *(continued)*

282

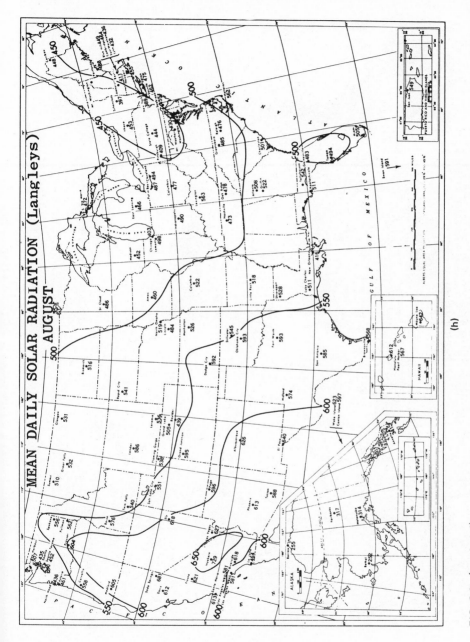

FIG. C.2 *(continued)*

283

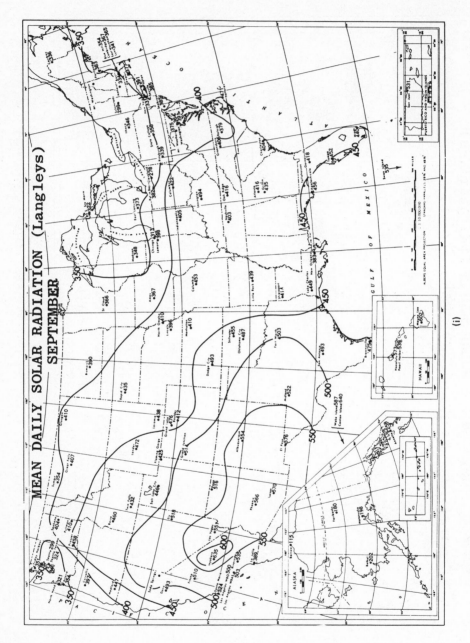

FIG. C.2 *(continued)*

284

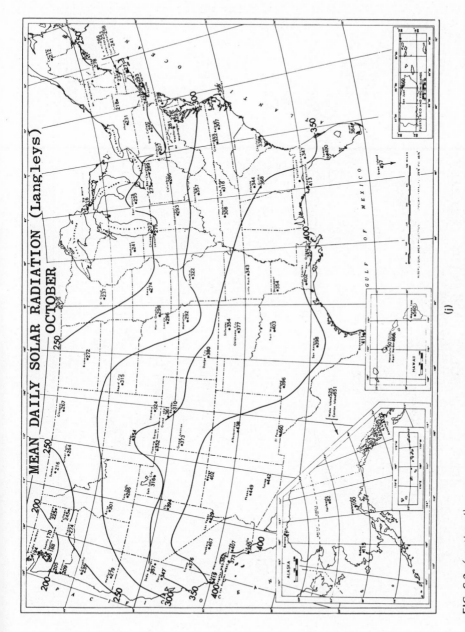

FIG. C.2 *(continued)*

285

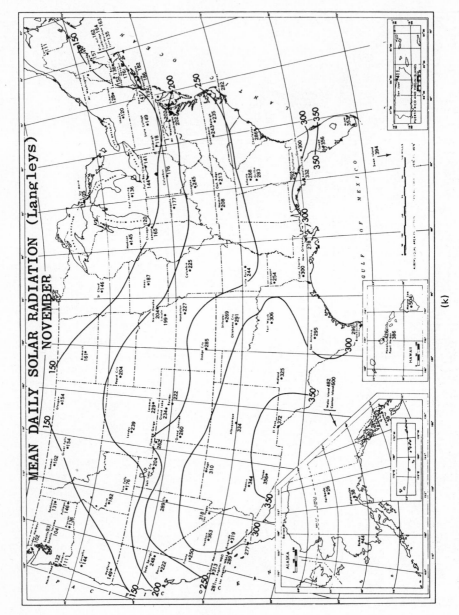

MEAN DAILY SOLAR RADIATION (Langleys)
NOVEMBER

FIG. C.2 *(continued)*

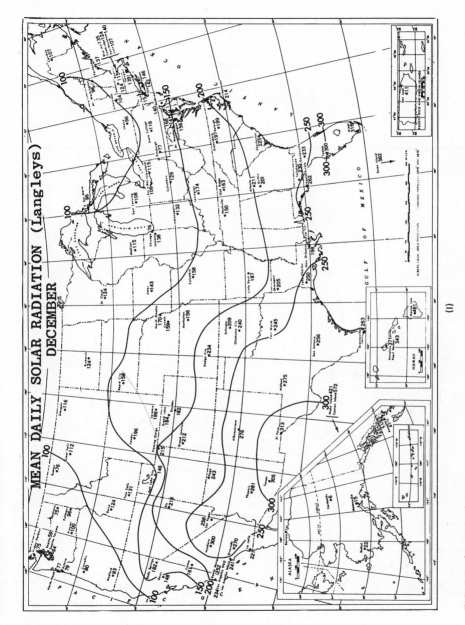

FIG. C.2 (continued)

287

LOCAL CLIMATOLOGICAL DATA

U.S. DEPARTMENT OF COMMERCE
NATIONAL OCEANIC AND ATMOSPHERIC ADMINISTRATION
ENVIRONMENTAL DATA SERVICE

DENVER, COLORADO
NAT WEATHER SERVICE FCST OFC
STAPLETON INTERNATIONAL AP
FEBRUARY 1973

Latitude 39° 45' N   Longitude 104° 52' W   Elevation (ground) 5283 ft.   Standard time used: MOUNTAIN   WBAN #23062

Date	Temp Max	Temp Min	Temp Avg	Dep. from normal	Avg dew point	Heating (7A)	Cooling (7B)	Weather types	Snow/ice on ground 05AM In.	Precip. Water equiv. In.	Snow, ice pellets In.	Avg station press. Elev 5332 ft m.s.l.	Wind Resultant dir.	Wind Resultant speed	Wind Avg speed mph	Fastest mile Speed mph	Fastest mile Direction	Sunshine Hours & tenths	Sunshine % of possible	Sky Sunrise-sunset	Sky Midnight-midnight	Date
1	39	20	30	2	15	35	0		1	0	0	24.65	03	2.5	6.5	14	N	9.3	92	3	3	1
2	48	16	32	0	12	33	0		1	0	0	24.77	19	9.6	9.6	14	S	10.2	100	0	5	2
3	49	25	37	7	17	28	0		1	0	0	24.65	20	6.5	8.1	13	SW	7.7	78	7	6	3
4	52	31	42	12	22	23	0		T	0	0	24.62	23	4.4	9.5	19	SE	7.7	75	0	0	4
5	52	27	40	10	22	25	0	1	T	0	0	24.73	17	6.7	9.5	11	SE	10.3	100	2	7	5
6	38	26	32	2	23	33	0	1 6 8	T	.03	.4	24.72	36	6.0	8.3	12	NE	6.3	61	0	0	6
7	30	12*	21	-9	13	44	0		T	0	0	24.86	05	5.4	6.0	17	N	0.5	0	10	10	7
8	29	11	18*	-12	10	47	0		T	0	0	24.90	18	4.2	7.5	09	S	10.5	100	1	2	8
9	44	25	37	-3	17	37	0		T	0	0	24.72	17	7.3	7.5	14	NE	8.9	100	8	7	9
10	46	29	36	11	20	29	0		T	0	0	24.52	17	4.7	4.7	17	N	10.4	85	5	8	10
11	51	30	42	-1	25	23	0		T	0	0	24.31	19	7.8	7.6	20	SE	8.5	80	9	8	11
12	54	24	41	10	18	24	0	1 8	T	.01	.2	24.30	16	5.3	5.3	16	NE	4.9	46	9	8	12
13	36	23	30	-1	18	35	0	1	T	0	0	24.72	07	5.8	11.8	12	NE	2.8	26	0	0	13
14	37	20	30	-1	15	35	0		T	0	0	24.75	16	6.3	8.1	09	NE	8.6	80	5	3	14
15	44	21	32	-1	14	33	0		T	0	0	24.80	13	5.1	5.2	17	SW	10.8	100	0	4	15
16	47	20	34	-2	18	31	0		T	0	0	24.78	16	5.5	8.6	11	NE	10.8	100	3	4	16
17	39	21	34	7	18	34	0		T	0	0	24.60	19	7.5	7.5	13	SW	9.7	89	5	7	17
18	52	25	39	-5	14	26	0		T	.10	2.2	24.85	13	3.2	10.4	20	E	9.9	94	10	7	18
19	58	23	41	9	19	24	0		2	.02	2.2	24.88	20	4.3	9.2	18	SW	10.9	100	8	5	19
20	50	24	37	4	16	28	0		1	0	0	24.97	18	1.8	12.2	18	N	10.3	97	4	6	20
21	55	27	41	8	21	24	0		T	0	0	24.60	17	2.4	7.5	10	NE	10.9	90	10	7	21
22	55	30	45	10	21	20	0		T	0	0	24.63	20	2.4	7.0	12	SW	10.4	94	6	7	22
23	57	32	44	12	19	21	0		T	0	0	24.75	17	4.3	7.1	13	SE	8.6	78	10	5	23
24	58	30	46*	10	25	19	0		T	0	0	24.63	21	6.8	4.3	13	N	10.5	94	6	6	24
25	58	29	44	10	25	21	0		T	0	0	24.67	35	7.3	6.8	14	SW	10.1	90	8	6	25
26	62*	32	44	10	27	21	0		T	0	0	24.56	15	5.2	6.9	15	SE	5.5	48	8	7	26
27	56																					27
28																						28

For the month:

	Temperature Max	Min	Avg	Dep.	Dew
Sum	1330	655			
Avg	47.5	23.4	35.5	4.0	18

Number of days — Maximum Temp: ≥90° 0, ≤32° 0 — Minimum Temp: ≤32° 28, ≤0° 2

Degree days Base 65: Heating Total 820, Dep. -118; Season to date Total 4742, Dep. 258. Cooling Total 0, Dep. 0

Number of days — Precipitation ≥.01 inch 4, ≥1.0 inch 1; Snow, ice pellets ≥1.0 inch 1; Thunderstorms 0; Heavy fog X 0; Clear 9; Partly cloudy 10; Cloudy 9

Precipitation Total .16, Dep. -0.53; Snow, ice pellets Total 3.0, Dep. 3.0

Greatest in 24 hours and dates — Precipitation .12 18-19; Snow, ice pellets 2.4 18-19

Greatest depth on ground of snow, ice pellets 2, date 19+

Pressure: Resultant dir. 17, Resultant speed 3.0, Avg speed 7.8, Fastest mile 20 SW, Date 19+

Sunshine: Total 240.4, Possible 299.7, % for month 80

Sky cover Tenths: Sunrise to sunset Sum 144 Avg 5.1; Midnight to midnight Sum 132 Avg 4.7

Weather types on dates of occurrence:
1 Fog
2 Heavy fog x
3 Thunderstorm
4 Ice pellets
5 Hail
6 Glaze
7 Duststorm
8 Smoke, Haze
9 Blowing snow

HOURLY PRECIPITATION (Water equivalent in inches)

Date	A. M. Hour ending at												P. M. Hour ending at												Date
	1	2	3	4	5	6	7	8	9	10	11	12	1	2	3	4	5	6	7	8	9	10	11	12	
1																									1
2																									2
3																									3
4																									4
5																									5
6																	T	T	T	T	T	T	T	T	6
7	T	T	T	T	T	T	T	T	T	.01	.01	.01	T	T	T	T	T	T	T	T	T				7
8																									8
9																									9
10																									10
11																									11
12																		T							12
13								T	T	T	.01	T	T							T	T	T	T	T	13
14	T	T	T	T	T	T	T																		14
15																									15
16																									16
17												T	T												17
18																				T	T	T	.04	.06	18
19	.02	T																							19
20																									20
21																									21
22																									22
23																									23
24																									24
25																									25
26																									26
27																									27
28																									28

- • Extreme temperatures for the month. May be the last of more than one occurrence.
- – Below zero temperature or negative departure from normal.
- ‡ ≥ 70° at Alaskan stations.
- + Also on an earlier date, or dates.
- X Heavy fog restricts visibility to ¼ mile or less.
- T In the Hourly Precipitation table and in columns 9, 10, and 11 indicates an amount too small to measure.

The season for degree days begins with July for heating and with January for cooling.

Data in columns 6, 12, 13, 14, and 15 are based on 8 observations per day at 3-hour intervals.

Wind directions are those from which the wind blows. Resultant wind is the vector sum of wind directions and speeds divided by the number of observations. Figures for directions are tens of degrees from true North; i.e., 09 = East, 18 = South, 27 = West, 36 = North, and 00 = Calm. When directions are in tens of degrees in Col. 17, entries in Col. 16 are fastest observed 1-minute speeds. If the / appears in Col. 17, speeds are gusts.

Any errors detected will be corrected and changes in summary data will be annotated in the annual summary.

Subscription Price: Local Climatological Data $2.00 per year including annual issue if published; 75¢ extra for foreign mailing. Single copy: 20¢ for monthly issue; 15¢ for annual issue. Make checks payable to Department of Commerce, NOAA; send payments and orders to: National Climatic Center, Federal Building, Asheville, N. C. 28801. Attn: Publications.

I certify that this is an official publication of the National Oceanic and Atmospheric Administration, and is compiled from records on file at the National Climatic Center, Asheville, North Carolina 28801.

William H. Haggard

Director, National Climatic Center

SUMMARY BY HOURS

Hour Local time	AVERAGES							Resultant wind	
	Sky cover Tenths	Station pressure In.	Temperature Air °F	Temperature Wet bulb °F	Temperature Dew Pt. °F	Relative humidity %	Wind speed m.p.h.	Direction	Speed m.p.h.
02	4	24.71	28	25	17	63	7.7	20	4.3
05	4	24.70	27	23	15	62	7.3	20	4.5
08	5	24.72	28	24	15	58	8.0	20	6.2
11	5	24.72	41	31	16	39	7.8	18	2.9
14	5	24.66	45	34	18	36	10.4	09	5.7
17	6	24.68	41	33	20	45	9.0	09	4.9
20	4	24.71	34	28	20	59	6.3	18	2.0
23	4	24.72	30	26	19	65	6.2	18	4.1

USCOMM — NOAA — ASHEVILLE 500

(a)

FIG. C.3 (*a*) U.S. Dept. of Commerce local climatological data; (*b*) observations at 3-hr intervals. (*Reproduced with permission from U.S. Dept. of Commerce.*)

OBSERVATIONS AT 3-HOUR INTERVALS

DAY 01

HOUR	SKY COVER Tenths	CEILING Hnds. of ft.	VISIBILITY WHOLE MILES	VISIBILITY 16ths MILE	WEATHER	AIR °F	WET BULB °F	DEW PT. °F	REL. HUM. %	WIND DIR	WIND SPEED Knots
02	7	UNL	20			29	25	16	58	36	8
05	3	UNL	20			26	23	15	63	36	6
08	2	UNL	60			25	22	14	63	00	0
11	2	UNL	60			36	28	12	37	01	5
14	8	UNL	60			39	30	16	39	02	10
17	4	UNL	60			33	28	18	54	13	6
20	1	UNL	20			28	25	17	63	18	7
23	0	UNL	20			22	20	13	68	17	3

DAY 02

HOUR	SKY COVER Tenths	CEILING Hnds. of ft.	VISIBILITY WHOLE MILES	VISIBILITY 16ths MILE	WEATHER	AIR °F	WET BULB °F	DEW PT. °F	REL. HUM. %	WIND DIR	WIND SPEED Knots
02	0	UNL	20			21	18	08	57	19	10
05	0	UNL	20			19	16	06	57	19	8
08	0	UNL	50			19	15	04	52	20	10
11	0	UNL	50			37	28	13	37	20	8
14	0	UNL	50			47	34	11	23	20	6
17	0	UNL	40			41	33	22	47	19	7
20	0	UNL	20			30	25	17	58	19	7
23	0	UNL	20			29	24	14	53	19	11

DAY 03

HOUR	SKY COVER Tenths	CEILING Hnds. of ft.	VISIBILITY WHOLE MILES	VISIBILITY 16ths MILE	WEATHER	AIR °F	WET BULB °F	DEW PT. °F	REL. HUM. %	WIND DIR	WIND SPEED Knots
02	0	UNL	20			27	22	11	51	19	7
05	0	UNL	20			29	23	10	45	19	10
08	6	UNL	50			26	22	11	53	22	8
11	8	250	50			43	33	18	37	20	7
14	7	250	25			47	36	22	37	01	5
17	8	250	50			43	35	24	47	21	3
20	8	200	20			35	30	23	62	20	6
23	5	UNL	20			36	30	19	50	19	10

DAY 04

HOUR	SKY COVER Tenths	CEILING Hnds. of ft.	VISIBILITY WHOLE MILES	VISIBILITY 16ths MILE	WEATHER	AIR °F	WET BULB °F	DEW PT. °F	REL. HUM. %	WIND DIR	WIND SPEED Knots
02	0	UNL	20			43	33	17	35	27	12
05	7	250	20			45	35	21	39	30	12
08	7	250	50			37	30	20	50	22	6
11	5	UNL	60			47	37	24	41	22	10
14	3	UNL	60			50	38	23	35	13	8
17	8	UNL	30			46	37	24	42	11	5
20	7	UNL	20			35	31	25	67	16	4
23	7	UNL	20			34	30	24	67	22	4

DAY 05

HOUR	SKY COVER Tenths	CEILING Hnds. of ft.	VISIBILITY WHOLE MILES	VISIBILITY 16ths MILE	WEATHER	AIR °F	WET BULB °F	DEW PT. °F	REL. HUM. %	WIND DIR	WIND SPEED Knots
02	3	UNL	20			31	28	23	72	23	6
05	2	UNL	20			29	26	20	69	20	6
08	2	UNL	40			31	27	20	64	20	8
11	0	UNL	60			43	35	25	49	20	8
14	0	UNL	60			51	39	22	32	08	9
17	7	UNL	50			44	36	24	45	12	10
20	0	UNL	20			36	31	23	59	19	8
23	0	UNL	20			34	29	21	59	15	11

DAY 06

HOUR	SKY COVER Tenths	CEILING Hnds. of ft.	VISIBILITY WHOLE MILES	VISIBILITY 16ths MILE	WEATHER	AIR °F	WET BULB °F	DEW PT. °F	REL. HUM. %	WIND DIR	WIND SPEED Knots
02	0	UNL	20			31	27	21	66	18	5
05	3	UNL	20			29	26	20	69	00	0
08	8	UNL	10			29	26	21	72	00	0
11	9	UNL	15			36	31	23	59	08	4
14	10	250	8			37	33	26	65	02	11
17	10	3	0	12	SFK	29	28	25	85	32	11
20	10	8	3		SFK	30	28	25	82	35	7
23	10	5	3		ZLSFK	30	28	25	82	02	4

DAY 07

HOUR	SKY COVER Tenths	CEILING Hnds. of ft.	VISIBILITY WHOLE MILES	VISIBILITY 16ths MILE	WEATHER	AIR °F	WET BULB °F	DEW PT. °F	REL. HUM. %	WIND DIR	WIND SPEED Knots
02	10	5	0	12	SFK	28	27	24	85	05	10
05	10	3	2		ZLSF	26	25	21	81	06	7
08	10	5	2		ZLSF	20	19	15	81	05	13
11	10	18	2		SF	17	16	12	81	05	8
14	10	14	3		SKH	18	16	09	68	04	8
17	10	13	0	12	SKH	17	15	09	71	04	7
20	10	95	10		SW	17	15	09	71	00	0
23	7	85	10			15	13	08	74	13	5

DAY 08

HOUR	SKY COVER Tenths	CEILING Hnds. of ft.	VISIBILITY WHOLE MILES	VISIBILITY 16ths MILE	WEATHER	AIR °F	WET BULB °F	DEW PT. °F	REL. HUM. %	WIND DIR	WIND SPEED Knots
02	9	95	15			14	11	01	56	12	7
05	6	UNL	20			12	10	03	67	22	3
08	1	UNL	15			12	10	01	61	23	5
11	0	UNL	10			22	18	05	48	17	6
14	0	UNL	10			26	21	10	51	20	4
17	0	UNL	20			27	20	03	35	13	7
20	0	UNL	15			21	18	09	60	18	8
23	0	UNL	20			16	14	06	64	19	6

DAY 09

HOUR	SKY COVER Tenths	CEILING Hnds. of ft.	VISIBILITY WHOLE MILES	VISIBILITY 16ths MILE	WEATHER	AIR °F	WET BULB °F	DEW PT. °F	REL. HUM. %	WIND DIR	WIND SPEED Knots
02	0	UNL	20			15	13	04	61	22	5
05	0	UNL	20			13	10	01	58	20	6
08	0	UNL	50			18	14	02	49	19	10
11	0	UNL	50			34	26	12	40	17	8
14	0	UNL	50			42	30	07	23	12	9
17	4	UNL	20			34	29	21	59	36	8
20	2	UNL	15			26	23	17	69	00	0
23	10	250	10			28	25	18	66	21	6

DAY 10

HOUR	SKY COVER Tenths	CEILING Hnds. of ft.	VISIBILITY WHOLE MILES	VISIBILITY 16ths MILE	WEATHER	AIR °F	WET BULB °F	DEW PT. °F	REL. HUM. %	WIND DIR	WIND SPEED Knots
02	10	250	10			27	24	17	66	17	4
05	6	250	20			25	22	15	66	00	0
08	3	UNL	50			26	23	15	63	00	0
11	7	UNL	50			42	32	16	35	17	11
14	10	300	50			44	33	15	31	15	15
17	9	250	50			44	34	19	37	15	7
20	4	UNL	20			37	30	19	48	19	12
23	7	250	20			34	28	19	54	17	4

DAY 11

HOUR	SKY COVER Tenths	CEILING Hnds. of ft.	VISIBILITY WHOLE MILES	VISIBILITY 16ths MILE	WEATHER	AIR °F	WET BULB °F	DEW PT. °F	REL. HUM. %	WIND DIR	WIND SPEED Knots
02	2	UNL	20			33	28	18	54	19	10
05	2	UNL	20			30	25	17	58	22	7
08	8	250	40			36	29	17	46	20	11
11	2	UNL	40			49	37	21	33	20	12
14	1	UNL	40			52	39	21	30	16	15
17	10	UNL	40			46	36	22	39	14	16
20	10	250	20			37	33	26	65	32	7
23	10	250	15			35	31	26	70	33	4

DAY 12

HOUR	SKY COVER Tenths	CEILING Hnds. of ft.	VISIBILITY WHOLE MILES	VISIBILITY 16ths MILE	WEATHER	AIR °F	WET BULB °F	DEW PT. °F	REL. HUM. %	WIND DIR	WIND SPEED Knots
02	10	250	20			34	30	25	70	22	5
05	8	200	20			30	27	23	75	20	5
08	10	120	15			32	29	25	75	25	5
11	8	UNL	15			42	35	25	51	13	3
14	8	200	20			48	37	22	36	08	8
17	10	80	20			41	35	28	60	00	0
20	3	UNL	20			36	32	27	70	16	7
23	7	80	20			32	29	25	75	10	4

DAY 13

HOUR	SKY COVER Tenths	CEILING Hnds. of ft.	VISIBILITY WHOLE MILES	VISIBILITY 16ths MILE	WEATHER	AIR °F	WET BULB °F	DEW PT. °F	REL. HUM. %	WIND DIR	WIND SPEED Knots
02	2	UNL	20			29	26	21	72	09	6
05	9	18	20			28	26	21	75	08	7
08	10	21	7		S	27	24	18	69	03	5
11	10	20	7		S	29	26	19	66	32	6
14	8	37	15			36	30	20	52	07	10
17	9	95	20			29	25	17	61	07	9
20	10	55	15		S	26	23	16	66	08	7
23	10	90	15		S	25	22	15	66	08	6

DAY 14

HOUR	SKY COVER Tenths	CEILING Hnds. of ft.	VISIBILITY WHOLE MILES	VISIBILITY 16ths MILE	WEATHER	AIR °F	WET BULB °F	DEW PT. °F	REL. HUM. %	WIND DIR	WIND SPEED Knots
02	10	37	10		S	24	21	15	68	09	6
05	10	30	10		S	23	20	13	65	13	6
08	9	180	10			25	22	15	66	14	4
11	6	UNL	15			31	25	15	51	09	6
14	6	UNL	15			34	28	19	54	12	8
17	4	UNL	20			33	29	21	61	03	6
20	0	UNL	15			26	24	19	75	00	0
23	0	UNL	15			25	23	18	75	00	0

DAY 15

HOUR	SKY COVER Tenths	CEILING Hnds. of ft.	VISIBILITY WHOLE MILES	VISIBILITY 16ths MILE	WEATHER	AIR °F	WET BULB °F	DEW PT. °F	REL. HUM. %	WIND DIR	WIND SPEED Knots
02	0	UNL	8			23	21	15	71	19	5
05	0	UNL	12			20	18	13	74	19	7
08	0	UNL	30			24	20	10	55	19	10
11	0	UNL	40			38	29	14	37	18	8
14	0	UNL	40			44	32	13	28	13	12
17	0	UNL	40			40	31	15	36	10	11
20	0	UNL	20			35	28	16	46	31	4
23	0	UNL	20			29	25	18	63	05	3

DAY 16

HOUR	SKY COVER Tenths	CEILING Hnds. of ft.	VISIBILITY WHOLE MILES	VISIBILITY 16ths MILE	WEATHER	AIR °F	WET BULB °F	DEW PT. °F	REL. HUM. %	WIND DIR	WIND SPEED Knots
02	0	UNL	20			23	21	16	74	00	0
05	0	UNL	20			22	20	13	68	20	6
08	0	UNL	50			24	20	10	55	20	5
11	0	UNL	20			36	28	12	37	11	5
14	0	UNL	20			40	31	16	38	07	8
17	0	UNL	20			37	30	19	48	01	8
20	0	UNL	20			32	27	18	56	17	3
23	0	UNL	12			26	24	19	75	21	4

DAY 17

HOUR	SKY COVER Tenths	CEILING Hnds. of ft.	VISIBILITY WHOLE MILES	VISIBILITY 16ths MILE	WEATHER	AIR °F	WET BULB °F	DEW PT. °F	REL. HUM. %	WIND DIR	WIND SPEED Knots
02	0	UNL	12			23	21	15	71	21	5
05	0	UNL	12			21	19	13	71	21	6
08	1	UNL	50			26	22	12	55	22	9
11	4	UNL	50			43	31	10	26	22	5
14	7	85	50			44	34	20	38	03	10
17	8	90	60			41	34	23	49	09	6
20	1	UNL	20			35	30	23	62	16	4
23	0	UNL	20			30	27	21	69	19	7

DAY 18

HOUR	SKY COVER Tenths	CEILING Hnds. of ft.	VISIBILITY WHOLE MILES	VISIBILITY 16ths MILE	WEATHER	AIR °F	WET BULB °F	DEW PT. °F	REL. HUM. %	WIND DIR	WIND SPEED Knots
02	0	UNL	20			29	25	18	63	19	7
05	0	UNL	20			26	23	15	63	18	8
08	0	UNL	60			31	26	16	54	20	9
11	1	UNL	60			47	35	16	29	32	4
14	7	UNL	60			51	38	19	28	09	15
17	7	UNL	60			46	35	19	34	10	8
20	10	110	20		SW	36	29	18	48	05	13
23	10	7	1		S	30	28	25	82	23	8

NOTES

CEILING COLUMN—

UNL indicates an unlimited ceiling.

WEATHER COLUMN—

- • Tornado
- T Thunderstorm
- Q Squall
- R Rain
- RW Rain showers
- ZR Freezing rain
- L Drizzle
- ZL Freezing drizzle
- S Snow
- SP Snow pellets
- IC Ice crystals
- SW Snow showers
- SG Snow grains
- IP Ice pellets
- A Hail
- F Fog
- IF Ice fog
- GF Ground fog
- BD Blowing dust
- BN Blowing sand
- BS Blowing snow
- BY Blowing spray
- K Smoke
- H Haze
- D Dust

WIND COLUMNS—

Directions are those from which the wind blows, indicated in tens of degrees from true North; i. e., 09 for East, 18 for South, 27 for West. Entry of 00 in the direction column indicates calm.

Speed is expressed in knots; multiply by 1.15 to convert to miles per hour.

DAY 19 DAY 20 DAY 21

DAY 22 DAY 23 DAY 24

DAY 25 DAY 26 DAY 27

DAY 28

ADDITIONAL DATA
Other observational data contained in records on file can be furnished at cost via microfilm, microfiche, or paper copies of the original records. Inquiries as to availability and costs should be addressed to: Director, National Climatic Center, Federal Building, Asheville, North Carolina 28801.

STATION: DENVER COLORADO   YEAR & MONTH: 73 02

(b)

FIG. C.3 *(continued)*

291

Nomenclature

A	Area, ft^2
A_{coll}	Collector area, ft^2
A_i	Area of the image of the sun on the collector of a concentrating collector, ft^2
A_{rf}	Roof area, ft^2
A_{wa}	Wall area, ft^2
$A_{wa,sh}$	Shaded wall area, ft^2
A_{wi}	Window area, ft^2
$A_{wi,sh}$	Shaded window area, ft^2
a_s	Solar-azimuth angle (positive west of south), deg
a_{wa}	Wall-azimuth angle (positive west of south), deg
B	Unpaid balance on a loan, $
C_{aux}	Average cost of auxiliary energy, $/MBtu
C_h	Additional annual cost of solar hardware, $
C_{hl}	Cost of heat energy loss, $
$C_{h,tot}$	Total additional, initial cost of solar system, $
C_{se}	Average annual cost of delivered solar energy, $/MBtu
C_{tot}	Total annual cost of delivered energy (solar + auxiliary), $
C_{wi1}	Total cost for a single-plate window, $
C_{wi2}	Total cost for a double-plate window, $
c	Angular coefficient for free convection from a surface, Btu/(hr)(ft^2)($^\circ$F$^{1.25}$)
c_a	Specific heat of air, Btu/(lb$_m$)($^\circ$F)
c_f	Energy transport-fluid heat capacity, Btu/(lb$_m$)($^\circ$F)
D	Flat-plate-collector intercover spacing, in. or ft
d_a	Diameter of the aperture area of a concentrating collector, in. or ft
d_{coll}	Depth of the collector, in. or ft
$E_{b,\lambda}$	Spectral emissive power, i.e., radiative heat flux in a small bandwidth interval centered at λ.
f_s	Fraction of heat demand delivered by solar system
F_{sh}	Decimal fraction of shaded portion of window (0.0, fully shaded; 1.0, unshaded)

f Outer-collector-cover heat loss ratio, i.e., the outer cover surface to non-outer-surface heat-loss rat'o

H_s Local solar-hour angle (measured from local solar noon, 1 hr = 15°), deg or rad

H_m Hour angle between sunrise and noon, deg or rad

h Enthalpy

h_c Coefficient of convection heat transfer from a surface to a fluid, Btu/(hr)(ft)(°F)

$h_{c,in}$ Coefficient of convection on the inside surface of a building, Btu/(hr)(ft^2)(°F)

$h_{c,out}$ Coefficient of convection on the outside surface of a building, Btu/(hr)(ft^2)(°F)

h_{c1} Coefficient of heat transfer from the outermost cover of a flat-plate collector to the environment, Btu/(hr)(ft^2)(°F)

I Insolation, i.e., the solar radiation incident on a surface, Btu/(hr)(ft^2)

I_b Beam, or direct, component of insolation at the surface of the earth, Btu/(hr)(ft^2)

$I_{b,coll}$ Component of beam, or direct insolation normal to the aperture of a collector, Btu/(hr)(ft^2)

I_{coll} Insolation absorbed by a collector surface, Btu/(hr)(ft^2)

I_h Total insolation on a horizontal surface, Btu/(hr)(ft^2)

$I_{h,b}$ Beam portion of horizontal total insolation, Btu/(hr)(ft^2)

$I_{h,d}$ Diffuse portion of horizontal total insolation, Btu/(hr)(ft^2)

I_r Diffuse insolation reflected from the ground, Btu/(hr)(ft^2)

I_{sc} Solar constant, i.e., the amount of solar energy received by a unit area of surface placed perpendicular to the sun outside the earth's atmosphere at the earth's mean distance from the sun (429 ±7), Btu/(hr)(ft^2)

i Incidence angle, deg and rad

i_{ann} Annual interest rate, \$/\$/yr

i_d Discount rate, \$/\$/yr

i_{eff} Effective interest rate, \$/\$/yr

K Storage rate of water per square foot of collector, lb$_m$/ft^2

k Thermal conductivity, Btu/(hr)(ft)(°F)

k_a Thermal conductivity of air, Btu/(hr)(ft)(°F)

k_g Thermal conductivity of glass, Btu/(hr)(ft)(°F)

L Latitude, deg

L_{coll} Latitude of collector (positive north of equator), deg

L_t Total thermal losses from a flat-plate collector, Btu/(hr)(ft^2)

m Air mass, i.e., the distance through which radiation travels from the outer edge of the earth's atmosphere to a

	recovery point on the earth divided by the distance radiation travels to a point on the equator at sea level when the sun is overhead
$\dot{m}$	Energy-transport-fluid flow rate per unit of collector area, $lb_m/(hr)(ft^2)$
$\dot{m}_a$	Infiltration/exfiltration rate, lb_m dry air/hr
$\dot{m}_{ab}$	Flow rate of absorbent, lb/hr
$\dot{m}_r$	Flow rate of refrigerant, lb/hr
$\dot{m}_s$	Flow rate of refrigerant–absorbent solution, lb/hr
n	Number of glass covers on a flat-plate collector
P	Principal, $
P_{ann}	Constant annual payments on a loan, $
P_0	Initial annual payment on a loan, $
p	Atmospheric pressure at the surface of the earth, $lb/in.^2$
Q_A	Heat flow from AAC absorber, Btu/hr
Q_C	Heat flow from AAC condenser, Btu/hr
Q_E	Heat flow into AAC evaporator, Btu/hr
Q_G	Heat flow into AAC generator, Btu/hr
Q_i	Heat flow through infiltration and exfiltration, Btu/hr
Q_{rf}	Heat flow through roof, $Btu/(hr)(ft^2)$
Q_{tot}	Total annual heat and hot-water load, MBtu
Q_w	Latent heat load, Btu/hr
Q_{wa}	Heat flow through unshaded walls, $Btu/(hr)(ft^2)$
$Q_{wa,sh}$	Heat flow through shaded walls, $Btu/(hr)(ft^2)$
Q_{wi}	Heat flow through unshaded windows, $Btu/(hr)(ft^2)$
$Q_{wi,sh}$	Heat flow through shaded windows, $Btu/(hr)(ft^2)$
q	Rate of heat flow (flux), $Btu/(hr)(ft^2)$
q_a	Rate of energy absorbed, $Btu/(hr)(ft^2)$
q_c	Rate of heat flow by convection, $Btu/(hr)(ft^2)$
q_f	Rate of useful energy transferred to the working fluid, $Btu/(hr)(ft^2)$
q_k	Rate of heat flow by conduction, $Btu/(hr)(ft^2)$
q_r	Rate of heat flow by radiation, $Btu/(hr)(ft^2)$
q_{rf}	Rate of heat loss through a roof, $Btu/(hr)(ft^2)$
q_{sky}	Rate of infrared radiation emitted by the sky, $Btu/(hr)(ft^2)$
q_{tot}	Total rate of heat loss through infiltration and/or ventilation air flow and through the walls, windows, and roof of a building, (i.e., the heat load,) $Btu/(°F)$ per day
q_v	Rate of heat loss by infiltration and/or ventilation air flow, Btu/hr
q_{wa}	Rate of heat loss through a wall or ceiling, $Btu/(hr)(ft^2)$
q_{wi}	Rate of heat loss through windows, $Btu/(hr)(ft^2)$

q_x	Rate of heat loss from walls, windows, ceilings, and roof, $Btu/(hr)(ft^2)$
R	Thermal resistance, $(ft^2)(hr)/Btu$
R_c	Thermal resistance to heat transfer by convection
$R_{c,in}$	Thermal resistance to heat transfer by convection inside a wall
$R_{c,out}$	Thermal resistance to convection outside a wall
R_{cr}	Thermal resistance to heat transfer by combined convection and radiation
R_k	Thermal resistance to heat transfer by conduction
$R_{k,a}$	Thermal resistance to heat transfer by conduction in an air gap
$R_{k,g}$	Thermal resistance to heat transfer by conduction in glass
R_r	Thermal resistance to heat transfer by radiation
r	Radius
r_a	Angle of refraction, deg
$r_{i,coll}$	Index of refraction in a collector cover or other transparent medium
T	Temperature, $°F$ or $°R$
T_{coll}	Average collector-surface temperature, $°F$ or $°R$
$T_{coll,in}$	Collector transport-fluid-inlet temperature, $°F$
$T_{coll,out}$	Collector transport-fluid-outlet temperature, $°F$
T_f	Fluid temperature, $°F$
T_g	Average outer-cover glass-surface temperature, $°F$ or $°R$
T_{in}	Building inside temperature, $°F$
T_{out}	Outside ambient temperature, $°F$
T_s	Surface temperature, $°F$
$T_{s,in}$	Interior-wall-surface temperature, $°F$
$T_{s,out}$	Exterior-wall-surface temperature, $°F$
T_{sky}	Equivalent sky temperature for radiation, $°R$
t	Expected lifetime of a solar system, yr
U	Overall heat transfer coefficient, $Btu/(hr)(ft^2)(°F)$
U_{rf}	Overall heat transfer coefficient for building roof (conduction, convection, and radiation), $Btu/(hr)(ft^2)(°F)$
U_{wa}	Overall heat transfer coefficient for building walls (conduction, convection, and radiation), $Btu/(hr)(ft^2)(°F)$
U_{wi}	Overall heat transfer coefficient for building windows (conduction, convection, and radiation), $Btu/(hr)(ft^2)(°F)$
v	Wind-speed velocity over collector surface, knots or mi/hr
W_{in}	Indoor humidity ratio, lb water/lb dry air
W_{out}	Outdoor humidity ratio, lb water/lb dry air
w	Building wall thickness, in. or ft
X_{ab}	Concentration of LiBr in absorbent, lb/lb of solution
X_b	Concentration of LiBr in refrigerant, lb/lb of solution

α	Angle of solar altitude above the true horizon, deg or rad
$\bar{\alpha}_s$	Absorptance of a surface for solar radiation
$\bar{\alpha}_{s,b}$	Surface absorptance of beam solar radiation
$\bar{\alpha}_{s,d}$	Surface absorptance of diffuse solar radiation
$\bar{\alpha}_{s,rf}$	Building-roof solar absorptance
$\bar{\alpha}_{sky}$	Surface absorptance of infrared sky radiation
$\bar{\alpha}_{s,wa}$	Exterior-wall solar absorptance
β	Collector tilt from horizontal, deg
ΔT_{wa}	Temperature difference between the average fluid tempera-ture in the collector and the outside ambient air tempera-ture, $^\circ$F
δ_s	Solar declination, deg
$\bar{\epsilon}_{ir}$	Infrared emittance of a surface; infrared emittance of collector absorber
$\bar{\epsilon}_{ir,c}$	Collector-absorber-plate infrared emittance (nonselective surface)
$\bar{\epsilon}_{ir,g}$	Collector-cover infrared emittance
η	Efficiency of a solar collector, i.e., the ratio of delivered heat to the incident solar radiation, %
η_b	Efficiency of a concentrator based on beam component of radiation, %
η_C	Efficiency of a Carnot power cycle, %
$\eta_{C,hp}$	Efficiency of a Carnot heat pump cycle, %
η_R	Efficiency of the Rankine Cycle, %
η_r	Relative efficiency, i.e., ratio of two collector efficiencies
λ	Wavelength of light, μm
λ_w	Latent heat of water vapor, Btu/lb
ρ	Density, lb_m/ft^3
$\bar{\rho}_{bn}$	Reflectance from a series of parallel glass plates calculated from the Fresnel equation for beam radiation
$\bar{\rho}_{b1}$	Reflectance from a glass surface calculated from the Fresnel equation for beam radiation
$\bar{\rho}_d$	Reflectance from a series of parallel glass plates for diffuse radiation
$\bar{\rho}_m$	Mirror reflectance
$\bar{\rho}_s$	Solar reflectance of a surface
σ	Stefan-Boltzmann constant (0.1713×10^{-8}), Btu/(hr)(ft^2)($^\circ$R^4)
$\bar{\tau}_{atm}$	Atmospheric transmittance for solar radiation
$\bar{\tau}_{b,wi}$	Window transmittance for beam insolation
$\bar{\tau}_{d,wi}$	Window transmittance for diffuse insolation
$\bar{\tau}_{r,wi}$	Window transmittance for ground-reflected insolation
$\bar{\tau}_s$	Solar transmittance of a medium
υ	Frequency of light
Ω	Rate of rotation of the earth $(\pi/12)$, rad/hr

CC   Cloud cover expressed as tenths of sky covered: $CC = 0$ indicates a clear sky; $CC = 10$ indicates the sky is fully covered with clouds

COP   Coefficient of performance

CR   Concentration ratio

CRF   Capital recovery factor

Conversion Factors

Given unit	X Conversion factor	= Converted unit
Length		
in.	0.0833	ft
ft	12.0	in.
in.	2.54	cm
meter (m)*	39.37	in.
cm	0.394	in.
μ	3.937×10^{-5}	in.
in.	2.54×10^4	μ
Mass		
lb	454	g
lb	0.454	kilogram (kg)*
kg*	2.205	lb
Energy and power		
ft-lb	1.29×10^{-3}	Btu
Btu	778	ft-lb
kcal	3.968	Btu
cal	4.186	joule (J)*
joule*	9.478×10^{-4}	Btu
Btu	1055.1	joule*
kw-hr	3413	Btu
Btu	2.93×10^{-4}	kw-hr
hp	2544	Btu/hr
hp	745.7	watt (W)*
Heat flux		
cal/(cm²)(sec)	13.272	Btu/(hr(ft²)
W/cm²	3171	Btu/(hr)(ft²)
Btu/(hr)(ft²)	3.154×10^{-4}	W/cm²
Btu/(hr)(ft²)	3.154	W/m² *
langley (ly)†	3.687	Btu/ft²
ly/min	221.2	Btu/(hr)(ft²)
Btu/hr	0.293	watt*
watt	3.413	Btu/hr
ton air conditioning	12,000	Btu/hr
Heat-transfer coefficient		
Btu/(hr)(ft²)(°F)	5.67	W/(m²)(°K)*
W/(m²)(°K)*	0.1761	Btu/(hr)(ft²)(°F)
W/(cm²)(°C)	1,761	Btu/(hr)(ft²)(°F)
cal/(hr)(cm²)(°C)	2.048	Btu/(hr)(ft²)(°F)
Thermal conductivity		
Btu/(hr)(ft)(°F)	1.731	W/(m)(°K)*
W/(m)(°K)*	0.5779	Btu/(hr)(ft)(°F)
W/(cm)(°C)	57.79	Btu/(hr)(ft)(°F)
cal/(min)(cm)(°C)	14,514	Btu/(hr)(ft)(°F)
cal/(sec)(cm)(°C)	241.9	Btu/(hr)(ft)(°F)

*Units followed by an asterisk are International Standard (SI) units.
†Langley is defined as 1 cal/cm².

	Given unit	Conversion factor	Converted unit
	Temperature		
	°F	+ 459.7	°R
	°R	− 459.7	°F
	°C	+ 273.1	°K*
	°K*	− 273.1	°C
	°C	$= \frac{5}{9}(°F - 32)$	
	°F	$= \frac{9}{5}°C + 32$	

*Units followed by an asterisk are International Standard (SI) units.

Energy Conservation and Passive Utilization of Solar Energy

Solar radiation impinging on a structure affects the structure in one of two ways: depending on whether the radiation strikes a wall or passes through a window, it can either be passively rejected or absorbed by the structure.

The architect can determine the effects of passive collection by his choice of

1. Window size, orientation, glass type, and shading features (louvers, overhangs, etc.)
2. Wall color
3. Wall area facing south compared with that facing east or west

The equations in Chap. 5 permit the performance of these types of solar-energy control devices to be estimated quantitatively. The tables in App. C provide guidelines to evaluate the absorptance of different wall surfaces.

The desirability of either passive solar energy gain or passive solar energy rejection depends upon the climate and time of year. For example, in the South, solar-heat rejection is important in summer to avoid high air-conditioning loads, while in the North solar-heat gain through windows during the day is important in winter as both a heating and lighting mode. It should be noted that covering the south wall of a house with glass in the North as a means of partial heating is not advised since the heat loss during 14 to 16 hr of darkness is greater than the possible gain from the winter sun during 8 to 10 hr of daylight. Careful use of passive solar devices can reduce the dollar cost of winter heating or summer cooling.

One example of a passive solar device is shown in Fig. F.1. Summer and winter sun positions can be used to determine the size

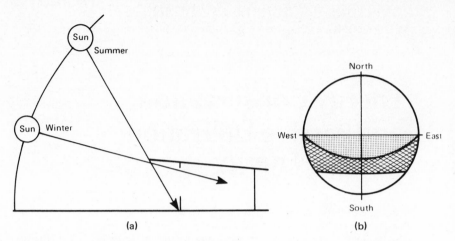

FIG. F.1 Example of summer/winter passive solar control. (*a*) Summer and winter sun positions; (*b*) sun path with shaded portion showing summer declination, crosshatched portion showing winter declination.

of overhangs on windows. With careful placement based on the summer and winter sun positions (see Fig. F.1(*a*)) the overhangs can be arranged both to block the sun in summer and to admit the sun in winter. In the sun-path diagram shown in Fig. F.1(*b*) the shaded portion represents the region of larger, positive declination, i.e., summer declination, and the crosshatched portion indicates winter declination. In his book *Design with Climate*,* Olgyay describes a number of passive measures to control the sun's effect on a building in various climatic zones of the world.

A necessary adjunct to successful passive or active use of solar energy is good energy-conservation practice. Figure F.2 presents a checklist of 37 energy-conservation techniques† with a ranking indicating the efficaciousness of each, depending on the nature of the climate. In his monograph, *Retrofitting Existing Housing for Energy Conservation*,‡ Peterson describes the method to be used in evaluating the marginal cost/benefit ratio for any energy conservation measure. Examples of the analysis and some results are also presented.

*Victor Olgyay, *Design with Climate*, Princeton University Press, Princeton, N.J., 1963.
†Figure F.2 is reproduced from "A Checklist of Energy Conservation Opportunities, Ranked in Priority According to Climatic Conditions," *AIA J.*, May, 1974, p. 34.
‡Stephen R. Petersen, *Retrofitting Existing Housing for Energy Conservation: An Economic Analysis*, U.S. Government Printing Office, C13.29:2/64, Washington, D.C., 1974.

A CHECKLIST OF ENERGY CONSERVATION OPPORTUNITIES, RANKED IN PRIORITY ACCORDING TO CLIMATIC CONDITIONS

This checklist of energy-saving opportunities, appended to the guidelines, includes some items that subsume others. Some seem to border on the obvious, yet many contemporaty buildings are testimony to the need for even seemingly obvious measures.

The items are ranked in priority and coded to the following climatic features: For winter, *A* indicates a heating season of 6,000 degree-days or more; *B* a heating season of 4,000 to 6,000 degree-days, and *C* 4,000 degree-days or less. The numeral *1* following these letters indicates sun 60 per cent of daylight time or more and wind nine miles per hour or more; *2* indicates the sun condition but not the wind condition; *3* indicates the wind condition without the sun condition, and *4* the absence of either condition.

For summer, the letter *D* indicates a cooling season of more than 1,500 hours at 80 degrees Fahrenheit; *E* 600 to 1,500 hours at the same temperature and *F* less than 600 hours. The numeral *1* indicates a dry climate of 60 per cent relative humidity or less and *2* indicates 60 per cent or more humidity.

Guidelines that are independent of climate are not rated in priority columns and are marked '*'.

SITE

	Priority			
	1	2	3	N/A
1. Use deciduous trees for their summer sun shading effects and wind break for buildings up to three stories.	A1 / D1	A2 / D2	A4 / E1	C4 / F
2. Use conifer trees for summer and winter sun shading and wind breaks.	C4 / D1	C1 / D2	C2 / E1	A2 / F
3. Cover exterior walls and/or roof with earth and planting to reduce heat transmission and solar gain.	A1 / D1	A2 / D2	A4 / E1	C4 / F
4. Shade walls and paved areas adjacent to building to reduce indoor/outdoor temperature differential.	C2 / D1	C1 / D2	C3 / E1	A2 / F
5. Reduce paved areas and use grass or other vegetation to reduce outdoor temperature buildup.	C2 / D1	C1 / D2	C3 / E1	A2 / F
6. Use ponds, water fountains, to reduce ambient outdoor air temperature around building.	C2 / D1	C1 / E1	C3 / D2	A4 / F
7. Collect rain water for use in building.	*			
8. Locate building on site to induce air flow effects for natural ventilation and cooling.	C2 / F	C1 / E1	C3 / E2	A4 / D2
9. Locate buildings to minimize wind effects on exterior surfaces.	A4 / F	A1 / E2	B4 / E1	C2 / D1
10. Select site with high air quality (least contaminated) to enhance natural ventilation.	C2 / F	C1 / E1	C3 / E2	A4 / D2
11. Select a site which has year-round ambient wet and dry bulb temperatures close to and somewhat lower than those desired within the occupied spaces.	*			

FIG. F.2 Energy conservation checklist. (Reproduced from the *AIA Journal* by permission of the publishers.)

| 12. Select a site that has topographical features and adjacent structures that provide breaks. | A4 | A1 | B4 | C3 |
| | F | E2 | E1 | D2 |

| 13. Select a site that has topographical features and adjacent structures that provide desirable shading. | C2 | C1 | B2 | A1 |
| | D2 | D1 | E2 | F |

| 14. Select site that allows optimum orientation and configuration to minimize yearly energy consumption. | * | | | |

| 15. Select site to reduce specular heat reflections from water. | C2 | C1 | B2 | A4 |
| | D2 | D1 | E2 | F |

| 16. Use sloping site to bury building partially or use earth berms to reduce heat transmission and solar radiation. | A4 | A1 | A3 | C2 |
| | D1 | D2 | E1 | F |

| 17. Select site that allows occupants to use public transport systems. | * | | | |

BUILDING

	Priority			
	1	2	3	N/A
1. Construct building with minimum exposed surface area to minimize heat transmission for a given enclosed volume.	A4	A1	A3	C2
	D1	D2	E1	F

| 2. Select building configuration to give minimum north wall to reduce heat losses. | A4 | A1 | A3 | C2 |

| 3. Select building configuration to give minimum south wall to reduce cooling load. | D1 | D2 | E1 | F |

| 4. Use building configuration and wall arrangement (horizontal and vertical sloping walls) to provide self shading and wind breaks. | A4 | A1 | B4 | C3 |
| | D1 | D2 | E1 | |

| 5. Locate insulation for walls and roofs and floors over garages at the exterior surface. | A4 | A3 | A1 | C2 |
| | D1 | D2 | E1 | F |

| 6. Construct exterior walls, roof and floors with high thermal mass with a goal of 100 pounds per cubic foot. | A4 | A1 | A3 | C3 |
| | D1 | D2 | E1 | F |

| 7. Select insulation to give a composite U factor from 0.06 when outdoor winter design temperatures are less than 10 degrees F. to 0.15 when outdoor design conditions are above 40 degrees F. | | | | |

| 8. Select U factors from 0.06 where sol-air temperatures are above 144 degrees F. up to a U volume of 0.3 with sol-air temperatures below 85 degrees F. | | | | |

| 9. Provide vapor barrier on the interior surface of exterior walls and roof of sufficient impermeability to provide condensation. | * | | | |

| 10. Use concrete slab-on-grade for ground floors. | A4 | A1 | A3 | C2 |
| | D1 | D2 | E1 | F |

FIG. F.2 (continued)

11. Avoid cracks and joints in building construction to reduce infiltration.	A4 D2	A1 E2	A3 D1	
12. Avoid thermal bridges through the exterior surfaces.	A4 D2	A1 D1	A3 E2	C3 F
13. Provide textured finish to external surfaces to increase external film coefficiency.	A4	A1	B4	C2
14. Provide solar control for the walls and roof in the same areas where similar solar control is desirable for glazing.	D2	D1	E2	A
15. Consider length and width aspects for rectangular buildings as well as other geometric forms in relationship to building height and interior and exterior floor areas to optimize energy conservation.	A4 D1	A1 D2	A3 E1	C2 F
16. To minimize heat gain in summer due to solar radiation, finish walls and roofs with a light-colored surface having a high emissivity.	D1	D2	E1	
17. To increase heat gain due to solar radiation on walls and roofs, use a dark-colored finish having a high absorptivity.	A1	A2	A4	C2
18. Reduce heat transmissions through roof by one or more of the following items:				
a. Insulation	A4 D1	A1 D2	A3 E1	C3 F
b. Reflective surfaces.	C2 D1	C1 D2	C3 E1	A4
c. Roof spray.	D1	E1	F	
d. Roof pond.	D1	E1	F	
e. Sod and planning.	A4 D1	A1 D2	A3 E1	C2 F
f. Equipment and equipment rooms located on the roof.	A4 D1	A1 D2	A3 E1	
g. Provide double roof and ventilate space between.	D1	D2	E1	F
19. Increase roof heat gain when reduction of heat loss in winter exceeds heat gain increase in summer:				
a. Use dark-colored surfaces.	A2	A1	B2	B1
b. Avoid shadows.	A2	A1	B2	B1
20. Insulate slab on grade with both vertical and horizontal perimeter insulation under slab.	A	B	C	
21. Reduce infiltration quantities by one or more of the following measures:				
a. Reduce building height.				
b. Use impermeable exterior surface materials.	A4	A1	A3	C4
c. Reduce crackage area around doors, windows, etc., to a minimum.	D2	E2	D1	F

FIG. F.2 (*continued*)

d. Provide all external doors with weather stripping.				
e. Where operable windows are used, provide them with sealing gaskets and cam latches.				
f. Locate building entrances on downwind side and provide wind break.				
g. Provide all entrances with vestibules; where vestibules are not used, provide revolving doors.				
h. Provide vestibules with self-closing weather-stripped doors to isolate them from the stairwells and elevator shafts.	A4 D2	A1 E2	A3 D1	C4 F
i. Seal all vertical shafts.				
j. Locate ventilation louvers on downwind side of building and provide wind breaks.				
k. Provide break at intermediate points of elevator shafts and stairwells for tall buildings.				
22. Provide wind protection by using fins, recesses, etc., for any exposed surface having a U value greater than 0.5.	A4	A1	B4	C2
23. Do not heat parking garages.	*			
24. Consider the amount of energy required for the protection of materials and their transport on a life-cycle energy usage.	*			
25. Consider the use of the insulation type which can be most efficiently applied to optimize the thermal resistance of the wall or roof; for example, some types of insulation are difficult to install without voids or shrinkage.	*			
26. Protect insulation from moisture originating outdoors, since volume decreases when wet. Use insulation with low water absorption and one which dries out quickly and regains its original thermal performance after being wet.	*			
27. Where sloping roofs are used, face them to south for greatest heat gain benefit in the wintertime.	A1	A2	B1	C4
28. To reduce heat loss from windows, consider one or more of the following:				
a. Use minimum ratio of window area to wall area.				
b. Use double glazing.				
c. Use triple glazing.				
d. Use double reflective glazing	A4	B4	C4	

FIG. F.2 (*continued*)

e. Use minimum percentage of the double glazing on the north wall.	A1	B1	C1	
	A2	B2	C2	
f. Manipulate east and west walls so that windows face south.	A3	B3	C3	
g. Allow direct sun on windows November through March.				
h. Avoid window frames that form a thermal bridge.				
i. Use operable thermal shutters which decrease the composite U value to 0.1.				

29. To reduce heat gains through windows, consider the following:

a. Use minimum ratio of window area to wall area.				
b. Use double glazing.				
c. Use triple glazing.	D1	E1	F	
d. Use double reflective glazing.	D2	E2	F	
e. Use minimum percentage of double glazing on the south wall.				
f. Shade windows from direct sun April through October.				

30. To take advantage of natural daylight within the building and reduce electrical energy consumption, consider the following:

a. Increase window size but do not exceed the point where yearly energy consumption, due to heat gains and losses, exceeds the saving made by using natural light.				
b. Locate windows high in wall to increase reflection from ceiling, but reduce glare effect on occupants.	C2	B2	A2	
	C1	B1	A1	
c. Control glare with translucent draperies operated by photo cells.	C3	B3	A3	
	C4	B4	A4	
d. Provide exterior shades that eliminate direct sunlight, but reflect light into occupied spaces.	F	E	D	
e. Slope vertical wall surfaces so that windows are self-shading and walls below act as light reflectors.				
f. Use clear glazing. Reflective or heat absorbing films reduce the quantity of natural light transmitted through the window.				

31. To allow the use of natural light in cold zones where heat losses are high energy users, consider operable thermal barriers.	A4	A1	B4	C3
32. Use permanently sealed windows to reduce infiltration in climatic zones where this is a large energy user.	A1	A4	B1	C3
	D1	D2	E1	F
33. Where codes of regulations require operable windows and infiltration is	A1	A4	B1	C3
	D1	D2	E1	F

FIG. F.2 (*continued*)

undesirable, use windows that close against a sealing gasket.

34. In climatic zones where outdoor air conditions are suitable for natural ventilation for a major part of the year, provide operable windows.

C2	C3	C1	A4
F	E1	E2	D2

35. In climate zones where outdoor air conditions are close to desired indoor conditions for a major portion of the year, consider the following:

    a. Adjust building orientation and configuration to take advantage of prevailing winds.

    b. Use operable windows to control ingress and egress of air through the building.

    c. Adjust the configuration of the building to allow natural cross ventilation through occupied spaces.

F	E1	E2	D2

    d. Use stack effect in vertical shafts, stairwells, etc., to promote natural air flow through the building.

PLANNING

	Priority		
1	2	3	N/A

1. Group services rooms as a buffer and locate at the north wall to reduce heat loss or the south wall to reduce heat gain, whichever is the greatest yearly energy user.

A4	A1	A3	C2
D1	D2	E1	F

2. Use corridors as heat transfer buffers and locate against external walls.

A4	A1	A3	C2
D1	D2	E1	F

3. Locate rooms with high process heat gain (computer rooms) against outside surfaces that have the highest exposure loss.

A4	A1	A3	C2

4. Landscaped open planning allows excess heat from interior spaces to transfer to perimeter spaces which have a heat loss. *

5. Rooms can be grouped in such a manner that the same ventilating air can be used more than once, by operating in cascade through spaces in decreasing order of priority, i.e., office-corridor-toilet. *

6. Reduced ceiling heights reduce the exposed surface area and the enclosed volume. They also increase illumination effectiveness. *

7. Increased density of occupants (less gross floor area per person) reduces the overall size of the building and yearly energy consumption per capita. *

8. Spaces of similar function located adjacent to each other on the same floor reduce the use of elevators. *

FIG. F.2 (*continued*)

9. Offices frequented by the general public located on the ground floor reduce elevator use.	*			
10. Equipment rooms located on the roof reduce unwanted heat gain and heat loss through the surface. They can also allow more direct duct and pipe runs reducing power requirements.	A4 D1	A3 D2	B4 E1	C2
11. Windows planned to make beneficial use of winter sunshine should be positioned to allow occupants the opportunity of moving out of the direct sun radiation.	*			
12. Deep ceiling voids allow the use of larger duct sizes with low pressure drop and reduce HVAC requirements.	*			
13. Processes that have temperature and humidity requirements different from normal physiological needs should be grouped together and served by one common system.	*			
14. Open planning allows more effective use of lighting fixtures. The reduced area of partitioned walls decreases the light absorption.	*			
15. Judicious use of reflective surfaces such as sloping white ceilings can enhance the effect of natural lighting and increase the yearly energy saved.	*			

VENTILATION AND INFILTRATION

	Priority			
	1	2	3	N/A
1. To minimize infiltration, balance mechanical ventilation so that supply air quantity equals or exceeds exhaust air quantity.	A D	B E	C F	
2. Take credit for infiltration as part of the outdoor air requirements for the building occupants and reduce mechanical ventilation accordingly.	A D	B E	C F	
3. Reduce C.F.M./occupant outdoor air requirements to the minimum considering the task they are performing, room volume and periods of occupancy.	A D	B E	C F	
4. If odor removal requires more than 2,000 cubic feet per minute exhaust and a corresponding introduction of outdoor air, consider recirculating through activated carbon filter.	C D	B E	C F	
5. Where outdoor conditions are close to but less than indoor conditions for major periods of the year, and the air is clean and free from offensive odors, consider the use	C2 F	C1 E1	C3 E2	A2 D2

FIG. F.2 *(continued)*

of natural ventilation when yearly energy trade-offs with other systems are favorable.

6. Exchange heat between outdoor air, intake and exhaust air by using heat pipes, thermal wheels, run-around systems, etc.

A	B	C
D	E	F

7. In areas subjected to high humidities, consider latent heat exchange in addition to sensible.

D2	E2	F

8. Provide selective ventilation as needed; i.e., 5 cubic feet per minute/occupant for general areas and increased volumes for areas of heavy smoking or odor control.

*		

9. Transfer air from "clean" areas to more contaminated areas (toilet rooms, heavy smoking areas) rather than supply fresh air to all areas regardless of function.

*		

10. Provide controls to shut down all air systems at night and weekends except when used for economizer cycle cooling.

A	B	C
D	E	F

11. Reduce the energy required to heat or cool ventilation air from outdoor conditions to interior design conditions by considering the following:

    a. Reduce indoor air temperature setting in winter and increase in summer.

A	B	C
D	E	F

    b. Provide outdoor air direct to perimeter of exhaust hoods in kitchens, laboratories, etc. Do not cool this air in summer or heat over 50 degrees F. in winter.

A	B	C
D	E	F

HEATING, VENTILATION AND AIRCONDITIONING

Priority			
1	2	3	N/A

1. Use outdoor air for sensible cooling whenever conditions permit and when recaptured heat cannot be stored.

*			

2. Use adiabatic saturation to reduce temperature of hot, dry air to extend the period of time when "free cooling" can be used.

D1	E1	F	

3. In the summer when the outdoor air temperature at night is lower than indoor temperature, use full outdoor air ventilation to remove excess heat and precool structure.

D1	E1	F	

4. In principle, select the air handling system which operates at the lowest possible air velocity and static pressure. Consider high pressure systems only when other trade-offs such as reduced building size are an overriding factor.

	*		

FIG. F.2 (*continued*)

310

5. To enhance the possibility of using waste heat from other systems, design air handling systems to circulate sufficient air to enable cooling loads to be met by a 60 degree F. air supply temperature and heating loads to be met by a 90 degree F. air temperature. | *

6. Design HVAC systems so that the maximum possible proportion of heat gain to a space can be treated as an equipment load, not as a room load. | *

7. Schedule air delivery so that exhaust from primary spaces (offices) can be used to heat or cool secondary spaces (corridors). | *

8. Exhaust air from center zone through the lighting fixtures and use this warmed exhaust air to heat perimeter zones. | *

9. Design HVAC systems so that they do not heat and cool air simultaneously. | *

10. To reduce fan horsepower, consider the following:
  a. Design duct systems for low pressure loss | *
  b. Use high efficiency fans. | *
  c. Use low pressure loss filters concommitant with contaminant removal. | *
  d. Use one common air coil for both heating and cooling. | *

11. Reduce or eliminate air leakage from duct work. | *

12. Limit the use of re-heat to a maximum of 10 percent of gross floor area and then only consider its use for areas that have atypical fluctuating internal loads, such as conference rooms. | *

13. Design chilled water systems to operate with as high a supply temperature as possible—suggested goal: 50 degrees F. (This allows higher suction temperatures at the chiller with increased operating efficiency.) | *

14. Use modular pumps to give varying flows that can match varying loads. | *

15. Select high efficiency pumps that match load. Do not oversize. | *

16. Design piping systems for low pressure loss and select routes and locate equipment to give shortest pipe runs. | *

17. Adopt as large a temperature differential as possible for chilled water systems and hot water heating systems. | *

18. Consider operating chillers in series to | *

FIG. F.2 (*continued*)

increase efficiency.

19. Select chillers that can operate over a wide range of condensing temperatures and then consider the following:

    a. Use double bundle condensers to capture waste heat at high condensing temperatures and use directly for heating or store for later use.   *

    b. When waste heat cannot be either used directly or stored, then operate chiller at lowest condensing temperature compatible with ambient outdoor conditions.   *

20. Consider chilled water storage systems to allow chillers to operate at night when condensing temperatures are lowest.   *

21. Consider the use of double bundle evaporators so that chillers can be used as heat pumps to upgrade stored heat for use in unoccupied periods.   *

22. Consider the use of gas or diesel engine drive for chillers and large items of ancilliary equipment and collect and use waste heat for absorption cooling, heating, and/or domestic hot water.   *

23. Locate cooling towers or evaporative coolers so that induced air movement can be used to provide or supplement garage exhaust ventilation.   *

24. Use modular boilers for heating and select units so that each module operates at optimum efficiency.   *

25. Extract waste heat from boiler flue gas by extending surface coils or heat pipes.   *

26. Select boilers that operate at the lowest practicable supply temperature while avoiding condensation within the furnaces.   *

27. Use unitary water/air heat pumps that transport heat energy from zone to zone via a common hydronic loop.   *

28. Consider the use of thermal storage in combination with unit heat pumps and a hydronic loop so that excess heat during the day can be captured and stored for use at night.   *

29. Consider the use of heat pumps both water/air and air/air if a continuing source of low-grade heat exists near the building, such as lake, river, etc.   *

30. Consider the direct use of solar energy via   *

FIG. F.2 (*continued*)

a system of collectors for heating in winter and absorption cooling in summer.

31. Minimize requirements for snow melting to those that are absolutely necessary and, where possible, use waste heat for this service.

32. Provide all outside air dampers with accurate position indicators and insure that dampers are air-tight when closed.

33. If electric heating is contemplated, consider the use of heat pumps in place of direct resistance heating; by comparison they consume one-third of the energy per unit output.

34. Consider the use of spot heating and/or cooling in spaces having large volume and low occupancy.

35. Use electric ignition in place of gas pilots for gas burners.

36. Consider the use of a total energy system if the life-cycle costs are favorable.

LIGHTING AND POWER

1.a. Use natural illumination in areas where effective when a net energy conservation gain is possible vis-a-vis heating and cooling loads.
    b. Provide exterior reflectors at windows for more effective internal illumination.

2. Consider a selective lighting system in regard to the following:
    a. Reduce the wattage required for each specific task by review of user needs and method of providing illumination.
    b. Consider only the amount of illumination required for the specific task considering the duration and character and user performance required as per design criteria.
    c. Group similar task together for optimum conservation of energy per floor.
    d. Design switch circuits to permit turning off unused and unnecessary light.
    e. Illuminate tasks with fixtures built into furniture and maintain low intensity lighting elsewhere.
    f. Consider the use of polarized lenses to improve quality of lighting at tasks.
    g. Provide timers to automatically turn off lights in remote or little-used areas.

Item	Priority			
	1 C F	2 B E	3 A D	N/A
31	*			
32	*			
33	*			
34	*			
35	*			
36	*			
1.b	*			
2.b	*			
2.c	*			
2.d	*			
2.e	*			
2.f	*			
2.g	*			

FIG. F.2 (*continued*)

	*				

h. Use multilevel ballasts to permit varying the lumen output for fixtures by adding or removing lamps when tasks are changed in location or requirements.

i. Arrange electrical systems to accommodate relocatable luminaires which can be removed to suit changing furniture layouts.

j. Consider the use of ballasts which can accommodate sodium metalhalide bulbs interchangeably with other lamps.

3. Consider the use of high frequency lighting to reduce wattage per lumen output. Additional benefits are reduced ballast heat loss into the room and longer lamp life.

4. Consider the use of landscape office planning to improve lighting efficiency. Approximately 25 percent less wattage per foot-candles on task for open planning versus partitions.

5. Consider the use of light colors for walls, floors and ceilings to increase reflectance, but avoid specular reflections.

6. Lower the ceilings or mounting height of luminaires to increase level of illumination with less wattage.

7. Consider dry heat-of-light systems to improve lamp performance and reduce heat gain to space.

8. Consider wet heat-of-light system to improve lamp performance and reduce heat gain to space and refrigeration load.

9. Use fixtures that give high contrast rendition factor at task.

10. Provide suggestions to GSA for analysis of tasks to increase use of high contrast material which requires less illumination.

11. Select furniture and interior appointments that do not have glossy surfaces or give specular reflections.

12. Use light spills from characteristic areas to illuminate noncharacteristic areas.

13. Consider use of greater contrast between tasks and background lighting, such as 8 to 1 and 10 to 1.

14. Consider washers and special illumination for features such as plants, murals, etc., in place of overhead space lighting to maintain proper contrast ratios.

15. For horizontal tasks or duties, consider fixtures whose main light component is

FIG. F.2 (*continued*)

oblique and then locate for maximum ESI footcandles on task.					
16. Consider using 250 watt mercury vapor lamps and metal-halide lamps in place of 500 watt incandescent lamps for special applications.	*				
17. Use lamps with higher lumens per watt input, such as:					
a. One 8-foot fluorescent lamp versus two 4-foot lamps.	*				
b. One 4-foot fluorescent lamp versus two 2-foot lamps.	*				
c. U-tube lamps versus two individual lamps.	*				
d. Fluorescent lamps in place of all incandescent lamps except for very close task lighting, such as at a typewriter paper holder.	*				
18. Use high utilization and maintenance factors in design calculations and instruct users to keep fixtures clean and change lamps earlier.	*				
19. Avoid decorative flood-lighting and display lighting.	*				
20. Direct exterior security lighting at entrances and avoid illuminating large areas adjacent to building.	*				
21. Consider switches activated by intruder devices rather than permanently lit security lighting.	*				
22. If already available, use street lighting for security purposes.					
23. Reduce lighting requirements for hazards by:					
a. Using light fixtures close to and focused on hazard.	*				
b. Increasing contrast of hazard; i.e., paint stair treads and risers white with black nosing.	*				
24. Consider the following methods of coping with code requirements:					
a. Obtaining variance from existing codes.	*				
b. Changing codes to just fulfill health and safety functions of lighting by varying the qualitative and quantitative requirements to specific application.	*				
25. Consider the use of a total energy system integrated with all other systems.	*				
26. Where steam is available, use turbine drive for large items of equipment.	*				
27. Use heat pumps in place of electric resistance heating and take advantage of the favorable coefficient of performance.	*				

FIG. F.2 *(continued)*

28. Match motor sizes to equipment shaft power requirements and select to operate at the most efficient point.	*				
29. Maintain power factor as close to unity as possible.	*				
30. Minimize power losses in distribution system by:					
a. Reducing length of cable runs.	*				
b. Increasing conductor size within limits indicated by life-cycle costing.	*				
c. Use high voltage distribution within the building.	*				
31. Match characteristics of electric motors to the characteristics of the driven machine.	*				
32. Design and select machinery to start in an unloaded condition to reduce starting torque requirements. (For example, start pumps against closed valves.)	*				
33. Use direct drive whenever possible to eliminate drive train losses.	*				
34. Use high efficiency transformers (these are good candidates for life-cycle costing).	*				
35. Use liquid-cooled transformers and captive waste heat for beneficial use in other systems.	*				
36. In canteen kitchens, use gas for cooking rather than electricity.	*				
37. Use conventional ovens rather than self-cleaning type.	*				

FIG. F.2 (*continued*)

Solar Rights in Colorado

June 28, 1973 a moratorium on the issuance of additional natural gas permits was declared by the City of Colorado Springs. Soon thereafter, Andrew Marshall, Mayor, initiated a unique community project to develop alternate energy sources for heating. Calling upon the Municipal Utilities, American Institute of Architects, The Home Builders Association, Society of Professional Engineers, and others, a project was initiated to develop solar energy in the Pikes Peak region. Phoenix of Colorado Springs, Inc., a Colorado nonprofit corporation dedicated to the development of new energy sources, was formed. Its first project is the development of a solar home, which appeared in the Parade of Homes in the summer of 1974, fully operational. The National Science Foundation will assist in the evaluation.

A collateral issue confronted by Phoenix has been solar rights. No established legal rights to radiation from the sun exists. A search for legal references unearthed no reported cases or zoning regulations in the United States specifically dealing with solar rights. The closest the law has come to granting solar rights is the old English doctrine of "Ancient lights". The doctrine gave an English landowner a right at law to receive the customary amount of light and air. American jurisdictions, however, have uniformly repudiated the doctrine, recognizing easements of light and air only if they are express or arise by prescription.

Although there are no specific laws available to deal with the question of solar rights at this time, there are legal principles and devices which may be applicable to the problems, either directly or by analogy, in the law of nuisance, easements, pollution and zoning. It is easiest to divide the problem into two sections: (1) protection of light coming from directly over the building lot; (2) protection of light coming to the building lot from an angle over other lots. The following is a discussion of the problems and some possible solutions to the preservation of solar rights.

I. LIGHT ABOVE THE LOT

It has traditionally been said that a land owner owns all the earth below the surface of his land and all the air space above.

Courts have drawn back a bit from that—one can't shoot down a plane for trespassing—but the gist of the statement is still true.

A. Injunction

If a neighbor builds in a way that encroaches on ones air space, an injunction for the removal of the encroachment is available. Similarly, when trees are located on X's property but have branches or roots extending into Y's land or air, Y may lop off the offending limbs (unless zoning or other regulations abrogate this right). *McCrann v. Town Plan & Zoning Com.*, 161 Conn. 65, 282 A.2d 900 (1971). The United States Supreme Court has recognized and defined the air space rights in a series of airport approach cases. Typical is Justice Douglas' opinion in *U.S. v. Causby,* 328 U.S. 256 (1945); which stated, "We have said that the airspace is a public highway. Yet is is obvious that if the landowner is to have full enjoyment of the land, he must have exclusive control of the immediate reaches of the enveloping atmosphere. Otherwise buildings could not be erected, trees could not be planted and even fences could not be run."

B. Environmental Protection Laws

The more difficult problem is smoke and pollution which originates elsewhere but settles stubbornly over one's lot. How much energy loss this could cause is a scientific problem, but it must be assumed that the conditions could become severe enough to warrant legal action. The first resort is probably to the state or local environmental protection laws. Unfortunately, that leaves one at the mercy of the law enforcement authorities, unless one can convince the state to allow private suits to enforce such laws. It is also possible that pollution could be within legal standards and still cause severe loss of solar energy.

C. Nuisance

The second resort, then, is nuisance law. But even if one can prove the pollution to be a nuisance, it may only be possible to collect damages. Obviously, without an injunction, protection of solar energy is virtually impossible in those circumstances. A case squarely presenting the problem is *Boomer v. Atlantic Cement Co.,* 26 N.Y.2d 219, 257 N.E.2d 870 (1970); which granted an injunction to private property owners against a polluting cement plant only until permanent damages could be determined, and thereafter the injunction was vacated.

II. LIGHT FROM AN ANGLE OVER OTHER LOTS

The most difficult situation to prevent is the loss of sunlight due to trees or construction on a neighbor's property. As noted at the beginning, generally in the United States "a landowner has no right to light and air from adjoining lands", and cannot recover damages where an adjoining land owner constructs improvements or otherwise interferes with the light and air previously coming over the property: *2 C.J.S.* "Adjoining Landowners" Paragraph 68 at 65. The following theories are intended to present some positive direction for future consideration.

A. Nuisance

One line of defense is a nuisance action. *Sundowner, Inc. v. King,* 95 Idaho 367, 509 P.2d 785 (1973), is an example of a successful nuisance action to maintain light and air. However, the case involved a "spite fence", and is typical in that it required a finding of malice and ill will on the part of the Defendant. Thus the action is very limited; it is difficult to imagine a jury finding malice in the planting of a tree on the building of a high rise complex.

There is a further negative corollary to the nuisance problem, in that the solar panels mounted on the building using solar energy are often of a highly reflective nature, and if not controlled, the reflected light can be of substantial nuisance to adjacent property owners and others in the reflective path. Such reflection could be the grounds for a nuisance action by adjacent property owners for injunctive relief and damages against the owner of the property on which a solar panel is located.

B. Easement

The second line of defense is prevention—acquisition of an easement. It has already been noted that no easement of light and air is implied at law. It is possible to acquire one by prescription, but the obstacles are formidable. How, for example, does one prove that the claim is notorious or adverse? Must the statutory prescription period race the growth of the neighbors trees?

A more satisfactory and more neighborly solution would be the purchase of a light and air easement from the adjoining landowners. This might involve some expense, but it could only involve a quid pro quo, a trading of easements.

C. Covenants

Easements are far more easily established when a large piece of land is being developed at once and this might constitute a third approach. Then the developer or original owner can place restrictive covenants in each deed which would give neighbors light and air easements over each other. The covenant itself could describe the restriction, or it could merely set up a controlling body which would rule on each situation as it arose. This latter possibility has been successfully used for architectural control in situations such as that presented in *Rhue v. Cheyenne Homes, Inc.,* 168 Colo. 6, 449 P.2d 361 (1969), wherein the Colorado Supreme Court upheld a covenant giving an architectural control committee the power of final approval for all construction plans and specifications.

D. Zoning

A fourth and final protection for light coming across neighbor's property might come from zoning and other regulations for the given region. Almost all zoning ordinances include height restrictions for residential zones. These could be expanded to provide restrictions on trees as well as buildings. Some problems exist. A neighbor usually won't have standing to enforce zoning restrictions absent exceptional damage. In the case of *Taliaferro v. Salyer,* 328 P.2d 799 (Cal. 1958), the Court stated, "In the exercise of the police power a local government can impose restrictions on the maximum height of buildings for the purpose of securing adequate sunlight to promote public health in general . . . however, a violation would not necessarily give a private individual a cause of action therefor. In order to state a cause of action based upon a violation of the building code, plaintiff must show that he has suffered some exceptional damage other than that suffered by the public generally. As plaintiff has no easement of light and air, he cannot complain in the absence of a statute. . .". Also, if the zoning violation is used to support a nuisance suit, the regulation must be clear that such violation is an absolute nuisance, both public and private, as a matter of law. See *Northwest Water Corp. v. Pennetta,* 479 P.2d 398 (Colo. App. 1970).

Perhaps a barrier to such rezoning would be the accusation that it seeks special privileges for a few property owners rather than preservation of the general welfare. *Trust Co. of Chicago v. City of Chicago*, 408 Ill. 91, 96 N.E.2d 499 (1951). But nothing seems more in the general interest than the development of new, clean sources of power, and it is believed that such zoning would be sustained by the Courts as an exercise of police power for the public good.

III. RECOMMENDATIONS TO ESTABLISH SOLAR RIGHTS

The overview presented does suggest the outlines of a scheme to assure solar rights to direct sunlight for those who wish to take advantage of this natural energy. The following steps should be considered:

A. Easements

When building a structure using solar energy, and no other protection of solar rights is available, secure easements of direct sunlight from adjoining landowners.

B. Covenants

When developing a large tract of land, place restrictive covenants on the property which would give neighbors light and air easements over each others lots. The covenant could describe the restriction, or could merely set up a controlling body which would rule on each situation as it arose.

C. Zoning

Zoning or rezoning could be initiated by governmental units with proper power and authority. Such a zoning code might approach the problem as follows:

1. Incorporate a generally stated protection of access to direct sunlight into the height and use restrictions.
2. Empower a board or panel to review particular applications of the protection. The power might include the power to order compensation and injunctive relief.
3. Give citizens the right to sue for enforcement.
4. Give citizens the right to sue to enforce antipollution laws.

We have taken solar rights for granted too long. With the energy shortage upon us, and solar energy the most plentiful and accessable source available, it is time to preserve solar rights for all who wish to use them. Only by early attention to the question will public interest and proprietary right develop compatibly.

Prepared by Sandy F. Kraemer, Attorney
for Phoenix of Colorado Springs, Inc., a
Colorado nonprofit corporation dedicated
to the development of new energy sources.

Typical Solar-conditioned Building Designs

Since the first 500 ft² solar-heated building was constructed at the Massachusetts Institute of Technology, Cambridge, in 1938, several dozen solar-heated, solar-cooled, and solar-heated and solar-cooled buildings have been constructed. Frequently, these buildings were constructed as engineering experiments; much of modern solar-conditioned building design is based on the data collected from these experiments.

Six modern building designs are summarized in this appendix to show the size and method of integration of solar components in practical building designs. The six designs include single-family dwellings and larger, commercial or institutional buildings; both air- and water-cooled collectors of focusing or nonfocusing design are illustrated in these examples. Each building design is summarized in tabular form, and floor plans or system schematics are provided for each design.

For a complete, detailed description of nearly every major solar-heated and solar-air-conditioned building in the United States through 1974 the reader is referred to the *1974 Solar Directory.* *

*Carolyn Pesko, Ed., "1974 Solar Directory," Environmental Action of Colorado, Denver, prepared under NSF Grant AER74-18503 A01, 1975.

PROPOSED OFFICES FOR THE
MASSACHUSETTS AUDUBON SOCIETY

Designer: Arthur D. Little, Inc.
                 Cambridge Seven Associates

Building: Office building, two-story (see Figs. H.1 and H.2)
                 Floor area: 8,000 ft^2
                 Heat-loss rate: Approx. 43,000 Btu/degree-day
                 Max. air-conditioning load: 17 tons

Location: Lincoln, Mass.
                 Latitude: 42°N

Collector: Type: Flat plate
                 Fluid: Water + ethylene glycol
                 Area: 3500 ft^2
                 Position: South roof
                 Tilt: 45°
                 Glazing: Double

Storage: Type: Water
                 Volume: 7,500 gal

Cooling: Full air conditioning, using a 15-ton lithium-bromide
                 absorption-cooling unit powered by the solar collector.

FIG. H.1 Proposed Massachusetts Audubon Society headquarters, Lincoln.

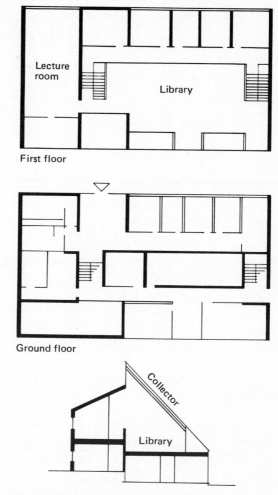

FIG. H.2 Proposed Massachusetts Audubon Society office: plans and sections.

System: Forced air

Performance: Est. solar contribution: 65 to 85 percent

DENVER, COLO. SOLAR-HEATED DWELLING

Designer: George O. G. Löf

Building: Residence, single-story (see Figs. H.3 and H.4)
Floor area: 2,050 ft²
Heat-loss rate: 32,000 Btu/degree-day

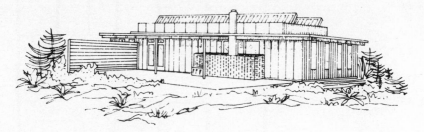

FIG. H.3 Denver solar-heated house.

Location: Denver, Colo.
                     Latitude: 40°N
                     Altitude: 5,280 ft
                     Degree-days: 6,283

Collector: Type: Flat plate (overlapping plates)
                     Fluid: Air
                     Area: 600 ft^2
                     Position: Mounted on flat roof
                     Tilt: 45°
                     Glazing: Half single, half double

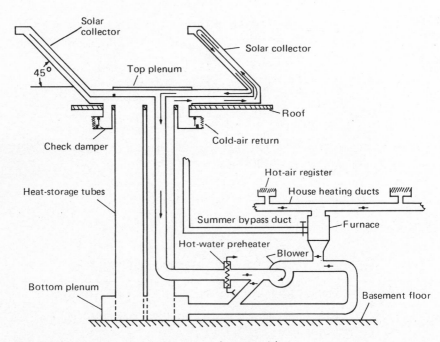

FIG. H.4 Denver solar space-heating system (not to scale).

Storage: Type: Crushed rock, 2.5-in. dia.
                Volume: 230 ft³, 11 tons
                Container: Two fiber drums
                Temperature: 140°F normal max

System: Forced air

Hot water: Preheated by thermosiphon air-to-water heat
                exchanger

Performance: Period: Winter 1959/1960 (5,700 degree days)
                Avg. collector efficiency: 34.6 percent
                Solar contribution: 25.7 percent
                See Table H.1 for annual performance summary for
                the winter of 1959/1960.

TORONTO SOLAR-HEATED DWELLING

Designer: F. C. Hooper, P.E.
                John Hix, Architect

Building: Three-story residence (Fig. H.5)
                Floor area: 2,500 ft²
                Heat load: 10,800 Btu/degree-day

Location: Toronto, Ont., Canada
                Latitude: 43.4°N

Collector: Type: Flat plate
                Fluid: Water + ethylene glycol
                Area: 850 ft²
                Position: South-facing roof
                Tilt: 60°

Storage: Type: Water
                Volume: 80,000 gal
                Location: Insulated, waterproof basement
                Capacity: Entire heating season

Performance: Designed to require *no* auxiliary heat sources

BOULDER, COLO. SOLAR-HEATED DWELLING

Designer: W. G. Steward (system inventor and engineer)
                C. Haertling, AIA

Building: Three-story residence (see Figs. H.6, H.7, and H.8)
                Floor area: 3400 ft²
                Heat-loss rate: 35,000 Btu/degree-day

TABLE H.1 Energy Balance of Denver Solar House Shown in Fig. H.0

Winter 1959–60: All values in million Btu

	September (18–30)	October	November	December	January	February	March	April	May	June (1–10)	Total
1. Total solar incidence on 45°, 600 ft² collector area	9.93	26.84	25.81	21.98	25.48	22.10	30.43	26.56	29.61	8.13	226.86
2. Total solar incidence available on 45°, 600-ft² collector area when collection cycles operated	6.94	20.16	17.55	15.67	17.05	13.38	22.03	19.94	22.60	6.00	161.33
3. Gross collected solar heat	*	*	*	*	5.99	4.33	9.08	9.16	8.86	2.40	*
4. Gross collector efficiency, percent	*	*	*	*	34.7	32.4	41.1	45.9	39.8	40.0	*
5. Useful collected heat	1.93	5.91	5.34	5.61	5.59	3.79	8.45	8.66	8.25	2.18	55.72
6. Net collector efficiency, percent	27.9	29.4	30.3	35.8	32.7	28.3	38.3	43.5	36.5	36.4	34.6
7. Solar heat absorbed by storage tubes	1.12	3.04	2.77	2.79	2.64	1.92	3.99	3.68	3.01	0.50	25.46
8. Storage-tube inventory	−0.03	0.07	−0.002	−0.008	0.017	−0.023	−0.008	0.07	0.17	−0.05	*
9. Solar heat absorbed by water pre-heater	0.11	0.35	0.30	0.32	0.27	0.21	0.51	0.72	0.87	0.29	3.95
10. Heat delivered by natural gas for house heating	3.19	12.26	19.65	22.16	26.90	28.10	17.45	7.09	4.78	0.25	141.83
11. Heat delivered by natural gas for water heating	0.67	1.49	1.79	2.01	1.84	2.68	2.60	3.23	3.24	0.88	20.43
12. Total heat load	5.79	19.66	26.78	29.78	34.33	34.57	28.50	18.98	16.27	3.31	217.98
13. Percent of useful collected heat absorbed by water preheater	5.7	5.91	5.63	5.7	4.84	5.55	6.04	8.2	10.56	13.4	7.09
14. Percent of total water-heating load supplied by solar energy	14.1	19.0	14.4	13.75	12.80	7.26	16.4	18.25	21.20	28.40	16.25
15. Percent of house heat load supplied by solar energy (including water preheating but excluding water heating)	37.5	32.3	21.4	20.2	17.2	11.87	32.6	55.3	63.4	89.6	28.20
16. Percent of house heat load supplied by solar energy (including both water pre-heating and heating)	37.0	30.2	19.9	18.0	16.25	10.95	29.6	45.8	50.7	65.8	25.7

*Not determined.

FIG. H.5 Toronto solar-heated dwelling.

Location: Boulder, Colo.
                Latitude: 40°N
                Altitude: 8,000 ft

Collector: Type: Fixed spherical concentrator (SRTA)
                Fluid: Dowtherm J or water
                Area: 755 ft²
                Position: South-facing roof, integral
                Tilt: 55°
                Reflecting surface: High-reflectance aluminum

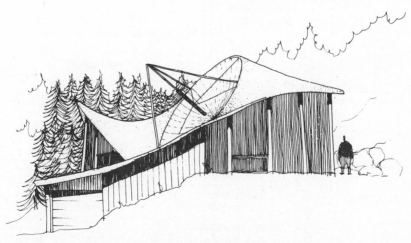

FIG. H.6 Architect's sketch of Dr. W. G. Steward's solar house.

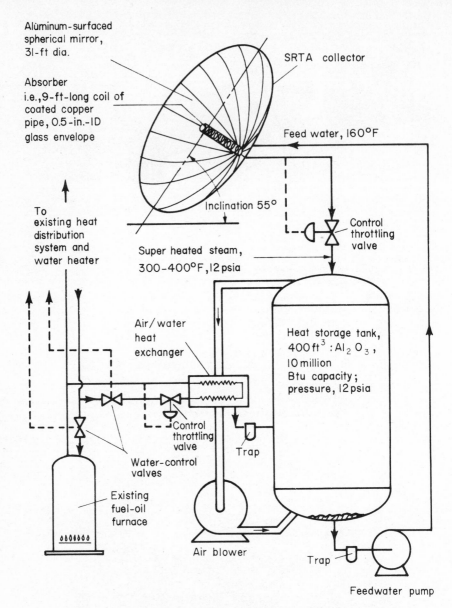

Aluminum-surfaced
spherical mirror,
31-ft dia.

Absorber
i.e., 9-ft-long coil of
coated copper
pipe, 0.5-in.-ID
glass envelope

SRTA collector

Feed water, 160°F

Inclination 55°

To
existing heat
distribution
system and
water heater

Super heated steam,
300-400°F, 12 psia

Control
throttling
valve

Air/water
heat
exchanger

Heat storage tank,
$400 \, ft^3 : Al_2O_3$,
10 million
Btu capacity;
pressure, 12 psia

Control
throttling
valve

Trap

Water-control
valves

Existing
fuel-oil
furnace

Air blower

Trap

Feedwater pump

FIG. H.7   SRTA home heating system, Steward house, Boulder, Colo.

Storage: Type: Aluminum-oxide gravel
                Volume: 400 ft³
                Location: Basement
                Container: Insulated (R-30) steel tank
                Temperature: 400°F max
                Capacity: 10 million Btu

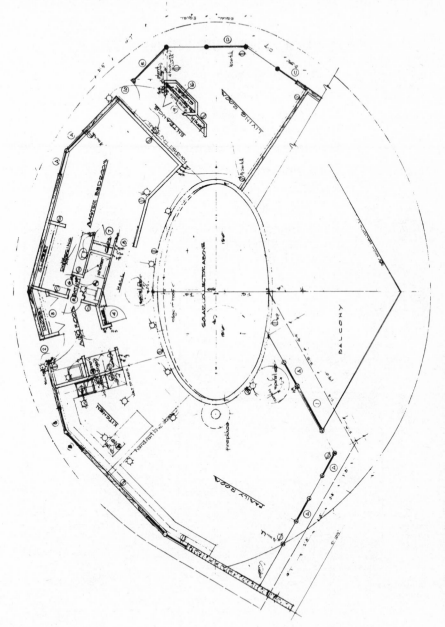

FIG. H.8 Main floor plan, Steward house, Boulder, Colo.

System: Baseboard convectors (hot water)

Performance: 80 to 90 percent of annual heating demand predicted

PHOENIX OF COLORADO SPRINGS, COLO.

Designer: D. M. Jardine, Engineer
               M. Lane, AIA

Building: Two-story (see Figs. H.9 and H.10)
               Floor area: 2,200 ft^2
               Heat-loss rate: 17,600 Btu/degree-day

Location: Colorado Springs, Colo.
               Latitude: 38.5°N
               Altitude: 4,000 ft

Collector: Type: Flat plate, double-glazed
               Fluid: Dowtherm J
               Area: 810 ft^2

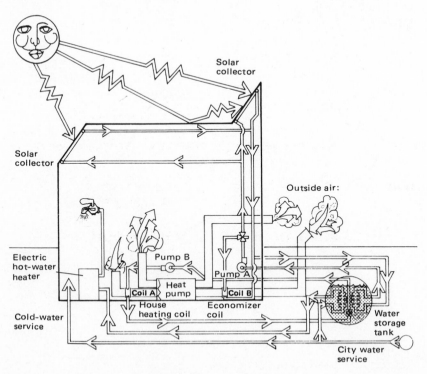

FIG. H.9 Schematic diagram of Phoenix solar-heated building, Colorado Springs, Colo.

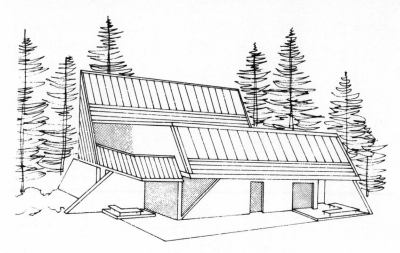

FIG. H.10   Phoenix solar-heated building, Colorado Springs, Colo.

	Position: Roof, integral Tilt: 55°
Storage:	Type: Water Volume: Variable, 8,000 gal. max Location: Buried beneath driveway Container: 9.5-ft dia. x 15-ft long uninsulated steel tank Temperature: 200°F max Capacity: 5 days carry through
System:	Solar-assisted, liquid-to-air heat pump; used for air conditioning in summer
Performance:	80 percent of annual heat requirement predicted

PYRAMIDAL OPTICS SOLAR HOUSE ADDITION

Designer:	G. Falbel, Wormser Scientific Corp.
Building:	Addition to the residence of Mr. Falbel (See Figs. H.11, H.12, and H.13) Floor area: 300 ft² Heat loss rate: Not known
Location:	Stamford, Conn. Latitude: 41.0°

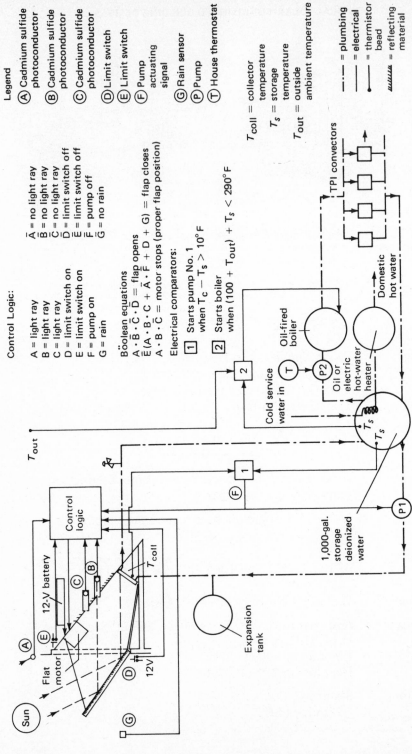

Legend

(A) Cadmium sulfide photoconductor
(B) Cadmium sulfide photoconductor
(C) Cadmium sulfide photoconductor
(D) Limit switch
(E) Limit switch
(F) Pump actuating signal
(G) Rain sensor
(P) Pump
(T) House thermostat

T_{coll} = collector temperature
T_s = storage temperature
T_{out} = outside ambient temperature

—··— = plumbing
——— = electrical
——• = thermistor bead
▨▨▨▨ = reflecting material

Control Logic:

A = light ray   Ā = no light ray
B = light ray   B̄ = no light ray
C = light ray   C̄ = no light ray
D = limit switch on D̄ = limit switch off
E = limit switch on Ē = limit switch off
F = pump on   F̄ = pump off
G = rain   Ḡ = no rain

Böolean equations

$A \cdot \bar{B} \cdot \bar{C} \cdot \bar{D}$ = flap opens
$\bar{E}(A \cdot B \cdot C + \bar{A} \cdot \bar{F} + D + G)$ = flap closes
$A \cdot B \cdot \bar{C}$ = motor stops (proper flap position)

Electrical comparators:

[1] Starts pump No. 1 when $T_c - T_s > 10°F$
[2] Starts boiler when $(100 + T_{out}) + T_s < 290°F$

TPI convectors

Domestic hot water

Oil-fired boiler

Cold service water in

Oil or electric hot-water heater

1,000-gal. storage deionized water

Expansion tank

Control logic

12-V battery

Flat motor

Sun

T_{coll}

T_{out}

T_s

FIG. H.11 Control system. (*From "Solar House Heating System Using Reflective Pyramid Optical Condensing System." Copyright 1974 by the Wormser Scientific Corporation. Used with permission.*)

(a)   (b)

FIG. H.12 Addition to Falbel residence, Stamford, Conn. (*From "Solar House Heating System Using Reflective Pyramid Optical Condensing System." Copyright* 1974 *by the Wormser Scientific Corporation. Used with permission.*)

Collector: Type: Pyramidal optics in concentration ratio of 1:3 in winter
                Fluid: Water + ethylene glycol
                Area: 32 ft² of collector
                Position: Interior of house

Storage: Water tank

Standby: Oil burner furnace

Performance: See Fig. H.14.

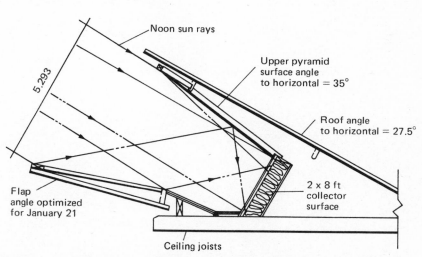

FIG. H.13 Arrangement of pyramidal optical concentration system. (*From "Solar House Heating System Using Reflective Pyramid Optical Condensing System." Copyright* 1974 *by the Wormser Scientific Corporation. Used with permission.*)

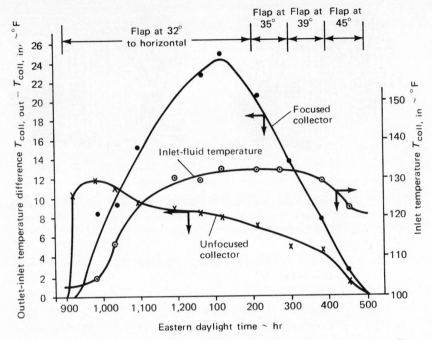

FIG. H.14 Comparison of performance between flat-plate and pyramidal optical collector systems. Data taken on Aug. 11, 1974, ambient temperature 75°F, cloudless sky all day. (*From "Solar House Heating System Using Reflective Pyramid Optical Condensing System." Copyright 1974 by the Wormser Scientific Corporation. Used with permission.*)

Bibliography

1. Aronin, J.: "Climate and Architecture," Reinhold Publishing Corp., 1953.
2. "ASHRAE Handbook of Fundamentals," American Society of Heating, Refrigerating and Air Conditioning Engineers, New York, 1974.
3. Brinkworth, B. J.: "Solar Energy for Man," Interscience Publishers, New York, 1972.
4. Daniels, F.: "Direct Use of the Sun's Energy," Yale University Press, New Haven, 1964.
5. Daniels, F., and J. A. Duffie: "Solar Energy Research," University of Wisconsin Press, Madison, 1961.
6. Duffie, J. A., and W. A. Beckman: "Solar Energy Thermal Processes," Interscience Publishers, New York, 1974.
7. Eberhard, John, Ed.: "Energy Conservation in Building Design," AIA Research Corp., Washington, D.C., 1974.
8. Pesko, Carolyn, Ed., "1974 Solar Energy Directory," Environmental Action of Colorado, Denver, prepared under NSF Grant AER 74-18503 AO1, 1975.
9. Federal Energy Administration, "Project Independence—Report of the Solar Energy Task Force," U.S. Gov't. Printing Off., Washington, D.C., 1974.
10. General Electric Space Div.: "Solar Heating and Cooling of Buildings (Phase 0)," Nat'l. Tech. Info. Serv., Springfield, Va., 1974.
11. General Services Administration: "Energy Conservation Design Guidelines for Office Buildings," AIA Research Corporation (see Chap. 11, prepared by Burt-Hill & Associates), Washington, D.C., 1975.
12. Halacy, D. S.: "The Coming Age of Solar Energy," Harper & Row Publishers, Inc., New York, 1973.
13. Hammond, A. L., et al.: Energy and The Future, American Association for the Advancement of Science, Washington, D.C., 1973.
14. Jordan, R. C., Ed.: Low Temperature Engineering Application of Solar Energy, ASHRAE, New York, 1967.
15. List, R. O.: "Smithsonian Meteorological Tables," Smithsonian Institution Press, Washington, D.C., 1971.
16. Mesarovic, M., and E. Pestel: "Mankind at the Turning Point—The Second Report to the Club of Rome," E. P. Dutton & Co., Inc., New York, 1974.
17. Pirsig, R.: Zen and the Art of Motorcycle Maintenance, William Morrow and Company, Inc., New York, 1974.
18. Rom, F., et al.: "Solar Energy to Meet the Nation's Energy Needs," Nat'l. Tech. Info. Serv., Springfield, Va., 1973.

19. Schumacher, E. F.: "Small is Beautiful—Economics as if People Mattered," Harper and Row, New York, 1973.
20. *Transition—A Report to the Oregon Energy Council,* Office of the Governor, Salem, Ore., 1975.
21. The Ford Foundation: "A Time to Choose—America's Energy Future," Ballinger Publishing Co., Cambridge, Mass., 1974.
22. Threlkeld, J. L.: "Thermal Environmental Engineering," Prentice-Hall, Inc., Englewood Cliffs, N.J., 1970.
23. TRW Systems Group: "Solar Heating and Cooling of Buildings (Phase 0)," Nat'l. Tech. Info. Serv., Springfield, Va., 1974.
24. Westinghouse Electric Corporation Special Systems: "Solar Heating and Cooling of Buildings (Phase 0)," Nat'l. Tech. Info. Serv., Springfield, Va., 1974.
25. Williams, J. R.: "Solar Energy Technology and Applications," Ann Arbor Science Publishers, Inc., Ann Arbor, Mich., 1974.

Index

solarheatingcool00krei

solarheatingcool00krei

solarheatingcool00krei